AF531361

TEXT BOOK
OF
APPLIED ZOOLOGY

TEXT BOOK OF APPLIED ZOOLOGY

(Vermiculture, Apiculture, Sericulture, Lac-culture, Agricultural Pests And Their Control)

By

Dr. Pradip V. Jabde

M.Sc., Ph.D., D.H.E., F.S.E.S.C.
Reader
Department of Zoology
Mahatma Gandhi Vidya Mandir's
Arts, Commerce and Science College
Manmad (Nasik)
(Maharashtra)

DISCOVERY PUBLISHING HOUSE
NEW DELHI-110002

First Published - 2005

Reprinted - 2023

ISBN: 978-81-7141-970-8

Text book of Applied Zoology

Published by:

DISCOVERY PUBLISHING HOUSE
4383/4B, Ansari Road, Darya Ganj
New Delhi-110 002 (India)
Phone: +91-11-23279245; 23253475; 43596065
E-mail: discoverybooksindia@gmail.com
discoverypublishinghouse@gmail.com
web: www.discoverypublishinggroup.com

Printed at:
Infinity Imaging Systems
Delhi

PREFACE

Recently Applied Zoology has been included in the national syllabus of U.G.C. as a compulsory course; this subject is taught in one form or another throughout the Indian Universities. All these applied fields are very vast and it is hard to cover up all aspects of each field. The information on these applied aspects are scattered in books, research papers, reports and pamphlets. Therefore, I decided to write a book on this important subject. The present work is an attempt to put the available information together in the form of a critical review. I had tried my level best to put the information in a simplified way with appropriate diagrams and tables. I did not claim perfection but if I get co-operation from the readers in the form of constructive suggestions, I hope, I would be able to improve it further in the next edition.

I received encouragement and help from Shri Prashant Hire, General Secretary, M.G. Vidya Mandir and Principal, Arts, Science and Commerce College, Manmad. I wish to record my gratitude to my wife sau. Shubhda and my kids Suraj and Shraddha, who fully helped in writing this book.

I am thanked to editor, staff, Discovery Publishing House for their co-operation for bringing forth this book in time.

Prof. Dr. P.V Jabde

CONTENTS

Preface

1. **Introduction** **1-8**

Nematology, Vermiculture, Parasitology, Applied Entomology, Pesticide Technology, Aquaculture, Pultry Science, Dairy Science, Biotechnology, Tissue Culture, Applied Embryology, Radiation Biology

2. **Vermiculture** **9-49**

History, Habitat Diversity, Ecology, Varieties of Earthworms, Types of Earthworms Suitable for Vermicomposting, Economic Importance of Earthworms, Harmful Activities of Worms, Methods of Vermicomposting, Basic Requirements, Preparation of Vermibed, Collection of Compost and Separation of Earthworms, Vermiwash, Method of Preparation, Effect of Vermiwash on Yield and Quality of Crops, Model Projects, Conditions of Margin Money Scheme

3. **Apiculture** **50-236**

History, Early Hives, Classification, Distribution, Colony Organisation, The Queen (Fertile Female), The Drones (Fertile Males), The Workers (Imperfectly Developed Females), Bee-hive or Comb, Caste Differentiation, Life History, Behaviour and Communication, External Characters of Bee, Bee Products, Bee Keeping Equipments, Bee Management, Bee Pollination, Bee Diseases and Enemies and Their Control, Pesticide Bee Poisoning, Economics of Bee-Keeping

4. **Sericulture** **237-333**

Introduction, Taxonomic Classification, External Morphology and Life Cycle, Cultivation and Harvesting of Mulberry, Silkworm Rearing, Diseases and Pests of Silkworms, Economics of Sericulture

5. **Lac Insect and Lac Culture** 334-346

History, Morphology, Fertilization, Natural Infection, Artificial Infection (Inoculation), Preservation of Brood Sac, Alteration of Lac Host or Brood Lac, Harvesting, Parasites and Predators of Lac Insect

6. **Agricultural Pests and Their Control** 347-494

Terminologies Used for Types of Damage to Crop Plants, Damaged Leaves, Damaged Flowers and Buds, Damaged Fruits and Seeds, Damage to Sown Seeds and Seedlings, Agricultural Pests, Principles of Pest Control and Practices, Integrated Pest Management (IPM), Pesticide Appliances

1

INTRODUCTION

To study Zoology, it is necessary to have a clear understanding of the organisation and functions of animal life around us; in other words, we must understand the great "variety of forms" in which the animal life expresses itself and its relationship with agriculture and other aspects of human welfare. Such a study alone can enable us to use to our advantage or to discard, destroy. The understanding of the "interrelationship of animal life with special reference to human life is, in scientific language, referred to as *Applied Zoology*. Or "science which is dealing with scientific application of zoological knowledge to the well-being of mankind." Applied zoology has its aim the manipulation of animals to man's advantage.

The sources of animal wealth have not been adequately tapped by man anywhere, much less to say in India. Tapping these sources is bound to add to the wealth and prosperity of men of all classes including agriculturists. It may not be necessary to deal with all the life processes of all the animal groups from practical point of view, but effort would be made to deal with comparatively more important animal groups right from the most primitive single-celled creatures, the protozoans, to the most highly evolved group, the mammals to which man himself belongs. This branch includes both benefits as well as harmfulness from related organisms.

It is known that the economy of India depends on agro-industry, which is still based on traditional equipments and methodology/and the agriculturists remain free without any much work at least half of the year and this is a major drawback. This problem could be solved by utilising life around us *i.e.,* to start industries based on them and he may utilise those off-season months.

BRANCHES OF APPLIED ZOOLOGY

This science of applied zoology has number of branches. Diagrammatically important branches are given below:

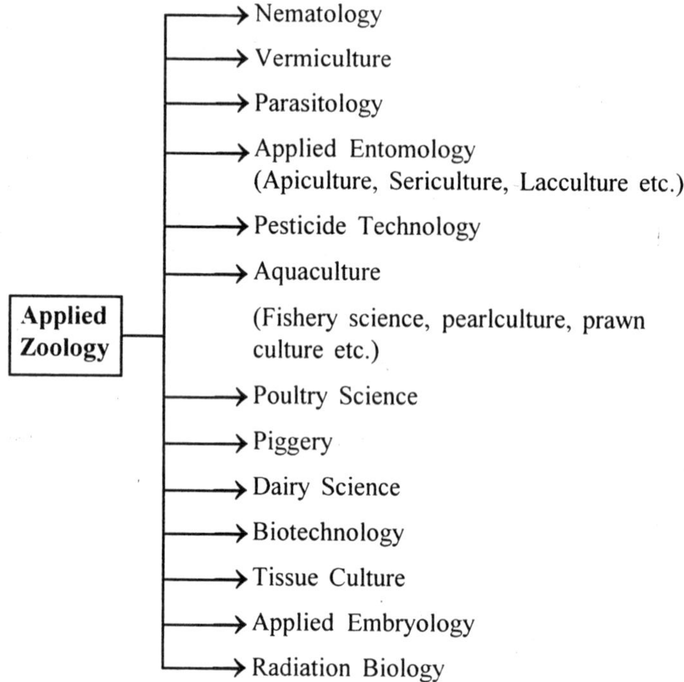

Nematology

Branch of Applied Zoology dealing with study of round worms of plants and animals and their control. *Ascaris, guinea worms, pinworm* are some of animal parasites; *root-knot nematode, root lesion nematode, citrum nematode* are some of plant parasitic nematodes.

Vermiculture

Vermiculture deals with the culture of soil burrowing worms could be used in production of natural fertilizers. Details will be discussed later on under Vermiculture.

Parasitology

The study of parasites in relation to man and related organisms is called parasitology. It is necessary to know the organisms causing diseases so that controlling measures could be managed properly, to maintain health.

Applied Entomology

Applied entomology has as its aim the manipulation of insects (both harmful and beneficial) to man's advantage. Successful manipulation of any organism depends on adequate biological understanding. Since time immemorial mankind has had to contend with insect pests which attack his cultivated plants, his domestic animals and his person. However, insects have also long provided man with valued products, in particular, honey, wax, silk, lac etc., so their actions are not entirely harmful. In addition, in more recent years, the beneficial roles of insects in facilitating ...ination of many cultivated plants and as natural enemies of pest ...s have been recognised. Some knowledge of insects, both harmful and beneficial, is therefore essential for anyone engaged in agriculture or horticulture.

Pesticide Technology

It is dealing with study and use of different pesticides (details are given under agricultural pests and their control) for controlling plant pests and diseases.

Aquaculture

Increase in production is one of the basic points under different plan periods of National policy. It has been remarked that science must help us to speedily improve production, natural and human resources can be profitably developed and equally shared, to create more employment and reduce drudgery (servile labour), to strengthen nation and reduce vulnerability. As per the scientific policy of Government of India, the need of our country is to utilise the resource potentials for enhancing the economic development of the country. This will result ultimately in increase of per capita income, per capita production, per capita consumption and better socio-economic condition of the people. This increase in per capita production and consumption will reduce malnutrition and vulnerability to diseases among people. The diet of an average Indian is very low in calories. Fish protein is supposed to be cheapest and of superior quality as it contains all the essential aminoacids for body building than plant proteins. Therefore, in the present context production of animal proteins from different water sources (fresh water and marine) is essential. Aquaculture includes all aspects of production of aquatic organisms (Bivalves, pearls, squids, fishes, prawns etc.) in captivity comprising either some or all stages of their life cycle, their live foods and the resultant marketable products in the habit of fresh, brackish and sea water. *Lagelar* (1956) has defined fishes are the most numerous of the vertebrates constituting above 40% of the vertebrate phylum. *Lone* (1988) has described aquaculture as an underwater agriculture.

Purpose of Aquaculture

(a) To increase production for per capita consumption and per capita income by which National Income will be higher.

(b) Ornamental purpose—like culture of anjel fish, black molly, red sword tail, gaurami etc.

(c) Sports and game purposes-like culture of trouts and mahaseers.

(d) Available natural water resource utilisation.

(e) Earning foreign exchange.

(f) Create employment opportunity.

(g) Utilisation of by-products of fish like issinglass, pearl-essence, liver oil, fish protein, fish glue etc.

(h) Controlling parasites like mosquito larvae, dengue by Larvicidal fishes (*Lebistes, Gambusia*).

(i) Utilisation of medicinal added value of fishery products.

Poultry Science

The poultry farming (rearing of varieties of birds) is a real profitable profession for Indian people which contribute more economy towards the India. It is transferred from a side business to the main business for rural as well as urban area. It becomes a fastest growing industry in India as it contains high proteins and fat. Birds form an important item of man's diet and are commonly sold even in the village markets for purpose of human consumption.

Dairy Science

Secretion of milk by the mammary glands of the mother is basically meant to nourish her young ones and not for commerce and trade. While other mammals depend only upon their own mother's milk, man, in addition, drinks milk of other mammals like that of cow, buffalo, uses the milk of other mammals as a source of additional nutrition for his offspring as well as for himself. In addition, he also uses mammalian milk for a variety of preparations ranging from curd, butter, ghee and cheese to sweets. To ensure the regular supply of milk man has domesticated a number of mammals which is called dairy. It is really distressing to note that in India adequate efforts have not been made to improve their lot and they are just reared as a means of livelihood either by the tribals or by certain other comparatively poorer classes of society. Most of these are maintained and reared almost free of cost without bothering about improving their breed. The only mammals which have received attention worth the name are cows and buffaloas. Even then much still remains to be done to meet the requirement of the growing population of the country.

Biotechnology

Biotechnology is a *multidisciplinary frontier area* of science. The whole world today is so much shaken up with this word that it has become a strong belief that this is "the field which will give solution to our "any problem." "Biotechnology, no surprise, is then variously described as "the last revolution of the current century", "the third wave in the evolution of human ambitions".

Biotechnology can be defined as "*The applications of scientific and engineering principles to the processing of material by biological agents to provide goods and services*". It involves the latest developments in the sciences of biology, microbiology, biochemistry, molecular biology, electronics, computers, physics, optics, photography, microscopy, instrumentation and practically most branches in engineering and technology. Biotechnology has the technical breadth and depth to change the industrial community of the 21st century because of its potentials:

1. to give products which were never available before;
2. to give products that are currently in short supply;
3. to give new methods which will reduce costs substantially;
4. to give safer, better quality products;
5. to give products which will use cheap raw materials which are plentily available but not used.

Today, the world is facing four major problems: malnutrition, diseases, energy scarcity and its high cost and environmental pollution. To overcome these problems has become the objective of biotechnology development. Biotechnology is applied today most vigorously in following major fields:

1. Medicine and pharmaceuticals.
2. Animal health, food industry.
3. Industrial (chemical manufacturing).
4. Agriculture.
5. Environmental protection.

Medical applications of biotechnology are really exciting and progress has been impressively fast. Many diverse aspects of disease prevention, diagnosis and cure are dealt with very effectively, by biotechnology. The areas in which biotechnology is working hard are:

1. Diseases caused by infective agents, including virus.

2. Diseases due to imbalance in body's natural chemistry.
3. Hereditary disorders.
4. Tumour and cancer therapy.
5. Organ transplantation (tissue culture technique).

Following are the newer approaches and some achievements of biotechnology in the field of medicine:

1. New and more powerful antibiotics.
2. Cheaper steroid drugs and hormones.
3. Improved more effective and safer vaccines for disease prevention.
4. DNA probe and monoclonal antibiotics in diagnostic kits.
5. Hormones, enzymes, interferon and some body proteins in therapy.
6. Monoclonal antibodies to reduce rejections in transplantation and to select a proper donor.
7. Liposomes, monoclonal antibodies for effective drug delivery, drug targeting.
8. Gene replacement therapy in hereditary genetic disorders.

The aim of biotechnology to the food industry is to produce food in abundance, with lowest cost, higher nutritive value, with greater variety, with more appealing in taste, from cheaper sources and food preservation over long periods.

Use of biotechnology in industries mainly in fermentation processes for products like vitamins, antibiotics, steroids, enzymes, alcohols, glycerol, organic acid and some new chemicals like plastics, polyacrylamide etc.

Environmental biotechnology is now a fast emerging frontier area of science. The techniques and methodologies adopted in municipal and industrial effluent treatment, water purification, composting and biogas fermentations were the earlier examples of environmental biotechnology. With the recent advances in molecular biotechnology development of new generation of fermentation and processing plants and skills available for introducing genetically engineered microbes, biotechnology has now opened new and unlimited vistas for protection of environment, water, land and air as well as faster regeneration of biomass to provide essential fuel and fodder.

The environmental biotechnological techniques and methods can be very effectively and usefully employed for pollution control and cleaning of major river systems. They can be very effective in supplementing the efforts being undertaken in programmes like the "Ganga Action Plan". The application of environmental biotechnology for pollution control in major rivers can be in the areas of:

1. primary and secondary treatment of waste water, sewage, municipal waste and industrial effluents before the discharge into the stream/river;
2. bioconversion of municipal and rural waste into biological fertilizers through composting;
3. weed and hyacinth control;
4. nitrogen (ammonia-nitrate) removal from fishery effluent;.
5. conversion of selected biomass species to organic acids and alcohols;
6. bioconversion of agricultural residues and waste into high value nutritious animal feed; and
7. preparation of anaerobic digester feed by blending biomass, sewage and fishery sludges.

Tissue Culture

The *in-vitro* (under artificial conditions, outside the living organism) culture of cells, tissues and organs of plants is an established technique of biotechnology. It is already being employed to propagate improved crops and ornamental plants. Such improvements can lead to inducing *genetic variation* and the selection, maintenance and propagation of desired variants.

In-vitro cultivation of animal tissue was first done in 1907. However, large scale culture of animal cells been done only during the last two decades or so. Initially this technique was used for the propagation of viruses for use as vaccines. In recent years the technique is also being used for production of proteins and monoclonal antibodies for therapeutical use. Examples are: Polio vaccine, modern rabies and measles vaccines, interferon (for the treatment of cancer), interleukins (for modulating the immune system).

The animal tissue culture technique's also being successfully employed for growing *in-vitro* skin cells taken from burns patient. An important use of this technique is in evaluating new drugs, and toxic compounds. With the use of this technique, alongwith conventional animal models, the evaluation of new drugs and biologicals can be done in much reduced time and cost. Similarly, this techniques are very extensively and successfully employed for use in the animal including vaccines for the cattle, growth hormones (somatotropin) for higher productivity (milk as well as body weight), diagnostics and therapeutics which are already in the market.

Applied Embryology

It is dealing with the study of artificial culture of embryo (test tube culture); for achieving higher productivity from animals, especially cattle, Embryo *Transfer Technology* (ETT) is a new advancement. This comprises of identification of an elite "*donor*" female and a sire both of high genetic quality. Either through normal breeding or by artificial insemination, the selected elite donor female is inseminated. 5 to 7 days after fertilisation, embryos are taken out from the elite donor female. From 4 to 10 embryos are normally collected at the same time. Each embryo is then transplanted into a normal indigenous or local (*deshi*) cow called the "recipient" or "carrier cow". The recipient can be any female having normal genitilia, *i.e.*, reproductive capability. Thus 4 to 10 normal 'deshi' recipient cows could be carrying (incubating) that many potential, offsprings of elite parents. As against this, under the normal circumstances, the elite donor mother would have been producing only one offspring after a pregnancy of 10-11 months.

Radiation Biology

It is dealing with use of electromagnetic radiations (X-rays, gamma-rays) and corpuscular radiations (α-particles, β-particles) (electrons and protons) for well-fare of humans.

Examples: Medical use in dental and embryo examination, use in TV, luminous dials, luminous markers etc.

2
VERMICULTURE

Wastes are the remains of anything that we use. They are also main pollutants of our environment. Wastes are of different types depending on their source. The deadliest of wastes is the fuel that has been used in nuclear power reactors. The wastes emanating from industries are mercilessly thrown into air and water bodies. Slaughterhouses are serving a social cause for the flies and pathogenic microbes, as breeding grounds for these disease-causing and disease-spreading agents. But probably the least take about type of waste is the one that comes out of every house, every day.

Domestic wastes (wastes originating from houses) includes leftovers of food, vegetable and fruit wastes, waste paper, old cloths, polythene bags, broken pieces of furnitures and scrap metal. This group of wastes can be broadly classified into two categories as *biodegradable* and *non-bio-degradable*. The biodegradable wastes comprise of decaying wastes such as vegetable matter, egg shells, paper, garden wastes such as dry leaves or woodchips. Biodegradable wastes are easily broken down by a class of microbes known as decomposers. The non-biodegradable domestic wastes include polythene, used furniture, styrofoam, plastic and rubber. Though wood is a biodegradable material, when it is used for making furniture, it is dressed with wood polish, warnish and paint, which make it difficult for the decomposers to feed on wood.

The domestic wastes that we discard everyday reach the local garbage dumps. From there they reach the main garbage dump of the city or town. Many a times, however, one often finds garbage carelessly dumped along the roadside or near the garbage dumps, by a callous public or the concerned authorities. Recently, there was a report which said that residents of an affluent locality in Malabar Hill in Mumbai dumped their household garbage into the Arabian sea, when they could afford their own garbage bins for the municipal corporations to collect garbages. Another report painfully observes the growth of a garbage mound at Ooty-one of the famous hill stations in India. This garbage dump is a grave threat to the people and the flora and fauna of the surrounding areas.

The decomposition, or stabilisation of organic matter by the action of microorganism has been taking place in nature since life first appeared on earth. "Composting" man's attempt to control and directly utilise the natural process for sanitary disposal and reclamation of organic waste material. The final product of composting is called "compost".

Vermiculture is the part of environmental biotechanology. Environmental biotechnology aims for bioconversion of organic wastes into resources which leads to protect the environment from getting-polluted. Environmental biotechnology involves harnessing diverse desired species and strains of bacteria to convert organic wastes into value added products. Bio-processing of waste biomass is then another way to stop non-conventional sources of energy.

When organic matter is decomposed in the presence of oxygen the process is called '*aerobic*'. Here living organisms (bacteria) which utilise oxygen, feed upon the organic matter and grow utilising the nitrogen (nitrogen fixing bacteria ex. *Rhizobium*), phosphorus, some of the carbon and other required nutrients. A great deal of heat is liberated during the conversion of carbon to CO_2 which destroy pathogenic bacteria. Aerobic bacteria can multiply very fast under ideal conditions of temperature, moisture (50-70%), pH-(6.5-7.5) and nutrients, doubling even in a short period of 20 minutes. Aim of speedy and effective waste management is to provide proper conditions for proliferation of aerobic bacteria so that they consume wastes voraciously.

For that purpose such bacterial strains are first isolated from the rest and allow to grow as a culture. The bacterial population is then scaled up from this purified mother culture by means of a sophisticated equipment called "*Fementer*" or "*Bioreactor*". Bioreactor creates ideal conditions of temperature, pH, moisture for speedy multiplication of desired bacteria. To achieve this one would literally need to construct a similar bioreactor inside the garbage, which is a difficult task.

Actually all animals are also bioreactors of some kind, because they regulate the temperature, moisture, pH level of their own bodies to ensure their own survival. The main task is now to find an animal whose bioreactor functions correspond to the conditions that suit aerobic bacteria and such suitable animal which is easily available is the *Earthworm*. The gut of an earthworm is a perfect natural bioreactor, which provides an ideal home for aerobic bacteria. Garbages in which earthworm lives is not a garbage for him but it is a habitat which he needs for getting nutritious food. Earthworm feeds more on bacteria present in organic matter and little organic matter (soil) (Fig. 2.1).

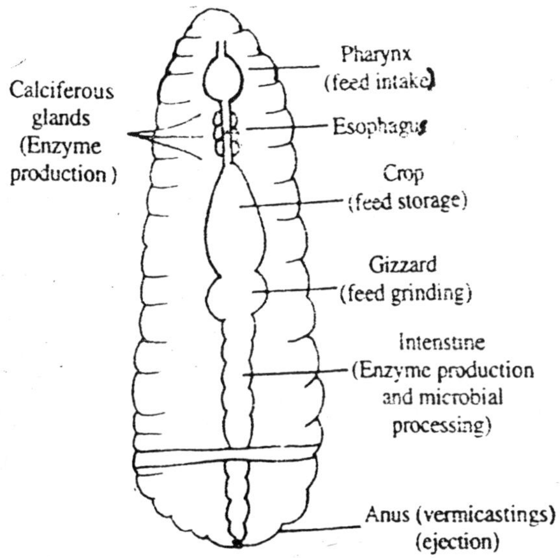

Fig. 2.1 : Gut of Earthworm.

History

Earthworm are known to inhabit the earth's soil since the precambrian period about 600 million years ago. The basic design of earthworms has not changed much over these millions of years and also does not very much between the species. Charles Darwin (1881) first studied the role of earthworms in breaking down of dead plant and animal residues in soil but earthworms have drawn the attention of philosophers and naturalists since ancient times because of their great importance in soil improvement. As early as 384 -322 B.C. the famous greek philosopher Aristotle described earthworms as "the intestine of earth". However, the zoological nomenclature of earthworms was adopted by *carolus Linnaeus* (1718) who listed two annelid species in his famous book entitled "systema Naturae" an oligochaete (*Lubricus terrestris*) and a polychaete (*Lumbricus marinus*). Existence of diversity in earthworms was brought to light by Savigny (1826) during his explorations of fauna of paris region in France. He described 20 species of Lubricidae from the area.

The oligochaete systematist *Michaelsen* named an earthworm genus after Wegener in honour of Wegener's work towards an understanding of the distributions of earthworms (Michelsen, 1933). The great upsurge of work on the morphology, histology and taxonomy of earthworms occurred in the late 19th and early 20th centuries. However, it was only in the last 25 years that interest in and research into ecology and biology of earthworms has peaked. Much of this work has been summarised by Edwards and Lofty (1977) in their book on the Biology and ecology of Earthworms by Satchell (1983) in *Earthworm Ecology*, from the Darwin to vermiculture and by Lee (1985) in his book on *Earthworms:* Their Ecology and Relationships with soil and land use.

Later on many of the research workers studied the mechanisms of conversion of organic matter into the humus by introducing vermiculture in the field. They have shown the importance of the earthworms for decomposition of organic wastes such a pig and cattle solids and sludge, wastes from layers, chickens, broilers, turkeys and ducks, horse manure and rabbit dropping, crop residues and city garbage. The various countries that are enguaged in the technology are. U.S.A., Mexico, West Germany, Italy, Holland, Switzerland, Austria, Japan, Phillipines, Thiland and very fast growing in India.

Habitat Diversity

Earthworms are found in all types of soils provided there is sufficient moisture and food supply. They have been recorded from forests, grasslands, cultivated fields, orchards, gardens, plant nurseries and green houses. A few species have been recorded to live under snow on high mountains (Julka, 1988). Most earthworms inhabit neutral soils, but they are also found in acidic conditions. A few small sized pieces are able to survive in desert and semi deserts, (Kubiana, 1953, Kollmannsperger, 1956). The occurrence of earthworms is related with physio-chemical conditions of the soil *i.e.*, temperature, moisture, pH, C/N ratio etc. Most species prefer soil with temperature of 10-35°C, moisture of 12-45%, pH about 7 and C/N ratio of 2 : 18.

Seasonal abundence of earthworms is affected by many ecological factors. In Indian subtropical climate, especially in the plains, earthworms are mainly active and abundant during the summer rains with maximum density occuring in September-October and minimum density in May-June. Agricultural management practices such as tillage, crop rotation, stubble retention, drainage, irrigation, time, fertilizer and slurry application, pesticide use and stocking rate can influence earthworm numbers and biomass.

Earthworms are simple, cylindrical, coelomate and segmented animals belonging to the order oligochoeta of the phylum Annelida and are characterized by the presence of setae in their body segments, the setae being modified in their genital segments called differently as the pencil setae. Or capulatoory setae. They lack specialized breathing devices. Respiratory exchange occurs through the body surface. (Fig. 2.2).

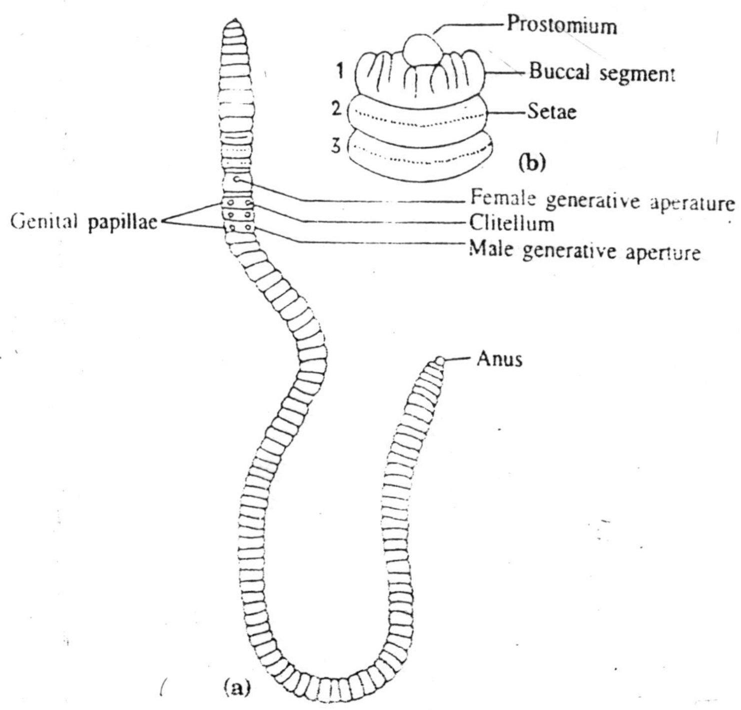

Fig. 2.2 (a) : Ventral view of an earthworm (b) Dorsal view of prostomium and first three segments.

Earthworms range from a few milimeters long to over 1 meter, but most of the common species are 5 to 10 cm. in length and hermaphroditic but self fertilization is avoided and a mutual exchange of sperms occurs between two worms during crossmating. Mature sperms and eggs are deposited in cocoons along with nutritive fluid. The cocoons are developed by clitellum, which is a conspicuous, girdle like structure near the anterior end of the body and its position may vary from 14th to 30th segment depending upon the species. The eggs are fertilized by the sperms within cocoon, when then slips off the worm and is deposited in or on the soil. The eggs hatch after about 3 weeks, each cocoon producing from 2 to 20 baby worms with an average of four. (Fig. 2.3).

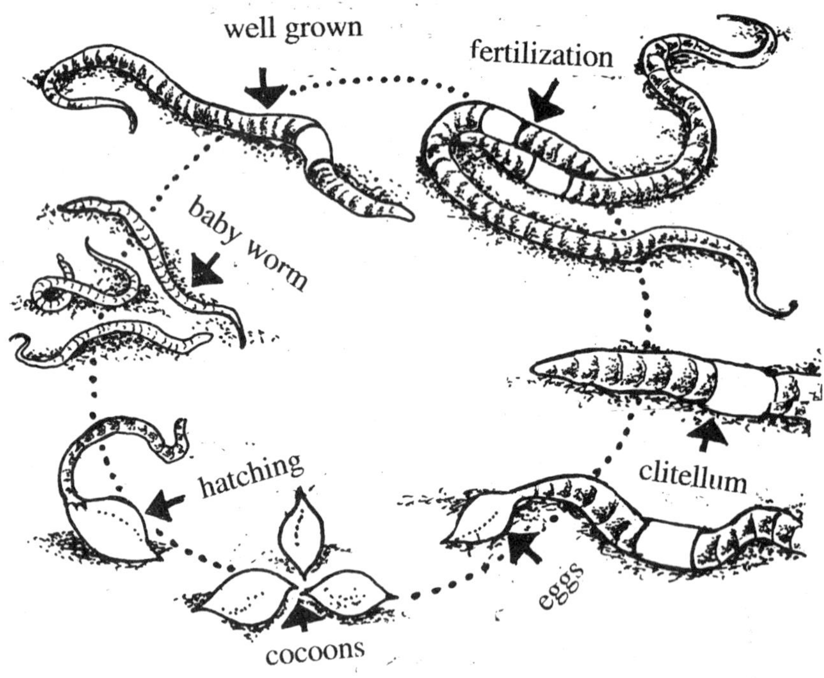

Fig. 2.3 : Life history.

Ecology

Earthworms are classified ecologically and according to their vertical distribution in soil biota as (1) Epigeic (2) Endogeic and (3) Anecic.

(1) **Epigeic:** These worms live in the soil surface of 3 to 10 cm. and feed on organic matter like leaf litter or animal excrements. They are very active and have high regenerative capacity within a short period of time. Normally they are richly pigmented worms. These have no effect on the soil structure as they generally cannot dig and they are efficient agents of communication and

Table 2.1 : Important earithworms showing their distribution through out the world.

Earthworm species	Weight adult worms	Temperature tolerance	Moisture toolerance	Feeding niche and virtical distribution	Active phase	Casting acivity	Distribution
1	2	3	4	5	6	7	8
Eudrilus eugeniae	1.5-2.5	18-35	20-40	Epigeric, surface living in organic matter	Throughout the year, No diapause	Non-burrower, surface casting,	Tropical Africa and South America
Eisenia foetida	0.3-0.7	15-30	20-40	Epigeic, Living organic matter	Throughout the years, No diapause	Non-burrower, surface casting, loose, granular	Temperate regions of Europe, North America, India
Perionyx excavatus	0.8-1.2	8-30	30-50	Epigeic, Living in organic matter	Throughout the year. No diapause	Non-burrower, surface casting, loose, granular	Temperate countries
Dichogaster caurgensis	0.07-0.12	20-25	20-30	Partially epigeic surface living in letter and soil with high humus content	July-October, No resting stage	Non-burrower, surface casting, loose, granular	South India at high altitudes and heavy rainfall area

Dichogaster bolaui	0.04-0.07	20-28	20-30	Partially epigeic, surface living in litter. Survive even in mineral soil with poor organic matter	July-October, No resting stage	None-burrower, surface casting, loose, granular	Tropical countries
Dichogaster affinis	0.04-0.07	20-28	20-30	Partially epigeic, surface living in mineral soil with poor organic matter	July-October, No resting stage	non-burrower, surface casting, loose, granular	Tropical countries
Drawida barwelli	0.2-0.5	20-30	40-50	Endogeic, 3-10cm living humus and mineral soil	August-November seasonal cessation	Surface casting, caste are thick long, mostely underground	Tropical countries shade essential for establishment
Lampito mauriti	0.8-1.5	18-28	20-40	Endogeic, 6-40cm living in humus and mineral soil	June-August seasonal cessation	Surface casting, caste loose and granuar complicated burrow	Plains cf Indian Penninsula
Octochaetona	1.2-2.0	20-28	30-50	Endogenic, 3-10 cm living	August-November		

fragmentation of leaf-litter, so classified as phytophagus earthworms.

(2) **Endogeic:** These worms live deep in the soil from 10-30 cm. and feed on the humic materials and mineral matter. They have very long life cycles with limited regeneration capacity and lightly pigmented. Due to feeding habit on humus also called Detritivores, and also feed on dead roots and other plant debris in the organic matter rich surface soil horizon. ex *Octochaetona serrata*, from Madras.

(3) **Anecic:** These worms can go very deep into soil upto 60 to 90 cm. and form complicated burrows for their movements. They plaster the burrows with their own excrements and mucus secretions. These are called geo-phytophagus earthworms. The external abiotic parameters and the poor nutritive resources of the soil system appear to be the controlling factors for earthworm population.

Varieties of Earthworms

In the phylum - Annelida there are about 1800 species of earthworms grouped into five families and distributed all over the world. The most common worms in North America, Europe and Asia belong to the family Lumbricidae which has about 220 species. Temperate soils have higher percentage of earthworms, which feed directly on organic matter, and low percentage of humus feeders. In tropical countries, the humus feeders predominate over the organic matter feeding worms. The *Lumbricides* and *Pheretima posthuma* distributed all over the world are called *pregrine.*

In the Indian subcontinent, the earthworms are well defined in 3 topographical divisions namely, Himalayan region, Indogangetic plain and Peninsular region. The Himalayan and Indo-gangetic plain show sparse earthworm population. The majority of earthworm species show endemicity in the permisular region. The important Indian species are *Plutellus (Acanthodrilidae), Perionyx, (Megascolicidae), Dichogaster (Ocnerodrilidae), Comarodrilus, Megascolex, Lennoscolex, Nelloscolex, Tonoscolex, Lampito mauritti, Hoplochaetella Khadalensis, Hoplochaetella kempi, Metaphire haulleti, Amynthas alexandri, Pontoscolex corethrums, Drawida Kanarensis etc.* Out of above and other only a few types are of commercial importance.

Overall the earthworms of the world are divided into following varieties.

(a) Night Crawlers

These are common to the northern states and may be picked from fields and lawns at night for commercial fish bait sale. Although, very popular with fishermen, they are not commonly raised on a commercial basis because they are not reproduce slowly and require a special production and control procedures.

(b) Field worms or Gardan worms

These make excellent fish boit and are often preferred by those who want a small number of worms for their own use. They are not prolific breeders, so are not recommended for commercial enterprises.

(c) Manure worms

These are also called wigglers or angleworms because of their squirming reactions when handled. They are particularly adopted to commercial production and are one of the two types most commonly grown by successful worm farmers.

(d) Red Worms

These are also another type of manure worms differing mainly in size and colour from their larger and darger cousins; and also adopted for commercial production.

Functionally earthworms may be classified into two groups humus formers and humus feeders. Humus farmers include surface dwelling worms which mainly feed on nearly 90% organic matter and 10% soil. They are generally red in colour. They are also known are detrivorous worms. Humus feeders includes deep burrowing worms which feed on 90% soil and 10% degraded organic matter. They are faint in colour and are useful in making the soil porous and distributing humus throughout the soil.

Types of Earthworms Suitable for Vermicomposting :

India has about 300 species of earthworms which are adopted to a range of vermiculture needs. The efficiency of vermicomposting system may depends upon the number and types of microorganisms present in the substrate and there are major three species are important and suitable for composting which are shown in Table 2.2 along with their life cycle, growth-rate, age for maturity, cocoon formation, incubation period, hatching capacity per cocoon etc.

Table 2.2 : Biology of earthworm species suitable for composting.

	Eisenia foetida	Eudrilus eugeniae	Perionyx excavatus
Duration of life cycle (days)	± 70	± 60	± 46
Growth rate (mg $worm^{-1}$ day^{-1})	7	12	3.5
Maximum body mass (mg individual worm)	1500	4294	600
Maturation attained at age (days)	± 50	± 40	± 21
Start of cocoon production (days)	± 55	± 46	± 24
Cocoon production ($worm^{-1}$ day^{-1})	0.35	1.3	1.1
Incubation period (days)	± 23	± 16.6	± 18.7
Mean number of hatching ($cocoon^{-1}$)	2.7	2.7	1.1
Number of hatching from one cocoon	1-9	1-5	1-3

(1) Eisenia foetida

It is a exotic species and commonly called *"European worm"* and considered suitable for vermicomposting because of its rapid growth rate, reproductive potential and occurrence in rich organic substrates in nature. It is shown experimentally that the manure samples inoculated with *E. foetida* decomposed more rapidly and showed a higher degree of humification than uninoculated samples.

(2) Eudrilus eugeniae

It is also a widely used exotic species introduced in India and is commonly called "African Night Crawler" and is distributed throughout world by earthworm growers as these are one of the best for vermi-composting and the production of protein meal. It has excellent growth and high conversion rate.

(3) Perionyx sansibaricus and P. excavatus

It is one of the best known indigenous (Indian) species for vermicomposting and found in the depth ranging from 3-8 cm., temperature range between 20-28°C and moisture content ranging between 20-40%. It is best suited in southern part of the country as summer temperature does not rise as high as in central and north India.

Economic Importance of Earthworms

Earthworms are familier to every one and their importance are globally realized that earthworms can do wonderful job in the management of different pedoecosystems. To a common man worms are rather insignificant animals which generally come out on the soil surface during rain. The role of earthworms in increasing soil fertility was however, know to even ancient farmers, but during the last 2 to 3 decades, the use of artificial and chemical fertilizers have faded their significance.. In recent years the farmers are once again realizing the benefits of these soil creature and are making all efforts to culture them. In addition to above accepted fact earthworms has following diversified potentials in different fields.

(a) Waste Decomposer :

continuous increase in population and industrialization created pollution of soil due to waste materials, which is a major hazard for maintenance of clean environment. The waste materials can be processed under vermicomposting programme for the stabilization of a variety of organic waste. Earthworms such as *E. Euginiae,* E. foetida and *P. excavatus* are easily adaptable to agricaltural wastes like sugarcane thrash, paper pulp, faecal matter of cow, sheep, horse, biogas sludge of poultry droppings, activated sludge etc. Processing of organic waste in large quantities has been carried out in Canada, Portland, Los Angeles in the United States, Newzealand , United Kingdom, Spain and now India also.

(b) Biofertilizer Manufacturer

Low level of soil fertility is one of the major problems enhancing the agricultural production in India. The low fertility of soil has increased the chemical fertilizers which in turn hit the small and marginal farmers. The organic wastes greater environmental problems and earthworms are probably one of the major contributors in the breakdown of organic materials. They have also the capacity to accumulate several inorganic substances including heavy metals. Earthworms ingest fresh or partially decomposed organic matter from the soil surface. The ingested organic matter is broken into small pieces by the grinding action of the earthworms gizzard. The fragmented organic matter is mixed with enzymes in the earthworm digestive tract and excreted as a colloidal humus in the form of cast either on the soil surface or in burrows depending on species. Their cast contain high concentration of organic material, slit, clay and cations such as iron, calcium, magnesium and potassium, Earthworms also release nitrogen into soil in their casts and in the mucus. The exchangeable cations such as Ca, Mg, Na, K, available P in wormcasts is more than in the surrounding soil. (Table-2.3).

Table 2.3 : Chemical composition of worm cast.

Sr. No.	Property	Value
1.	pH	7.1
2.	Electrical conductivity (mhos/cm)	1.0
3.	Organic carbon (%)	4.6
4.	Nitrogen (%)	0.5
5.	Phosphorus as P_2O_5 (%)	1.2
6.	Potassium as K_2O (%)	0.1

Earthworms dead tissues are added to the soil as nitrogenous fertilizers. Earthworms casting are also used as casing layer for mushroom cultivation. It has been demonstrated that four tons of vermicompost can replace the use of forty tons of cow manure per hectare by giving 36% higher yield. Vermicomposting offer to be the most promising which help to process wastes, simultaneously giving biofertilizer for agricultural and horticultural uses.

The recent development of vermicomposting at Pune (Maharashtra) Institute found burrowing earthworms, *Pheretima elongata,* especially efficient in breaking down the toughest of organic waste like sugarcane thrash. The work at BHU, Varanasi supported by United States, India Fund (WSPP) Project has identified *Amynthus morrisi,* Dichogaster *bolani* and *Perionyx sansibaricus* as potential decomposers. Now the solid waste planners and environmental engineers are looking with great interest at the potentials of worms to convert wastes into assets.

(c) Land Reclaimer and Water Conservation

In many parts of world, agricultural productivity is lowered by decreased soil fertility or poor physical conditions. Uncultivated land may be reclaimed for agricultural use. Earthworms play a significant role in managements and reclamation of degraded pedoecosystem and enrich the barren soil with nutrients. The introduction of earthworms population in the exhausted soils make it more compact and its poor structure changes into deep friable top soil, this in turn is beneficial for plant growth and crop yield.

A valuable by-product of waste decomposition by the earthworms is water, which is produced upto the cxtent of 60% of weight by (dry) organic wastes. This water is released slowly and transported by earthworms to the root zone of the plants.

Decomposition → Water $\xrightarrow{\text{via}}$ Earthworm → Root zone.
(by product)

Earthworms can soak in upto 120 mm rainfall in an hour without earthworms, the soil may soak in hardly upto 10 mm rainfall per hour. Since the peak rainfall rarely exceeds 75 mm per hour any where in the world, earthworms ensure ground water rechange and prevent runoff causing soil erosion and flash floods. *Lumbricus terrestris, Aporrectodea caliginosa* are specialist at creating channels deep into the soil that increase the capacity of plant to grow deep roots. The plants can then take up more water and nutrients than in surface soil establishing worms as a part of sustainable form management. In Indian conditions *Polypheretima elongata* is suitable for soil reclamation. This species produces more mucus which binds available nutrients. It helps in nutrient restoration and moisture conservation.

The rehabilitation of open-cast mining sites and industrial spoil heaps may be dobe by introducing earthworms. The presence of earthworms rapidly improves water balance, organic matter availability and microbial activity, and hence, facilitates the establishment and growth of vegetation. The use of earthworms for soil and land reclamation is likely to increase in developed, developing and undeveloped countries as the problem of population and land use become more acute.

(d) Earthworms as Protein Source :

Earthworms contain 70-80% protein on a dry weight basis and is of a high quality with good balance of essential amino acids, specially rich in lysine. The amino acid composition is for superior to snail and fish meat. The presence of essential amino acids in earthworms tissue are sufficient in order to fulfill the recommendations of FAO and WHO Particularly in terms of lysine, methionine, cysteine, and tyrosine, all of which are important components of animal feed. The species of *Perionyx, Eudrillas, Metaphire* and *Eisenia* are good for biomass production. *Lampito mauritii* has high protein contents in their body tissue and they are suitable as fish bait, poultry and Fish feed. Not only protein but high content of nitrogen and fat have also been reported in this species. Scientists have shown that earthworms are better animal protein than local fish meal and also shown the improved growth rate of fishes when fed on dry or fresh worms. A small white earthworm *(Enchytraeus albidus)* is often grown in soil and used to feed aquarium fish and small laboratory animals.

(e) As Vitamin Source :

The vitamins which are important and present in the earthworms are niacin, riboflavin (B_2), Pantothenic acid, folic acid, biotin (B-Complex), thiamine (B_1), Pyridoxine (B_6), Vitamin (B_{12}). Among these niacin and B_{12} are of significant value. Niacin is a valuable constituent of animal feeds.

(f) Earthworms as Drug Source

In India, the paste of dried worms are prepared for curing diseases in Unani system, of medicine. It has been used in treating wounds, chronic boils, piles, sore throat, hernia and impotency when applied externally. When used internally they are useful in curing, chronic cough, diphtheria, Jaundice, rheumatic pains, tuberculosis, bronchitis, facial paralysis and impotency. They have been used in folk medicine to treat fever, pyorrhoea and small pox and for hair growth and explosion of stones from the bladder in Iran. It has been shown that antipyretic activity is due to all cis-5, 8, 11, 14-ico satetraenoic acid (archiclonic acid) and all cis-5,8,11, 14,17- ico satetraaenoic acid.

There is a agreeing historical as well recent evidence of the value of earthworms in curing rheumatism. In Japan four kinds of drugs including antibioses, aphrodisease (Sex Stimulant) antipyretia and antidots. In USA universities like California the earthworms are being utilized in medicinal research to study the regeneration mechanism, cancer, and birth control.

(g) Earthworms as Natural Detoxicant for Pollution Control :

Many workers have described earthworms as bioindicator of soil contamination with heavy metals and pesticides, other agricultural chemicals, acidrain etc. The special features of heavy metal chemicals are strong attraction to biological tissue but their slow elimination from biological systems. The accumulation of toxic chemicals in earthworms tissue is ecologically important, because they are the important component in the food chain of several species of birds and mammals. They form food for wood cock, Robins and reptiles. This is very important from the point of minimisation of soil pollution.

Earthworms have the ability to take up into their tissue a number of unwanted organic acid, inorganic chemicals, so pollutants can be detected from earthworms tissues which provide a good tool to test the level of pollution including industrial one. Ireland (1975) observed that earthworm *Drawida rubida* living in heavy metals contaminated acid soil can increase the amount of available Zn and Cu by excretion in the faeces with little or no effect on the amount of available lead.

Earthworms can process household garbage, city refuse, sewage sludge and wastes from paper, wool, and food industries. They dominate the soil invertebrate biomass in many ecosystems of the world. European Economic Community (E.E.C.) and FAO and many environmental pollution committees recommended earthworms as one of the five key indicator organisms for ecotoxicological testing of the toxicity of industrial chemical.

(h) Earthworms as Bait

Fish capturing is not only a option of food source but also a very interesting entertainment. As we know in our rural areas and far from costal region of sea, the utilization of earthworms is used as very good fish bait on large scale as compared to other baits due to its easy availability in every climate and pedoecosystems of tropical, subtropical and temperate regions, as well as easy to rear and culture at all wide scale without extra burden of money, equipments, space and at low tech level. Additionally with use of earthworms and fish bait, simultaneously we can manage organic wastes.

(i) Vermicompost and Crop Production :

Recent studies have reported that the vermicompost not only helped to protect fertility of the soil but also boosted productivity depending on crops, season and other factors and also enhanced the quality of end products. Continual use of vermicompost over the year has also resulted in reduction of pests and disease problems. Palanisamy in 1996 studied the effect of mixing 10 to 20% worm cast in the upper 30 cm. of a sandy soil on the growth parameters and yield of wheat. He observed 40% increase in yield of wheat due to mixing of cast in 30 cm. soil. Similarly, as increase in shoot length and total dry matter was also noted. When soyabean seeds cv. Co-1 were pelleted with vermicompost 50 g/Kg seed., The pod yield was increased by 16%. Recent studies have observed that application of earthworm compost to wheat increased the plant height, number of tillers, numbers of leaves, early earheading, earhead length and dry matter per plant than control.

Jambhekar (1990) reported that application of vermicompost increased the available N.P. and K content in soil. There was a significant increase in the up take of N.P. and K by grape. *Tomati et. al., (1990)* reported increase in protein synthesis in raddish than grown in the presence of worm cast. It is reported that integrated use of recommended dose of fertilizers and vermicompost resulted into maximum up take of major nutrients such as N, P, Ca, Mg and S in comparison to application of either recommended dose of fertilizers or vermicompost alone. Dhoparkar (2001) compared the effect of various organic manures including vermicompost with chemical fertilizers on the yield and quality parameters of Alphonso mango in lateritic soil. He has reported significant improvement in quality parameter of mango pulp due to vermicompost application.

HARMFUL ACTIVITIES OF WORMS

In some cases, the earthworms became nuisance in following way:

(1) Soil Erosion

Exceptionally, their borrows may cause loss of water by seepage from ditches in irrigated lands. Their castings on sloping lands tend to be washed away by rain and thus contribute to soil erosion, through to a lesser extent.

(2) As Disease Carrier

Certain species live as external parasites of frog and man. Sometimes, they bury in the car cases of buried animals and bring the disease-germs to the surface, where they may infect other animals. The earthworms are said to serve as intermediate hosts in the transmission of some parasites, such as the tapeworm (*Amoebotania sphenoides*) and the gapeworm (*Syngamus*) of chiken and lung nematode (*metastrongylus elongatus*) of pigs. The latter parasite carries a virus which together with a bacterium, causes hog influenza or swine influenza.

(3) Damage to Plants

They may cause harm to young and tender plants by eating them up. Some species become pests of plants. *Pheretima elongata* is suspected of damaging the roots of the Betel-vine (*Piper Betel*) in Coimbatore. *Malabaria padudicola* and *Aphanascus oryzivorus* are said to injure the roots of paddy in mababar. A species *Perionyx* damages cardamon stems grown on the Anamalai Hills in South India.

(4) The habit of bringing sub-soil to the surface interferes with the course of the golf balls on the putting greens of golf courses. To get rid those greens of this annoyance they are sometimes sprinkled with a lime and sulphur mixture or any other poison that is harmless to green.

Methods of Vermicomposting

As mentioned earlier that earthworms have dynamic potentials and can do wonderful jobs for man and biosphere. In the treatment of organic waste, surface feeding earthworms function as a shredding machine, breaking up large lumps of the material as they ingest it mixing the material and increasing the surface area to allow better aeration and drying. This inturn, stimulates the activities of decomposing micro-organisms. The earthworms gut is a stable environment. The end result should be a fine stable odour free material containing most of the nutrients from the the original waste *i.e.,* a negative asset into a profit, by producing useful materials and at the same time minimizing environmental pollution. The vermicomposted materials are then used as bioferitilizers. Developed countries are earning millions of dollars from vermicomposting programme. Vermiculture is possible on,

(1) Small Scale in suitable containers or specially designed boxes for processing kitchen and garden wastes, and (2) Commercial or large scale in farm or municipal garbage etc. Both the technologies are simple and can easily adoptable even in rural areas.

(1) Small Scale Vermiculture Technique

The vermiculture boxes or containers are made up of light weight materials like plastic, tin, wood etc. Which could be easily handled and carried from one place to another. Size always vary according to need but measures, 50 cm. in length, 35 cm. in width and 15-20 cm. in depth. the bottom of the box is provided with a few holes of 50 mm. diameter. A plastic window screen is placed on the inside bottom with a burlop or jute cloth lining on top of the screened sides before the culture medium is added. This prevents the culture medium from sticking to the

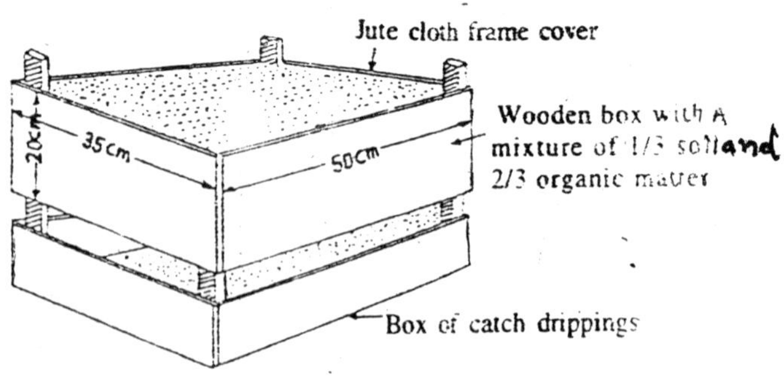

Fig. 2.4 : A design of vermiculture wooden box.

box and escape of worms through the holes but allows the excess of water to drain. Top of the box is covered with jute cloth frame before inoculation of medium and earthworms. (Fig. 2.4 and 2.5).

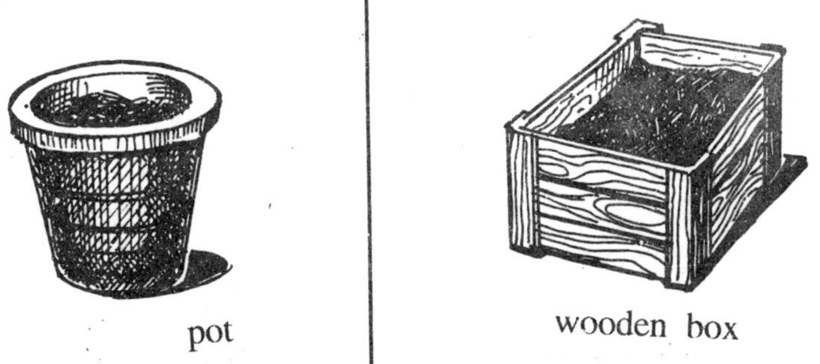

Fig. 2.5 : Preparation of vermicompost on small scale.

Recycling of Kitchen Waste

A-Demonstration

(a) Preparation of Vermiculture Bed

- Make a drainage hole on the lower side of a plastic bucket saving 18-20 litres capacity.
- Put 3 inches of soil layer.
- Put 1 inch of the vermiculture layer.
- Again put 1 inch of soil.
- Sprinkle a layer of cowdung.
- Cover it with grass cuttings or leaves to a thickness of 3 inches.
- Moisten the entire systems gently by sprinkling water, which has been stored in a tub for three days (storage removes chlorine from water).

- As grass and leaves dissolve, add more of the same and continue to maintain 3 inches of this mulch layer.
- Do not allow the system to become soggy due to excess water.
- The system will take 6 to 8 weeks for the earthworms to hatch and stabilise.

(b) Treatment of Kitchen waste using Vermiculture bed

- Spread the kitchen waste on the vermiculture bed.
- Initially, put only waste vegetable greens (no cooked food waste) on the vermiculture bed.
- Rock dust or black sand should be sprinkled every day with the food waste.
- Lime should be sprinkled with fruit peels and non-vegetable
- Bad smell indicates overloading. Inserts, flies indicate acidity. Allow the vermiculture bed to be airy. If the vermiculture bed is functioning properly, it will not smell.
- Stir the top occasionally.

  The entire contentscan be used as manure within 3 months of treatment.
- Set up a second vermiculture bed using a small portion of the above manure as vermiculture.

(2) Large Scale Commercial Vermiculture Farming

Under improper intensive farming and technologies the use of chemical fertilizers and pesticides in the soil is increasing . The soil becomes dead and unproductive as a result the production level goes on decreasing. Organic farming in the form of vermicompost obtained from the earthworms is one way to overcome the problems of low productivity. The earthworms enhance the decomposition rate of organic waste and improve the biological activities in the soil. The production of vermicompost from any organic wastes *i.e.,* agricultural waste, city garbage, industrial waste by using earthworms and its utilization in agriculture is one of the most economic ways in keeping the soils alive for sustainable production/productivity.

Fig. 2.6 : Production of vermicompost in a shed.

Fig. 2.7 : Vermicompost unit.

A Shed for Vermicoposting

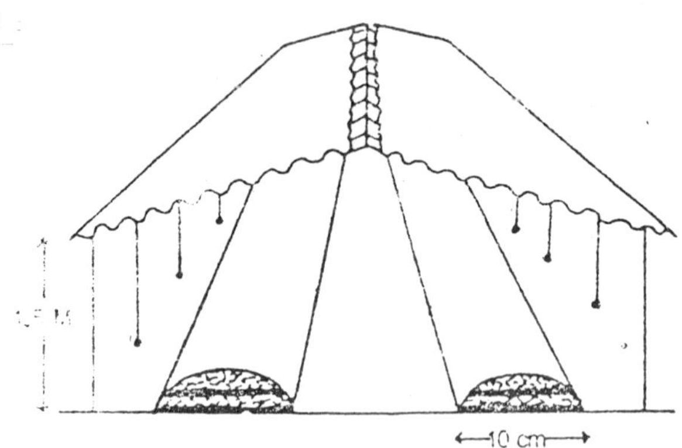

Fig. 2.8

Basic requirements

To produce vermicompost enough earthworm population is required and their multiplication on large scale is essential. To achieve the goal of economic multiplication of earthworms it is necessary to fulfill the following basic requirements :

(a) Selection of suitable species

As written earlier under heading "wastes decomposer" 3 species are important for vermicomposting are (1) *Eisinia foetida* (2) *Eudrillus euginiae* and (3) *Perionyx excavitis.*

(b) Suitable and adequate food

Well decomposed food of any organic waste in adequate quantity having C:N ratio less than 20 is essential for earthworm breeding. The dung of cattle, sheep, horses, pigs or dropping of poultry birds and vegetable waste from the ideal feed for the worms. Cattle dung can be fed as such whereas the other dung material or vegetable wastes are to be mixed in equal quantity with cattle dung for the feed acceptability. Wheat bran, gram bran and vegetable waste when added to dung in 10:1:1:1 ratio would enhance the biomass production considerably. A regular supply of 4.0 tones 1 day or 30 tonnes 1 week of organic waste is required at the site.

(c) Adequate moisture

Earthworms cannot survive without moisture in the feed. Earthworm contains 85% water in their body and hence it is the basic need. The water conservation mechanisms are poorly developed in earthworms. Respiration is done through body wall, which helps in keeping the body moist and much water is lost from the body through semisolid excreta. More than 35% water must be present in the earthworm feed for proper growth.

(d) Suitable temperature

The temperature limit of the earthworm feed should be in the range between 0°C to 35°C and best suitable range with 20°-30°C. The high temperature 735°C results into dessication of the body and moisture stress and temperature below 0°C stops earthworm activities.

(e) Protection from light

The earthworms are nocturnal in habit. They are injured and may be killed by exposure to light are affected by ultraviolet wavelength.

(f) Suitable pH (Hydrogen ion concentration)

For effective multiplication of earthworms pH of the feeding material should be at neutral level *i.e.*, 7-0 (Neutral). The earthworm population is severally affected if the pH of feed material is < 4 and > 9.

(g) Location

The site for multiplication should be shaded and on sloppy land, so that there are least chances for accumulation of rain water during rain season. For the best results erection of shade made either of wood, bamboo or permanent shade tin. High raised sheds of size (12' × 10' × 20'). Two sheds to accommodate 150-175 tons of waste. The earthworms can be multiplied very well n Pit, Pot and raised beds or on heap of 2' height filled with ready food of decomposed organic waste.

(h) Operation site

A small piece of land admeasuring around 2000sq.ft. is required for production and storage units to produce 500 tones of vermicompost per annum.

(i) Transport

Covered trucks / tractor to transport the organic waste from different parts of the village.

(j) Civil Work

(1) Pipe lines with sprayers for watering beds.

(2) Exhaust fans for removing smell for proper air circulation and cooling inside the sheds.

(3) Vermicompost drying yard (12' × 6' × 1') cement platform with a thatched or high-density interwoven nylon net roof for drying the vermicompost.

(4) Store room for processed vermicompost.

(5) House for the watch and ward.

(6) Power supply to run the motors and lighting.

(7) Bore wells for water supply.

(8) **Machinery:** (1) *Mechanical Choppers :* Crushers and mixers are required as shredding and crushing of organic waste brings down the volume of the material. It also enhances the feeding rate by earthworms using of mixer help in producing nomogenious material saves time and labour input. (2) *Mechanical Sieves :* Density gradient motorized roller drum services is recognised for sieving the vermicompost to separate small earthworms and cocoons.

(9) **Man Power :** (1) *Supervisors :* For overall supervision of work. (2) Labourers.

(10) **Purchase of earthworms :** For processing 125 to 150 tonnes of organic waste per month to yield 50-55 tonnes of dry vermicompost 1000 Kg. of earthworms or 10 to 12 lacs adult earthworms are required. The work can be initiated with 7 lacs of earthworms that work on organic waste.

Preparation of Vermibed (Fig. 2.9)

Vermibed is prepared under sheds either in the form of pit, pot, or raised beds or heaps; which measures about 6 Ft length, 3 Ft wide and 2 Ft deep. Keep a 2 Ft distance between two pits as a footpath for spreading the wastes, spraying the water, collection of vermicompost etc. Cut the two sides of pit at 45^0, so that there is no chance for soil to fall inside the pit while working. Fill the pit with ample of water so that non essential animals like ants, insects etc. Start coming out side. Then add pieces of brick to prepare a layer of 2-inches and sprinkle the water. On second day add 2 inch layer of sand and dry leaves or hay and sprinkle the water. On third day, add cowdung compost and soil with addition of water so that vermibed is prepared at the level of earth. Now add a layer of about 6 inches of partially decomposed wastes, soil and cowdung and prepare a heap like bed with addition of water. Now vermibed is ready for inoculation of earthworms. On next day add 8-10 earthworms per sqaure feet in vermibed.

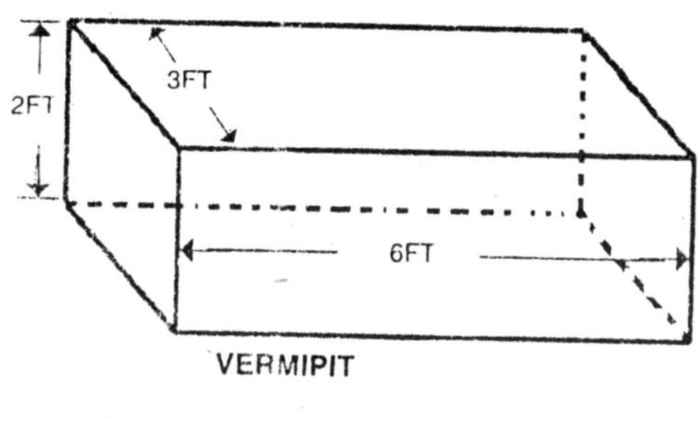

VERMIPIT

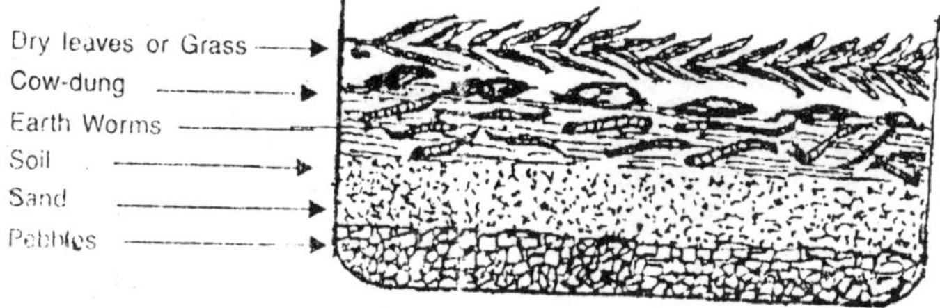

FILLING OF VERMIPIT

Fig. 2.9 : Preparation of Vermibed.

Collection of compost and separation of earthworms :

When earthworms are released into waste mix (vermibed), they start feeding from the top layer of the waste and make their way downwards into the medium. They have the tendency to move away from their own excrements. Within a span of 1 to 2 months earthworm feed actively, assimilate only 5 to 10% for their growth and the rest is excreted as loose granular rice shaped pellets as vermicastings on the surface.

Stop to sprinkle the water 3-4 days before harvesting, due to this upper layer become dry and in search of moisture they go in deep layers of vermibed. Before disturbing the bed, The excrements and left over feed should be removed. It is advisable to wear postmortem gloves while disturbing the bed to avoid causing injury to worms. The worms are harvested by hand sorting. Small conical heaps can be made and left for a few hours. Within 6 to 8 hrs time earthworms move down and settle at the bottom of the heap as a cluster. This way it is very easy to separate the earthworms and the compost. The separated worms can be added to the next fresh material for continuation of the next cycle. The compost (castings of earthworms) is passed through 3 to 4 mm. mesh sieve to separate the unfed material along with cocoons and young earthworms, which can be added to the fresh material. After drying the compost in sheds/shed till it is semi-dry can be stored in sacs until marketing or further use. (Fig 2.10).

Fig. 2.10 : Harvesting of Vermicompost.

Several institutions like the Sanjeevan Vidyalaya at Panchgani, Maharashtra have began to use virus vermiculture technology to solve their dual problems of savage disposal and water shortage (Fig. 2.11). The sewage from the school hostel is fed to a vermicomposting plant where earthworms convert it to vermicomposting, releasing clean water in the process. The vermicastings obtained are used by the school to green its surroundings, while the water produced suffices for gardening, flushing and other such operations, making considerable saving in water bill. The central pollution board is now taking effective steps to promote this effective method of sewage treatment.

Another project of converting trash into cash is going on at Surushi resort at famous will station, Mahabaleshwar, Maharashtra. A daily dose of left over is fed into a digester (vermifilter). The decomposed product is dried and sold as manure. (Fig. 2.12).

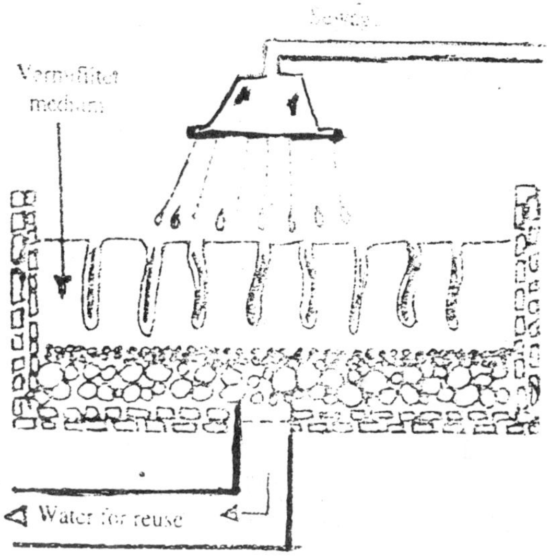

Fig. 2.11 : Sewage Treatment Plant.

Another example can be found at venkateshwara hatcheries, pune (maharashtra), where a vermicuture facility has successfully begun processing poultry residues of 10,000 birds amounting to over 4 tons daily are converted by earthworms in special bins to produce vermicastings. Now-a-days bioferfilizers are directly sold ex. Sujala is a biofertilizer consisting of specific culture of beneficial bacteria fixed on vermicastings produced by specific species of earthworms and is sold by Bhavalkar Earthworm Research Institute (BERI), Pune, Maharashtra.

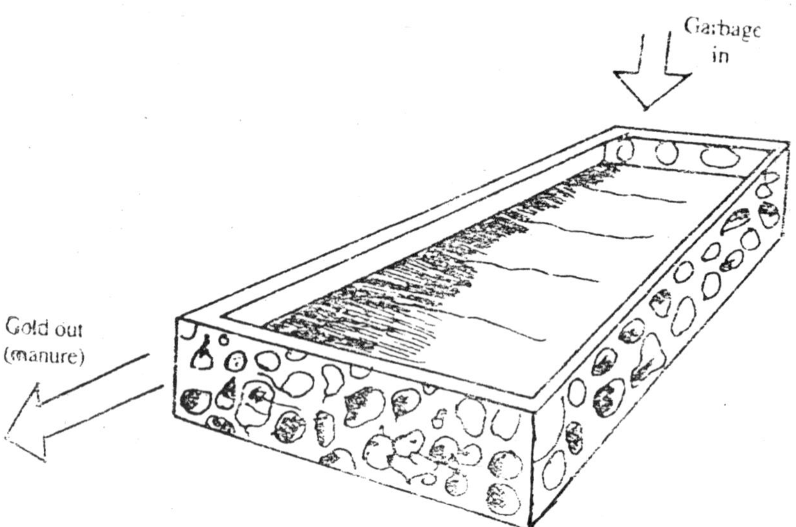

Fig. 2.12 : Formation of Manure.

Vermiwash

Vermiwash is a liquid collected after the passage of water through a column of vermibed. This vermiwash is very useful as a foliar spray to enhance the plant growth and yield and to check development of disease. It a collection of excretory products and much secretions of earthworms, along with micro-nutrients from the soil organic molecules. Vermiwash if collected properly is a clear and transparent, pale yellow coloured fluid. The utility of vermiwash in aquatic productivity was found to be immense but its potential as a biocides, either singly or when mixed with botanical pesticide, was under investigation. The vermiwash complex was observed to be efficient in raising nurseries, lawns and orchids.

Method of Preparation

For preparation of vermiwash take a barrel with capacity of 10 tc 100 liters with a outlet for collection of vermiwash and in which the different layers were placed from base to top as shown in Fig. 2.13. prepare a layer of about 25% length of washed and broken pieces of bricks, then add a layer of again 25% of washed thick sand, than on that add a layer mixture of cowdung soil and hay on the top. Sprinkle the water on bed and let it go through outlet and this procedure should be continued for 2 to 3 days so that extra wastes were washed out. Then introduce the well grown earthworms (40-50 sq.ft.) in side soil layer and then add 10% layer of cowdung and may and cover it by filter paper or muslin cloth. The vermiwash was collected from the bottom of the barrel after sprinkling water through perforated mud or metal pot of 5 litre capacity from the above. It is suggested to use the vermiwash either as it is or dilution with water or 10% cow's urine. The physico-chemical characterstics of vermiwash are as follows—pH 6.9 dissolved O_2-1.14 ppm, alkalinity-70.00 ppm, chloride-110.00 ppm, sulfates-177.00 ppm, inorganic phosphate - 50.9 mg/l, ammonical nitrogen - 2.00 ppm, potassium 69.00, and sodium-122.00 ppm. Grappelli *et al.* (1985) noted growth regulators content in worm costings as follows : Gibberellins (GA3) - 2.75 mg/g, cytokinins (IPA) - 1.05 mg/g and auxins (IAA) - 3.80 mg/g.

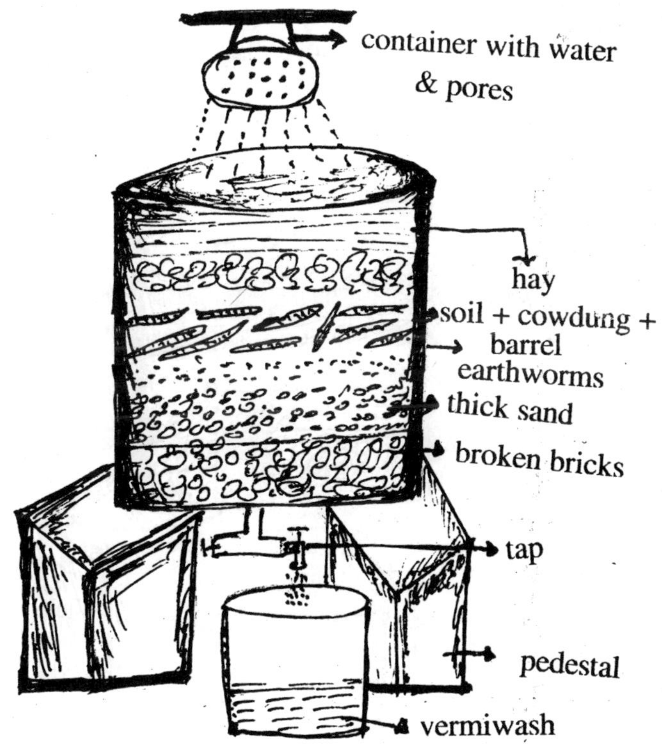

Fig. 2.13 : Vermiwash Unit.

Effect of Vermiwash on Yield and Quality of Crops

Recent findings showed that coelomic fluid from earthworm body had antibacterial properties. Studies on effect of spraying of vermiwash on vegetables indicated that quality and quantity of yield improved markedly, even the foliage turned dense green in 2 to 3 days. *Karuna et al.*, (1999) worked out two different concentrations of vermiwash of *Eudrilus eugeniae* and tested as spray to study its effect on anthuriums which has capture a international market, it was found that the tested lower concentration of the vermiwash *i.e.*, 50% was most effective in inducing vegetative growth like number of suckers, length and breadth of leaves and length of petiole and even initiated early flowering in plants. Balam (2000) studied the biopesticidal properties of vermiwash prepared from cowdung and vegetable wastes by inoculating *E. foetida* species of earthworms in the laboratory and the effect of said vermiwash on the powderty vmildew disease in cowpea, vermiwash at 20% and 30% inhibited the mycelial growth of pothogenic test fungi. It is shown that there is 75.14% disease control of powdery mildew when treatment of vermicompost and vermiwash was given. The next effective treatment was vermicompost + vermiwash + 10% cow urine which showed 73.37% disease control.

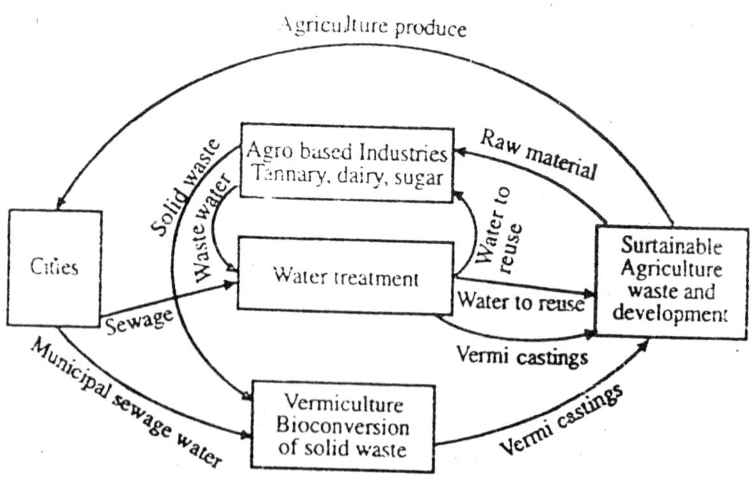

Fig. 2.14 : Vermiculture Biotechnology.

Model Projects

To help establish projects by entrepreneurs/ Institution/societies/ model projects on vermiculture are enumerated in the table 4, which will throw much light on the projects and its profitability. Beneficiary institutions/individuals are welcome to come up with their proposals by considering local needs and can submit projects to bank on vermicompost method, under margin money scheme of khadi and village Industries commission.

CONDITIONS OF MARGIN MONEY SCHEME

(1) Eligible Projects

The scheme is applicable to all viable village industry projects set up in rural areas.

(2) Eligible Borrowers

The eligible agencies under the scheme (i) Individuals (rural artisans / entrepreneurs) for projects upto Rs. 10 lakhs and (ii) Institution, co-operative societies, trusts registered with KVIC/ KVIB for projects upto Rs. 25 lakhs.

(3) Margin Money

25% of the project cost upto Rs. 10.00 lakhs will be provided by KVIC as margin money by way of backend subsidy, for projects above Rs. 10.00 lakhs and upto Rs. 25 lakhs rate of margin money will be 25% of Rs. 10.00 lakhs plus 10% of the remaining cost of the project.

In case of weaker section beneficiary viz. SC/ST/OBC/WOMEN/ PHYSICALLY Handicapped/ Ex- Servicemen and minority community beneficiary Institutions and for Hill, Border and tribal areas North Eastern Region, Sikkim, Andaman Nicobar Inlands, Lakshadweep, margin money grant will be at the rate of 30% of the project cost upto Rs. 10.00 lakhs and above this amount upto Rs. 25 lakhs will be 10% of the remaining cost of the project.

(4) Borrower's Contribution

under the scheme the borrower is required to invest his own contribution of 10% of the project cost. In case of sc/st and other weaker section borrowers, the contribution will be 5% of the project cost.

(5) Quantum of Loan

Banks will sanction 90% of the project cost in case of general category borrowers and 95% of the project cast in case of weather section beneficiary. Institutions and disburse full amount of the loan.

(6) Adjustment of Margin Money

KVIC will place a lumpsum deposit of margin money in advance with the corporate office of each bank or a nodal Branch designated by the banks in savings banks account in the name of KVIC.

After sanction of the credit facility by the bank branches, eligible amount of margin money will be kept in term deposit of 2 years in the name of the borrower at the bank branch which sactions the loan to the borrowers which will be first available for adjustment by banks to the borrower's loan account after a period of two years from the date to disbursement of loan. Banks would furnish quaterly progress report of adjustment of margin money directly to KVIC. For further details contact nearest nationalised Bank.

Table 2.4 : Showing projects and profitability from vermicompost unit, source : KVIC

Model Scheme-I (Vermicompost) (Individual)

I. PROJECT DETAILS				
1. Name of the Project	:	Vermicompost (Bio-Manure)		
2. Name of unit (Ind./Instt./Co-op	:	Individual		
3. Annual Capacity	:	500 M. Tonnes/Year		
4. Name of the product	:	Vermicompost (Bio-Manure)		
5. No. of working days	:	300 days		
6. Estimated area of work place	:	1900 Sq. Ft.		
7. No. of persons/Employment	:	20 Persons		
8. No. of cycle/Turnover	:	4 Cycles in a year		
II. COST OF PROJECT				
(a) Land : Own				
(b) (i) Shed and Office: 1400 Sq., ft. @ Rs. 250 per sq.ft.	:	3,50,000/-		
(ii) Godown 500 sq.ft. @ Rs. 250 per sq.ft.	:	1,25,000/-	4,75,000/-	
(c) Machinery/Equipments				
(iii) Shovel for mixing 1	:	25,000/-		

(iv) Sieving machine 1	:	25,000/-		
(v) Cutter & Blender 1	:	50,000/-		
(vi) Sewing machine 2	:	15,000/-		
(vii) Breeder Box 150 @ Rs. 400 each	:	60,000/-		
(viii) Mini Tractor	:	1,50,000/-	3,25,000/-	
(d) Working capital) for one cycle	:	1,28,000/-	1,28,000/-	9,28,000/-

III. MEANS OF FINANCE

(a) Own fund. 10%	:	92,800/-		
(b) Capital expenditure loan	:	7,20,000/-		
(c) Working capital for one cycle	:	1,15.200/-	9,28,000/-	9,28,000/-

IV. WORKING OF ECONOMICS

A. Raw Material (per annum)

1. Beeding material 120 tonnes @ 250/- per tonne	:	30,000/-	
2. Agriculture/Municipal/ Kitchen waste 800 tonnes @ Rs 200/- tonne	:	1,60,000/-	
3. Cow dung 350 tonnes @ Rs. 300 per tonne	:	1,05,000/-	2,95,000/-

B. Salary/Wages (Workers)

(i) Wages

1. 5 Skilled workers @ Rs. 80 per day	:	1,20,000/-	
2. 12 Semi Skilled worker cum helpers @ Rs. 40 per day	:	1,44,000/-	2,64,000/-

(ii) Salary (Administrative/ Managerial staff)

1. 1 Manager @ Rs. 3.000/- p.m.	:	36,000/-	
2. 1 Driver cum Assistant @ Rs. 2,500/- p.m	:	30,000/-	

3. 1 Watchman @ Rs. 2.000/- p.m.	:	24,000/-	90,000/-
C. (i) Overhead Charges			
(a) Interest on C.E. @ 14% p.a.	:	73,000/-	
(b) Interest on W.C. @ Rs. 14% p.a.	:	12,000/-	85,000/-
(ii) Insurance @ 1%			
(a) Stock in hand of raw materials	:	2,950/-	
(b) Capital Expenditure	:	8,000/-	10,950-
(iv) Depreciation			
(a) Building @ 5%	:	23,750/-	
(b) Machinery and equipment @ 10%	:	32,500/-	56,250/-
(v) Indirect Expenses			
(a) Freight and octroi	:	10,000/-	
(b) Packing and forwarding charges	:	25,000/-	
(c) Office expenses such as postage, printing, stationary etc.	:	10,000/-	
(d) Publicity	:	15,000/-	
(e) Power/Electricity	:	10,000/-	
(f) Transport	:	30,000/-	
(g) Selling Exp. (Comm.) @ 5% on sale	:	75,000/-	
(h) Repair and Maintenance	:	23,000/-	
(i) Other exp.	:	12,000/-	2,10,000/-

D. WORKING CAPITAL REQUIREMENT

	100% Requirement	70% Required	80% Required	90% Required
I. Fixed Working Capital				
(i) Salaries	90,000/-	90,000/-	90,000/-	90,000/-
(ii) Office expenses	15,000/-	15,000/-	15,000/-	15,000/-

(iii) Repairs	23,000/-	23,000/-	23,000/-	23,000/-
(iv) Interest on C.E. @ 14% p.a.	75,000/-	73,000/-	67,000/-	57,000/-
(v) Insurance @ 1%	10,950/-	10,950/-	10,950/-	11,000/-
(vi) Publicity	25,000/-	25,000/-	25,000/-	25,000/-
(vii) Depreciation	56,250/-	56,250/-	56,250/-	56,000/-
(vii) Others	–	–	–	–
Total Fixed Working Capital	2,47,200/-	2,47,200/-	2,41,200/-	2,31,000/-

II. Variable Working Capital

	100% Require-ment	70% Require-ment	80% Require-ment	90% Require-ment
(i) Wages	2,64,000/-	1,84,800/-	2,11,200/-	2,37,600/-
(ii) Interest on Working capital (14%)	12,000/-	12,000/-	10,000/-	8,000/-
(iii) Raw material chemical, fuel, Electricity etc.	2,95,000/-	2,06,500/-	2,36,000/-	5,500/-
(iv) Packing material	25,0007-	17,500/-	20,000/-	22,500/-
(v) Freight	10,000/-	7,000/-	8,000/-	9,000/-
(vi) Transport	30,000/-	21,000/-	24,000/-	27,000/-
(vii) Power	10,000/-	7,000/-	8,000/-	9,000/-
(viii) Selling Expanses	75,000/-	52,500/	60,000/-	67,000/-
(ix) Others	12,000/-	8,400/-	9,600/-	10,800/-
Total Variable Working Capital	7,33,000/-	5,16,700/-	5,86,800/-	6,56,400/-

E. COST ANALYSIS (COST OF PRODUCTION)

	@ 100% capacity utilization	@ 70% capacity utilization	@ 80% capacity utilization	@ 90% capacity utilization
Fixed costs	2,47,200/-	2,47,200/-	2,41,200/-	2,31,000/-
Variable costs	7,33,000/-	5,16,700/-	5,86,800/-	6,56,400/-
Total cost of production	9,80,200/-	7,63,900	8,28,000/-	8,87400/-

F. SALES

Sales realization 360 Tonnes @ 3,000 per cent (excluding sales commission)	15,00,000/-	10,50,000/-	12,00,000/-	13,50,000/-
Net Profit	5,19,800/-	2,86,100/-	3,72,000/-	4,62,600/-
Profit/Sale %ages	34%	27%	31%	34%

G. REQUIREMENT OF FUNDS

Capital Expenditure Loan	7,20,000/-
Working Capital Loan	1,15,200/-
Total	8,35,200/-

BEP = $\frac{12.50 \times 2.47}{12.50-6.85}$ = 5.46 BEP/Safe = 43%

The unit has capacity to pay back the loan in 7 years.

Cash Accrued			
Profit	2.86	3.72	4.63
Interest	0.85	0.77	0.65
Depreciation	0.56	0.56	0.56
Total	4.27	5.05	5.84
Capital Expenditure	0.38	0.76	0.76
Working Capital	0.12	0.12	0.12
Interest	0.85	0.77	0.65
Total	1.35	1.65	2.63
DSCR	3.16	3.06	2.22
Average		2.81	

Model Scheme-II (Vermicompost) (Individual)

I. PROJECT DETAILS

1. Name of the Project : Bio-Manure
2. Name of unit (Ind./Instt./Co-op.) with address : Institution/Co-op. Society
3. Annual Capacity : 1000 M. Tonnes per year

4.	Name of the product	:	Vermicompost (Bio-Manure)
5.	No. of working days	:	300 days
6.	Estimated area of work place	:	4000 Sq. Ft.
7.	No. of persons/Employment	:	41 Persons
8.	No. of cycle/Turnover	:	4 Cycles in a year

II. COST OF PROJECT

(a) Land	:	Own		
(b) (i) Shed and Office: 3000 Sq., ft. @ Rs. 250 per sq.ft	:	7.50.000/-		
(ii) Godown 1000 sq.ft. @ Rs. 250 per sq.ft.	:	2,50,000/-	4,75,000/-	
(c) Machinery/Equipments				
(iii) Shovel for mixing 1	:	60,000/-		
(iv) Sieving machine 1	:	60,000/-		
(v) Cutter & Blender 1	:	60,000/-		
(vi) Sewing machine 2	:	20,000/-		
(vii) Breeder Box 150 @ Rs. 400 each	:	2,00,000/-		
(viii) Mini Tractor	:	6,00,000/-	10,00,000/-	
(d) Working capita) for one cycle	:	2,70,000/-	2,70,000/-	22,70,000/-
*Total Cost of Project (a+b+c+d)				

III. MEANS OF FINANCE

(a) Own fund. 10%	:	2,27,000/-		
(b) Capital expenditure loan	:	18,00,000/-		
(c) Working capita) for one cycle	:	2,43,000/-	22,70,000/-	22,70,000/-

IV. WORKING OF ECONOMICS at 100% capacity :

A. Raw Material (per annum)

1.	Beeding material 300 tonnes @ 250/- per tonne	:	75,000/-

2. Agriculture/Municipal/ Kitchen waste 1500 tonnes @ Rs 200/- per tonne	:	3,00,000/-	
3. Cow ung 750 tonnes @ Rs.300 er tonne	:	2,25,000/-	6,00,000/-
B. Salary/Wages (Workers)			
(i) Wages			
1. 10 Skilled workers @ Rs. 80 per day	:	2,40,000/-	
2. 25 Semi Skilled worker cum helpers @ Rs. 50 per day	:	3,75,000/-	6,15,000/-
(ii) Salary (Administrative/ Managerial staff)			
1. 1 Manager @ Rs. 4,000/- p.m.	:	36.000/-	
2. 2 Accounts Clerk @ Rs. 2500/-	:	30.000/-	
3. 2 Driver cum Assistt @ Rs. 2,500/- p.m.	:	60,000/-	
4. 1 Watchman	:	24.000/-	
5. 1 Peon @ Rs. 2000/- p.m.	:	24.000/-	1,86,000/-
C. (i) Overhead Charges			
(a) Interest on C.E. @ 14% p.a.	:	2,05,000/-	
(b) Interest on W.C. @ Rs. 14% p.a.	:	28,000/-	2,33,000/-
(ii) Insurance @ 1%			
(a) Stock in hand of raw materials	:	6,750/-	
(b) Capital Expenditure	:	20,000/-	26,750/-
(iv) Depreciation			
(a) Building @ 5%	:	50,000/-	
(b) Machinery and equipment @ 10%	:	1,00,000/-	1,50,000/-

(v) Indirect Expenses			
(a) Freight and octroi	:	25,000/-	
(b) Packing and forwarding charges	:	50,000/-	
(c) Office expenses such as postage, printing, stationary etc.	:	25,000/-	
(d) Publicity	:	50,000/-	
(e) Power/Electricity	:	20,000/-	
(f) Transport	:	30,000/-	
(g) Selling Exp. (Comm.) @ 5% on sale	:	1,50,000/-	
(h) Repair	:	50,000/-	
(i) Other exp.	:	24,000/-	4,24,000/-

D. WORKING CAPITAL REQUIREMENT

	100% Requirement	70% Required	80% Required	90% Required
I. Fixed Working Capital				
(i) Salaries	1,86,000/-	1,86,000/-	1,86,000/-	1,86,000/-
(ii) Office expenses	25,000/-	25,000/-	25,000/-	25,000/-
(iii) Interest on C.E @ 14% p.a.	2,05,000/-	2,05,000/-	1,90,000/-	1,60,000/-
(iv) Insurance @ 1%.	27,000/-	27,000/-	27,000/-	27,000/-
(v) Publicity	50,000/-	50,000/-	50,000/-	50,000/-
(vi) Repairs	50,000/-	50,000/-	50,000/-	50,000/-
(vii) Depreciation	1,50,000/-	1,50,000/-	1,50,000/-	1,50,000/-
(viii) Others	24,000/-	24,000/-	24,000/-	24,000/-
Total Fixed Working Capital	7,17,000/-	7,17,000/-	7,02,000/-	6,72,000/-

II. Variable Working Capital

	100% Require-ment	70% Require-ment	80% Require-ment	90% Require-ment
(i) Wages	6,15,000/-	4,30,000/-	4,92,000/-	5,53,000/-
(ii) Interest on Working capital (14%)	28,000/-	28,000/-	24,000/-	20,000/-
(iii) Raw materia chemical, fuel, Electricity etc.	6,00,000/-	4,20,000/-	4,80,000/-	5,40,000/-
(iv) Power	20,0007-	14,000/-	16,000/-	18,000/-
(v) Packing material	50,000/-	35,000/-	40,000/-	45,000/-
(vi) Freight and Octroi	25,000/-	17,000/-	20,000/-	22,000/-
(vii) Transport	30,000/-	21,000/	24,000/-	27,000/-
(viii) Selling Expanses	1,50,000/-	1,05,000/-	1,20,000/-	1,35,000/-
(ix) Others	-	-	-	-
Total Variable Working Capital	15,18,000/-	10,70,000/-	12,16,000/-	13,60,000/-

E. COST ANALYSIS (COST OF PRODUCTION)

	@ 100% capacity utilization	@ 70% capacity utilization	@ 80% capacity utilization	@ 90% capacity utilization
Fixed costs	7,17,000/-	2,47,200/-	7,02,000/-	6,72,000/-
Variable costs	15,18,000/-	5,16,700/-	12,16,000/-	13,60,000/-
Total cost of prod.	22,35,000/-	17,87,000/-	19,18,000/-	20,32,000/-

F. SALES	100%	70%	80%	90%
Sales realization 1000 Tonnes @ 3,000 per tonne (excluding sales commission)	30,00,000/-	21,00,000/-	24,00,000/-	27,00,000/-
Net Profit	7,65,000/-	3,13,000/-	4,82,000/-	6,68,000/-
Profit/Sale%ages	25%	15%	19%	22%

G. REQUIREMENT OF FUNDS

Capital Expenditure Loan	18,00,000/-
Working Capital Loan	2,43,000/-
Total	20,43,000/-

$$BEP = 14.51 = \frac{30 \times 7.17}{30\text{-}15.18} = \frac{215.10}{14.82}$$

The unit has capacity to pay back the loan in 7 years.

BEW/Sales = 48%

Cash Accrued				
Profit	3.13	4.82	6.69	
Interest	2.33	2.14	1.80	
Depreciation		1.50	1.50	1.50
	Total	6.96	8.46	9.98
Capital Expenditure		1.09	2.18	2.18
Working Capital		0.29	0.29	0.25
Interest	2.33	2.14	1.80	—
	Total	3.71	4.61	4.23
	DSCR	1.87	1.83	2.36
	Average	2.02		

3

APICULTURE

Honey bee is one of the very few domesticated insects, which are accompanying man from several thousands of years. Honey was the only sweetening agent throughout the world before entry of canesugar and beetsugar. Rearing of honey bees on a commercial scale is called *Apiculture*. Apiculture (bee-keeping) is popular throughout the world and becoming popular in India not only because of the production of honey and wax but also because bees act as good pollinating agents for cultivated plants. The amount of honey produced by these bees is negligible, so culturing of honeybees on commercial basis may not be suitable, but in nature their role as a pollinating agent is more important.

Now-a-days beekeeping is an ideal agro-based subsidary enterprise providing supplementary / major income to the people in rural areas. A few colonies can be maintained in the courtyard of the house and the honey thus produced will add to the quality of diet of the family. This profession has proved to be a very good side business which with little efforts adds considerable income to inprove the socioeconomic states of the family. Now-a-days this enterprise is developing as a major industry and many entrepreneurs are taking it up on commercial scale.

History

It seems that man has taken honey from bees since he first set foot on earth. The oldest records of this are in prehistoric caves paintings, a notable number of which are found throughout Spain. One at Cuevas de la Arana in Valencia shows a man climbing up a cliff to rob a swarm of wild bees. It is estimated that it is around 15,000 years old. Honey became an important commodity throughout the ancient civilised world— Assyria, Babylon, Egypt, India, Persia, Greece and Rome-all made extensive references to honey in their art and literature. Long before the Roman invasion, in Britain, honey was much used for cooking and baking in those pre-sugar days, and almost importantly for making alcoholic ale or mead drinks.

Ancient Germany was another major honey country; and upon the introduction of Christianity honey production went into overtime there in order to keep up with the demand for church candles. In ancient times, it appears the honeybee *Apis mellifera* was non-existent in America and Australia. The Americas had smaller stigless bees such as *Apis Trigonae* and *Meliponae*, and the Indians cultivated them to produce a thinner style of honey. *A. mellifera* honeybees were brought from the old world to the new by Spanish, Dutch and English settlers at the end of the 16th century. In some areas this was a mistake, for they unsurprisingly robbed the sugarcane from the plantations in places like Cuba and Barbados. The planters did their best to annihilate them, and meanwhile they spread across the United States.

They swarmed from the Atlantic to the Pacific with the west proving a real paradise for these nectar seeking insects which only reached California in the mid of 19th Century. Mostly they were wild swarms which made their nests in forests, and occasionally in abandoned houses.

Early Hives

The ancient Greeks and Egyptians used hive made of pottery, shaped like giants thimbles and about the size of a modern skep. These would have been laid lengthways with entrance covers, probably stored in the walls of houses. Similar hives can still bee seen in use in some countries.

The beekeepers of an ancient Rome used all kinds of hives mentioned in their literature-long hives; hives made from cork bark; hives made from wooden boards; woven wicker hives; dung hives; earthenware hives, brick hives and the aforementioned "transparent stone" hives.

In India, with little success, the first attempt for bee-keeping was made in Bengal in the year 1882 and then Punjab during the year 1883-1884. Later, government of India collected the information on bee-keeping from different states and published the data in 1883. Later, in 1907 the work on bee-keeping was carried out in Pusa by Entamologists, and by Ghosh, Sir Louis Dane (Punjab) and others. As a result, Beekeepers association of Punjab was established with head office at Simla (1916). Later on, in 1938, All India Bee-keeper Association was established by beekeepers, and publishing and arranging different conferences and exhibitions at various centres of the country. The Indian Council of Agricultural Research co-ordinates the activities of Research workers associated in this field. After independence, a commission was established in 1953 called *Khadi* and *Village Industries Commission* to manage bee-keeping in modern organised way.

In 1956 this board was reconstituted under the industry ministry, having khadi ad village boards at state level. Thus, beekeeping got a good support from these boards. Some states like J and K, Karnataka, UP and HP established Departments of Beekeeping under the state ministry of Agriculture, KVIC established a central bee research institute (CBRI) at pune during 1962, with overall development of beekeeping as its mandate. Subsequently, CBRI also established some regional centres in various parts of the country. The apiculture research in the right earnest started when Indians council of agriculture research (ICAR) started funding the beekeeping projects in the states, central institution and other organizations.

On the recommendations of the national commission on agriculture (1976) are "All India coordinated project on honey bee research and training" was launched by ICAR in 1981 with CBRTI, pune are its main centre and at present, its headquarters has been shifted to CCS Haryana agriculture University, Hissar, which has sponsored many projects at different centres and several state. Some voluntary centres and several state agricultural universities (SAUS) are also engaged in beekeeping research and training. During 1943 Ministry of agriculture, deptt. of agriculture and cooperation laid special emphasis on beekeeping as are important component of the total programmes of the ministry and started a "National scheme on the development of beekeeping for Increasing crop productivity". Under this scheme, beekeeping research and development projects are sectioned to vatious SAUS/agriculture developments, govt. and Non-govt. organizations. Thus we have a good network of research and development centres in the country.

Presently there are about 6 lakhs bee colonies, established with financial aid from government. Technical guidance and supervision is afforded to thousands of bee-keepers through network of trained and experienced field supervisors. Development of scientific bee-keeping in our country through, the apiary industrial sector is exclusively done by the Khadi and Village Industries Commission. Suitable guidance and assistance is also made available to bee -keepers' societies for processing, grading and packing of honey for marketing.

The State Khadi and Village Industries Board and Khadi and Village Industry Commission are involved in research and extension of bee-keeping in Maharashtra State. The research centres are located at Mahabaleshwar, Pune, Latur, and Nasik, and their work is to educate farmers and aspiring bee-keepers and organise the different types of training programmes; not only this but centres are also providing necessary equipments and bee-boxes. At present Maharashtra has 15,000 bee colonies and about 2,000 bee-keepers and 4-co-operative societies which are engaged in this field.

CLASSIFICATION

Phylum—Arthropoda

This includes bilaterally symmetrical and metamerically segmented animals. Body usually covered by tough, chitinous exoskeleton. The segmented body bears paired and jointed appendages. Body is divided into head, thorax (cephalothorax) and abdomen. Body movements can be effected by sclerites with thin, soft and articular membranes between two sclerites. Blood vascular system is open type. Respiration with the help of gills, or trachea or book lungs, and with the aid of copper containing respiratory pigment haemocyanin. Excretion occurs by green glands or by Malpighian tubules. Sexes are generally separate with sexual dimorphism. Fertilisation internal and development includes metamorphosis.

Class—Insecta (Hexapoda)

They are air-breathing, mostly terrestrial and aerial and rarely aquatic. Body is divided into head, thorax and abdomen. Head with different types of mouth parts adapted for chewing, biting, piercing, sucking, siphoning or sponging type. Thorax made up of 3 segment, each-bearing a pair of jointed legs and two pairs of wings. Abdomen is divided into 7-11 segments. Respiration by trachae. Excretory organs are Malpighian tubules. Sexes separate.

Sub-class—Pterygota (Metabola)

Wings are usually present. Abdomen has no appendages except genitalia and cerci.

Metamorphosis simple or complex.

Division—Endopterygota

Wings develop internally. Metamorphosis complete including pupal stage.

Order—Hymenoptera (Membrane wings)

This includes social or parasitic insects. Mouth parts modified for biting and licking. Two pairs of wings, with reduced venation. Organs of defence are stings. Worker possesses stings and is absent in drones. In queen the sting is vestigial. Larvae are legless or apodus. Pupae are exarate (legs are not appressed to body). Polyembryony and parthenogenesis.

Sub-order—Clistogastron

A distinct constriction between head, thorax and abdomen. Thorax and abdomen are connected by petiole. *Ex.* ants, wasps and bees (all with stings).

Super family—Apoidea

Solitary and social bees. There is caste differentiation. Female antenna has 12 segments while male antenna has 13 segments.

Family- Apidae.

Bees are of moderate sized, females (except for parasitic and robber species and queens of the highly eusocial forms) carry pollen in a corbicula or pollen basket on each hind tibia.

Sub-family—Apinae

These are minute to moderate sized bees. All species live in perennial, highly eusocial colonies.

Tribe-Apini

This tribe, the true or stinging honey bees was originally restricted to Eurasia and Africa, but one species, the common honeybee, has been introduced into all parts of the world.

Genus—Apis

Species; dorsata, florea, indica, mellifera, cerana indica

Out of above species in India 3 species of honey bees viz. *A dorsata, A. florea and A. cerana indica occur, whereas A. mellifera* is distributed in Europe and Western countries. The *A. mellifera which originally occurred* in natural abodes in Europe was subsequently taken by man to various countries like America, Australia etc.

Apis dorsata (Rock bee)

It is the largest Indian species and commonly called giant or rock bee. It is distributed to Indo-Malayan region : to the west, occurs not further than the Indus river, avoiding the Xerotherm coasts of the Persian Gulf. To, the East, however, all the Philippine islands are included, which even crosses the Wallace line as far as the Kei Islands east of Timor. The giant bee is reported to occur in altitudes up to 1000m to 1500-1700m or, during migration, even upto 2000m in different regions. It seems to be common in areas where food and nesting sites are available.

Fig. 3.1 : Bee-tree in Srilanka with ten clearly visible dorsata **nests.**

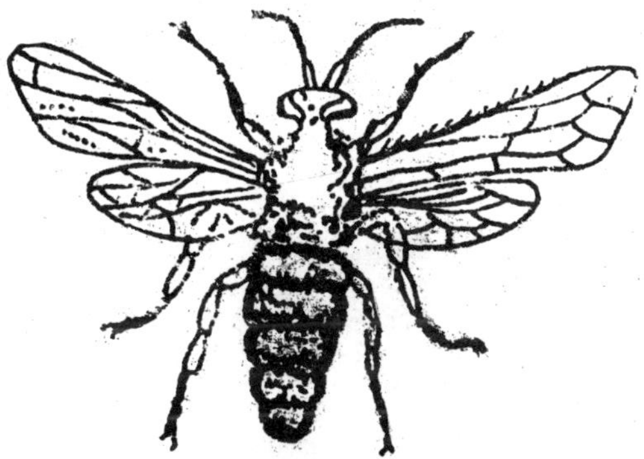

Fig. 3.2 : ***A. dorsata.***

Each colony of this bee is in the form of a single huge open comb on faces of overhanging rocks or under branches of large trees. The bees are very good honey gatherers, a single comb yielding about 25kg of honey and about 1-2 kg of beeswax. On account of their open air habit, their migratory habit (nomadie) and due to their irritability (attacking behaviour) disturbed rock bees cannot be kept in hives and cannot be reared successfully. Then also, some beekeepers are trying to keep them in movable frame hives.

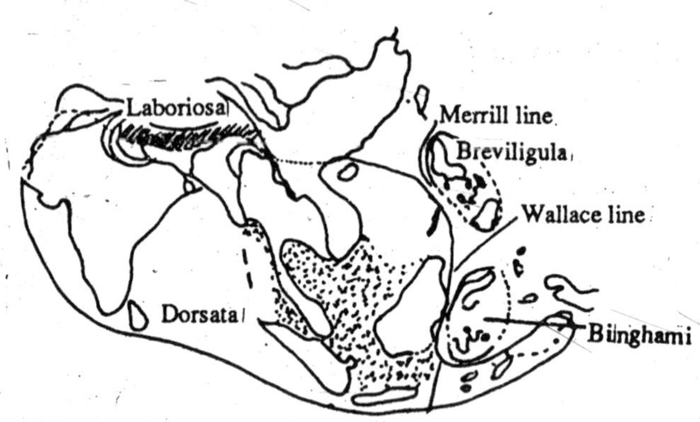

Fig. 3.3 : Distribution of *A. dorsata* in S. Asia.

Apis Florea (Garden bee)

The little bee, builds isolated small combs in open situations (branches of tree, walls and ceilings). These bees migrate frequently. Honey is stored in the top portion of the comb which may become 5 to 6.27 cm thick at the end of the honey flow season. The queen is golden brown in colour, drones are black with smoky grey hairs. These are poor

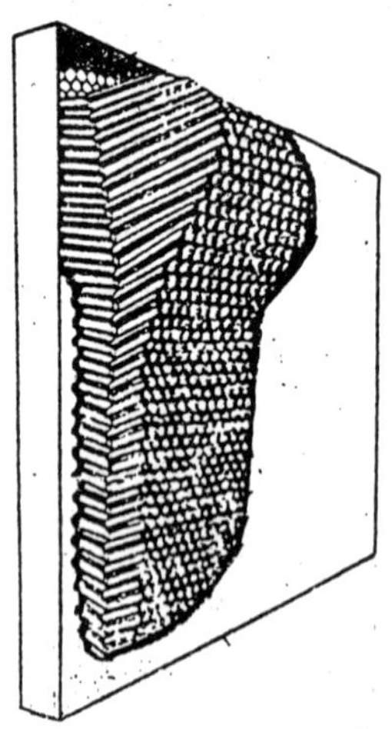

Fig. 3.4 : Florea comb attached to a vertical wall.

gatherers of honey, gathering about 1-2 pounds of honey *i.e.,* it is not economically important species but honey is medicinally used.

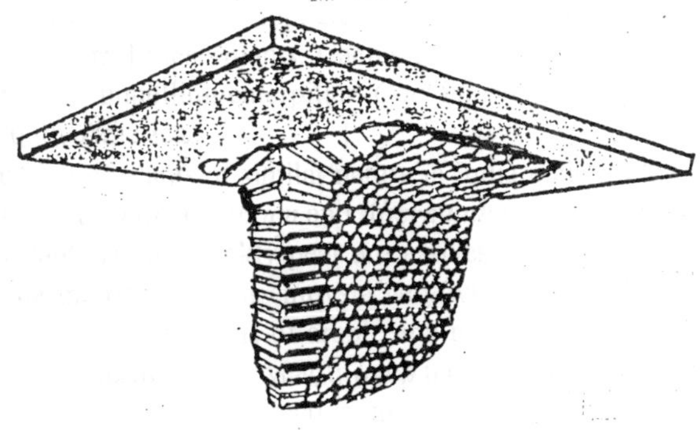

Fig. 3.5 : *Florea* comb attached to a horizontal ceiling.

Fig. 3.6 : *Apis florea.*

Distribution

A. florea is the honeybee of the low lands of South Asia. The main habitat of the species is Pakistan, India, Sri Lanka, Thailand, Indochina, Malaysia, parts of Indonesia and Palawan in altitude upto 500 m. In India it is restricted to Bengal, Assam, Madras, Malabar and Central India.

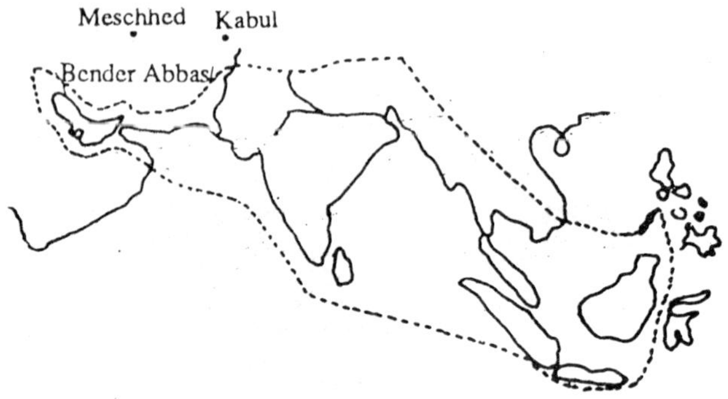

Fig. 3.7 : Distribution map of *A. florea*.

Apis mellifera (Domesticated bee) (Fig. 3.8)

Apis mellifera is similar in its habits to *A. indica*. Presently this bee has been successfully introduced in India. It makes its nest in the concealed spaces and build parallel combs. It maintains a prolific queen, swarms less, has gentle temperament, and has good nectar collecting capacity. The bees are well adapted to movable frame hives. A single comb yields 45 to 181.4 kg. The best yield recorded from a colony is 453.5 kg from U. S. A. A beekeeper of South Africa in 1949 reported a yield of 959.1kg from a multi-queen colony.

Fig. 3.8 : *A. mellifera*.

Distribution

It is commonly called as European bee, as it is found in Europe, Africa, U. S. S. R. It is artificially domesticated in America and Australia. Due to systematic breeding, it has developed into number of sub-species. A sub-species *A. mellifera adansonni* is most ferrocious.

Apis cerana indica (Domesticated bee)

It is distributed in China, Japan, Malaya and all over the India; so it is commonly called as Indian honey bee. This is locally named as mahun, darohia or mauna. There are two varieties (a) Hill and (b) Plain type. Hill type is more prevalent in Jammu and Kashmir and found at higher altitudes upto the height of 45,000 feet. Worker bees are larger and darker in colour. The plain type is seen in Kanya Kumari and other regions of India. Worker bees are comparatively smaller and appear yellow coloured. The Hill type variety makes 22 worker brood cells to 4 linear inches and plain type makes 24 and 25 worker brood cells to measure 4 linear inches.

Both these types make series of parallel combs. They construct their combs parallel to the direction of the entrance in plains and right angles to their entrance in the high altitudes. The modification is seen in hind-legs of worker to form pollen basket to collect the pollen. These are mild in their nature and therefore easy to handle. They are having tendency to heavy swarming, absconding, robbing and developing large number of laying workers. These bees are most productive as they are good gatherers of nectar. A colony yields about 3.55 to 4.5 kgs annually of honey at higher altitudes and 1.3 to 2.2 kg in the plains.

Fig. 3.9 : ***Apis indica.***

COLONY ORGANISATION

Biologists use the word *social* in a variety of ways. Ecologists may speak of a society consisting of all the organisms of a certain area of

the social relationships between members of a sexual pair. All these meanings, however, indicate not only interactions among individuals of a group, but interactions that produce effects qualitatively different from the mere summation of the independent activities of the individuals. If a number of organisms are close together, yet do not influence one another, one may speak of the group as an aggregation but not as a society. Social organisms, live as groups of adults of different generations, with cooperative activity and with different individuals performing different functions necessary for the group.

The honeybee colony presents a very well stabilised system of living, which has given an incentive to the human society to organise their living. A colony of bees is a small city in itself. The bee colony is polymorphic, comprising their varieties or castes : *queen, workers* and *drones*.

The Queen (Fertile female)

Queen is the only perfectly developed female and is the mother of entire colony. She is largest in size and can be easily recognised. Normally there is only one queen in each colony. Her sole function is to produce and lay eggs for future generation. She is unable to produce wax or to gather pollen or nectar. The queen is raised in a special big cell and is fed with a special food known as 'Royal jelly'. A queen can lay about 1500 eggs daily and as many as a million eggs in her life-span of 3-4 years. She mates on wing, with the drone bee after 2-3 days of her emergence. The mother queen can lay both fertilised and unfertilised eggs—the fertilised eggs result in the development of workers and queen depending upon the type of food the developing larvae are fed and the unfertilised eggs develop into males. Additionally, queen provides the cohesive force which keeps the thousands of workers members of the colony together as a social unit. This cohesive property of the queen is due to production of queen substance secreted from mandibular glands. This queen substance contain a scent of fatty and g-oxodecenoic acid which attracts workers only over a very short distance. During food-sharing 'queen substance' is probably transmitted from bee to bee. Therefore the interchange of food is one of the factor responsible for the maintenance of colony cohesion.

Fig. 3.10 : Colony members of a hive.

The Drones (Fertile Males)

They are considerably larger and stouter than the workers and have greatly enlarged eyes which cover most of the surface of the head. The tongue is not developed for gathering nectar, although, a drone can feed himself on honey from the cells. Drones never do any work in the hive and lack the wax-producing glands and the pollen-collecting apparatus. The biological function of drones is to fertilise queens. Drones take about 24 days to develop from the eggs. Drones have extremely good vision, which is most important during the nuptial flight, when they have to follow the rapidly flying queen. They appear most plentifully in the early summer at swarming time. As soon as the fertilised queen starts laying eggs the drones are either driven out or killed by the workers as their services are no longer required in the colony.

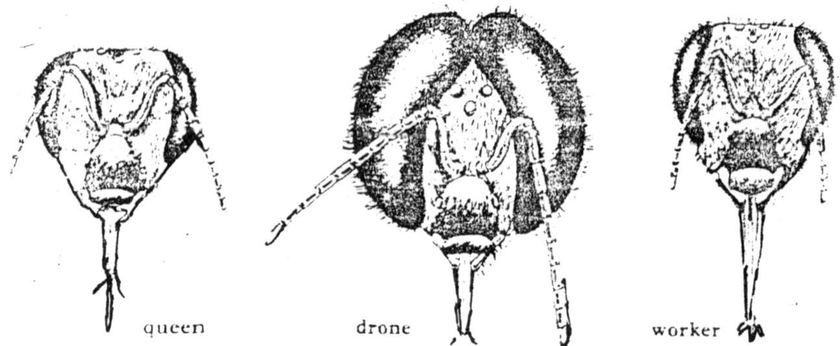

Fig. 3.11 : Anatomical differences between the queen, worker and drone.

The Workers (Imperfectly Developed Females)

These are most numerous and small bees but with well- developed and powerful wings. The workers are sterile females unable to reproduce but possess all the maternal instincts. They are raised in cells known as worker's cells. It takes about 21 days for the development of a worker from egg to adult. Workers practically sacrifice themselves for the wellbeing of the community *i.e.,* perform various duties, for that they possess number of modified special organs like pollen baskets, wax glands, scent glands etc. During the first half of the life of the worker, she performs indoor duties like secretion of royal jelly, nourishment of brood feeding the queen and attending to her, construction of honey comb by secreting wax, comb cleaning, ventilating or cooling the hive and guarding the hive, evaporating nectar and honey storage.

The second half life period of worker is engaged in outdoor duties like collection of nectar, pollen, propolis and water *i.e.,* foraging duty.

The bees manage their affairs on a more sound and organised basis. They have the perfect co-operation and unity of action. Their scheme of the division of labour is the best in the world. Every worker does her job and does it without being directed by any superior, for there is no unemployment nor there is any under-employment or old-age pension. The population of the bee colony is well and carefully regulated to the season and amount of work to be done. This means of course a perfect scheme of birth control. When a bad weather or season comes in, the bee colony stops raising more babies. When there is shortage of food or there is starvation the control bees will dump the half-grown babies (*i.e.,* larvae) out of the colony. If any one of the full grown youngsters is crippled, sick or not fully developed, they too meet the same fate. If they don't go out voluntarily they are kicked out. This is quite natural because to feed the unborn and those that cannot work might mean that all will starve. Everyone works, except the father (drone) and even he is ruthlessly turned out when his services are no more needed. Queen lays all the eggs and if she falls down on the job, she too is kicked out and another queen takes her place. Their slogan is "efficiency and work. If you do not work or cannot keep up your end of job in that bee colony you shall not eat." Therefore the lame and inefficiency die so that others may live. This is the political economy of the bee colony.

Bee-hive or Comb

The comb of honey bee is composed of hexagonal cells built up horizontally on each side from a mid-rib or septum. The comb is made up of bee wax secreted by the worker bees. Generally honey is stored in the upper most cells, next pollen and further down are the worker

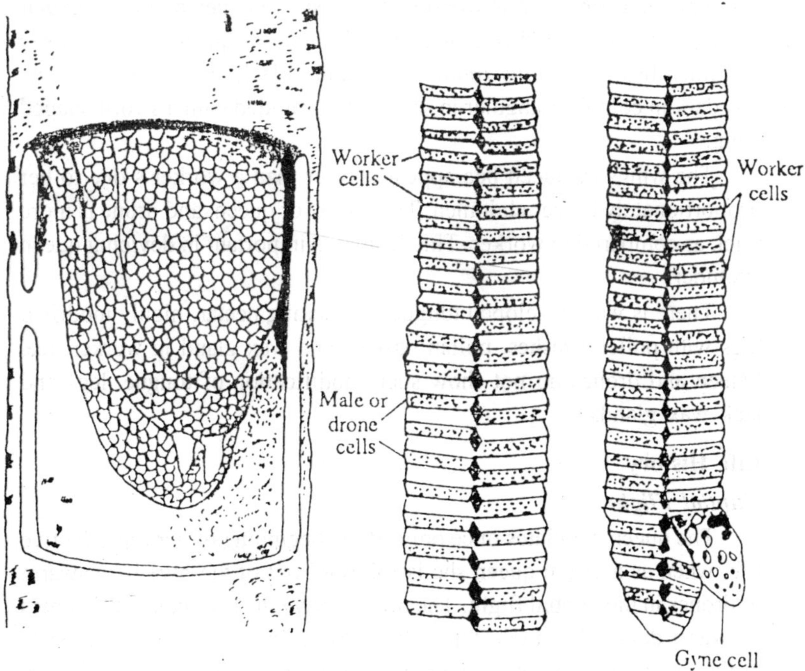

Fig. 3.12 : Nest of the common honey bee, *Apis mellifera.*

brood cells followed by drone brood and queen cells which are built along the lower edge of the comb. Number of cells in a honey bee nest varies greatly but there are commonly many thousands, perhaps 1,000,00 in a strong colony of *A. mellifera.*

Caste Differentiation

Eggs of different castes *i.e.,* queen, worker and drone appear similar in size and appearance. The eggs are laid in the brood cells provided for rearing the different castes. Similarly the grubs of all castes look similar but the queen and drone grubs are larger than the worker grubs during the later half of their development. The pupae of drone, worker and queen can be distinguished by examining their eyes. The eyes meet over the head in drone but in worker they are far apart. In addition to this, in brood comb the worker pupa has a flat capping with a dull and dry appearance. The drone pupa has a convex capping with a hole in the centre. During honey flow period the cap may be covered by a further layer of wax. Drone has black rectangular abdomen with a blunt free end without a sting. While worker has striped abdomen and is triangular in shape and has a barbed sting. The abdomen of queen of *Apis florea* is golden brown, in *Apis India* it is shining black and in *Apis dorsata* it is dark in colour, triangular in shape and provided with a sword-shaped sting.

The wings of queen bee appear shorter than a worker's due to her proportionately longer abdomen. The thorax of queen, and drone is more thick as compared to worker. This feature is important in devicing queen excluder.

There is well developed tongue for sucking nectar, highly modified legs with pollen brushes, pollen baskets and wax gland in worker bee. Queen and drones do not show such modification of tongue, legs and lack wax glands.

Life History

Nuptial Flight

The queen copulates once only, at the beginning of her reproductive life, when as a virgin queen she is followed out of the hive by a swarm of drones in the nuptial flight. The drones leave the hive in large numbers on warm days. The single function of drones is to fertilise the queens, of their own, or some other, colony. Fertilisation takes place only in the air. The male is always killed in the act of fertilisation, since he can eject the sperm only by generating great pressure in his abdomen with aid of muscles and fluid pressure of blood.

Development and Caste Determination

The young fertilised queen begins to oviposit three or four days after mating. The eggs are placed in the bottom of newly cleaned and polished cells and hatch in 3 days into minute white, legless grubs, or larvae. Fertilised eggs are laid in worker or queen cells while unfertilised eggs are laid in drone cells. Workers maintain a constant temperature of 35°C in the colony. This they do either by crowding or fanning the cells. Caste determination in honey bees is dependent not only on fertilisation or otherwise of the egg (hence chromosome number) but also on the diet which the grub receives. All grubs receive the same food for the first two days, a substance known as "royal jelly", which is secreted by pharyngeal glands opening into the mouths of the workers. It is not produced by the queen or by drones. The grubs destined to produce queens receive this food throughout their life in abundant quantity. After the third day, the food given to the developing workers and drones however contains an increasing proportion of honey. The royal jelly is reputedly rich in vitamin B and proteins and leads to queen development, while worker larvae are fed primarily on food rich in carbohydrates (honey).

The production of a sterile worker caste is an example of intraspecific control and cooperation exerted through polymorphism. The presence or absence of the queen determines both the behaviour and the sexual development of the workers. Loss of the queen leads to changes that tend to increase the production of sexually active females. This control is exerted through the release from the queen of a secretion called *queen substance*. Queen substance inhibits sexual development of workers that ingest it and regulates their building activities.

Cells of mature larvae are capped by the workers, and the larvae then spin their cocoons and pupate. Transformation to the adult takes about 12 days. Thus the development of a worker bee from egg to adult takes 21 days. Drones require 24 days and queens only 16. These times seem to be related both to the richness and quantity of the diet received, and to the relative sizes of the three castes.

The worker bee, on reaching maturity, cuts out the cell cap and crawls out. For a time it is relatively inactive. Then each worker bee undertakes a number of successive activities according to its age, as demonstrated by experiments of Rosch with marked, bees in a glass-walled observation hive. During the active summer months the workers live five or six weeks and the succession of duties is as follows:

Age in days	Duty
0-3	Cleaning brood-cells and keeping brood warm.
3-6	Feeding older grubs.
6-14	Feeding younger grubs and the queen.
14-18	Secreting wax, comb-building, cleansing hive.
18-20	Guarding entrance of hive.
20-40	Orientation and foraging.

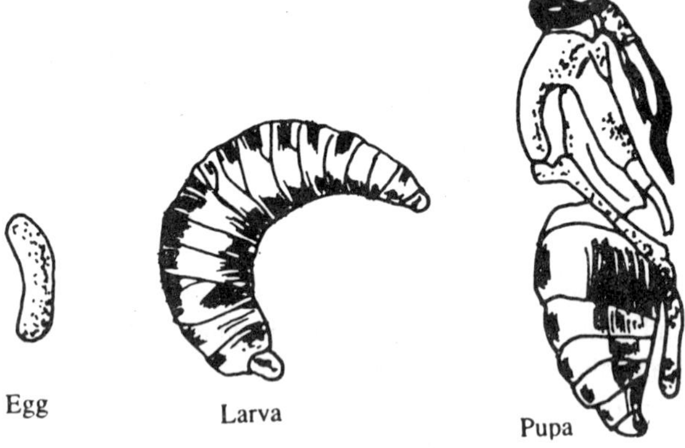

Fig. 3.13 : (a) Life Stages in Apis indica (isolated).

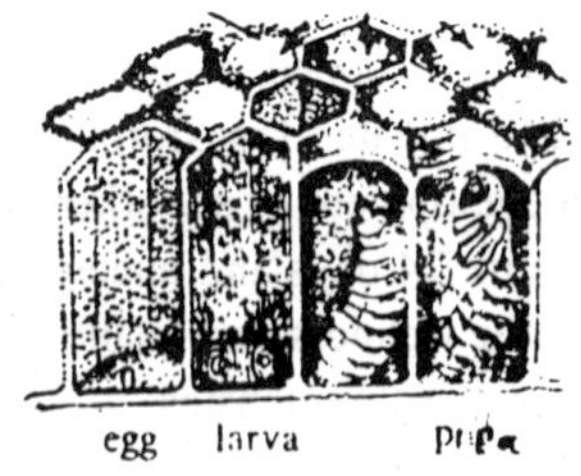

Fig. 3.13 : (b)

From about the 20th day onwards, the worker bee-takes short flights to familiarise itself with the surroundings of the hive. For this purpose they memorise the colours, positions and relative distances of landmarks, and the odours produced by the scent glands of the other bees. Further, bees are able to orient themselves by the sun, using it as a compass to establish their position and having also a sense of time which enables them to allow for the movement of the sun during their absence from the hive. Also their eyes are sensitive to polarised light so that they can orient themselves by a patch of blue sky, even if the sun itself is concealed.

(A)

BEHAVIOUR AND COMMUNICATION

Swarming

In highly eusocial forms like honeybees, if the reproductive rate exceeds the death rate of workers, a colony will grow larger and larger until the stage is reached when, probably due to social factors or to the limited space available in the nest, the Colony divides or multiplies. Swarming is a characteristic of all species of honey bees, since it is the method by which the colonies reproduce. Swarming is a type of movement of a queen and workers that establish a new colony. There are very simple reasons why bees choose to swarm :

1. They are overcrowded.
2. The queen has run out of laying space.
3. As queen gets older and the hive gets bigger, the queen's pheromone does not spread adequately through the hive. The bees does not reach assume they have no queen, and start to build queen cells in preparation for swarming. It is a inherent tendency of honeybees.

In *Apis* the division of the mother colony into two is abrupt and complete. Swarming in temperate climate usually occurs in late spring at a time of plentiful food supplies. This makes biological sense, for the swarm must establish itself, build combs, store food, and rear young workers, all before the onset of dry (flowerless) or cold weather. In warmer places, December to January and in colder regions March to April are the swarming periods. The honeybee swarm, consists not only of a queen but also of a large number of worker bees and few drones as well.

When the "swarming impulse" of a colony of honey bees has been aroused, a sort of communicative behaviours are performed, these are as follows:

1. Reduction in the inhibitory pheromone production by the queen, releasing workers to construct gene (queen) cells.

2. Reduced willingness of house bees to accept food from foragers, presumably stimulating nest searching by the former foragers.
3. Buzzing runs, which say in effect "let's go", when the swarm leaves the parental hive and again when it leaves the place where it has hung awaiting a decision about the new nesting place.
4. *The dances concerning nesting sites.* The first swarm which comes out of a hive is called the "*prime swarm or first swarm.*" Generally old queen accompanies such swarm, unless she dies due to accident. The swarms which are seen emerging after prime swarm are called as "*after swarms*". In all after swarms, virgin queen is present with them with majority of young bees. On the day of swarming queen becomes restless and stops egg laying and roams over the colony. She communicates the message to the whole colony. The bees (workers) accompanying the queen will usually fill themselves with honey. After a short time some bees come out in the air with their heads towards the hive. These bees swiftly fly in and out and thus persuade the other bees of the colony to follow them. Finally a sudden violent agitation commences in the colony resulting in the mass of bees, buzzing and whirling around in a circle restlessly. Then queen joins this group and a swarm leaves the parent colony.

Absconding

Absconding is a behaviour in which there is departure of a whole colony of highly eusocial bees for a new nest site. Absconding is sociophysiologically very like swarming, except that all or nearly all the bees leave instead of some staying behind with a young queen. Absconding is a type of defence to avoid unfavourable conditions of hive such as destruction of the combs by wax moths, termites, ants or other predators and lack of nectar producing flowers. Tropical forms of Apis are quite capable of migrating, and at certain seasons abscond from their nesting sites and establish themselves elsewhere. Thus the African race *A. mellifera adansonii* absconds when there is a shortage of either food or water, and tropical forms of Apis do likewise.

Bee Communication

Communication in bees is nothing but the sending of signals that influence the behaviour, development of others, usually in the same colony. The biological significance of communication with respect to food supplies in highly eusocial bees is considerable.

In *Apis*, communications concerning the distance and direction of food sources is in part by means of the famous dances first described by Karlvon Frisch, whose great work on the subject explains the dance communication in elaborate detail. The mechanism serves not only to alert or recruit potential forages and get these new forages to food supplies that have been discovered by other members of the colony, but also serves to get them there in approximately the right numbers to exploit those supplies. Honeybee dances appear to communicate information for vector orientation and sometimes locality knowledge by odour. Probably all species of *Apis* can increase the attractiveness of food sources to recruits by supplementing the natural floral odours with their own odours. At very large and concentrated food sources, rarely at flowers, workers of *A. mellifera* that have already made one or more trips to the food often expose Nasanov's gland, (an odour or pheromone producing gland occupies approximately the position of a wax gland on the apical abdominal tergum of worker) and liberate its secretion (geraniol, nerolic and geranic acids and citral) present in small amounts but an important part of the attractant. While hovering (lingering near) over the food or usually on alighting beside it, they raise the abdomen and expose the gland openings. Such scent making is especially important for odourless sources of food or for water. This type of communication is called *olfactory communications*.

Dance Communication

Karl von frisch showed that when bees returning to hive with nectar or pollen they quickly convey this message of food source to other bees with the help of special body movements, which is called dancing. The dances are not performed in flight; the bees walk or run over the surfaces of the combs of cells, that is, they dance on vertical surfaces in the dark of the nest, surfaces also supporting thousands of other bees. The dances themselves are small and mostly more or less round in shape.

The Round Dance

This type of dance indicates a food source (nectar) only a short distance away and that contains no distance or direction information. Before the round dance starts, a returned successful forager enters the hive and promptly offers her crop contents (nectar) to other bees waiting about on the combs. Many of these are young workers who accept nectar from returning foragers and either pass it on or process it into honey and place it in cells where it will be stored. These workers are not yet ready to forage. Others, however, are older workers, some of which are in the proper state to forage. Such bees not only accept bits of syrup from the returned forager but follow her on the combs, often being attracted to her by odour. She now dances her dance.

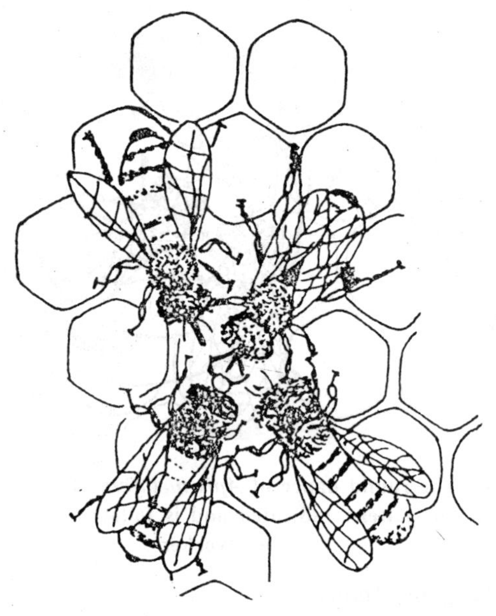

Fig. 3.14 : Liquid transfer in *Apis mellifera*. The returned forager (lower left) is giving nectar to three other workers.

In the round dance the bee runs in a small circle, reversing and going in the opposite direction after every turn or two, or sometimes after less than a full turn. The dance may stop after one or two reversals or may go on through twenty or more, after which it stops abruptly, often to be resumed once or twice more by same bee at the same place or elsewhere in the nest. Young bees seem repelled by the dance and move away, cleaning a little space for it, while some bees of foraging age are attracted, touch the dancer, and may be recruited to the food source about which she is dancing. After a short period of cleaning herself, removing pollen if she carried any, and feeding, the forager then leaves the nest again.

Those bees that were stimulated by the dancer touch her with their antennae, often trailing after her as she dances or cutting across the circle to catch up with her. Recruited foragers leave the dancer, clean themselves, take honey on food needed for the prospective trip, and then, usually within a minute, leave the hive.

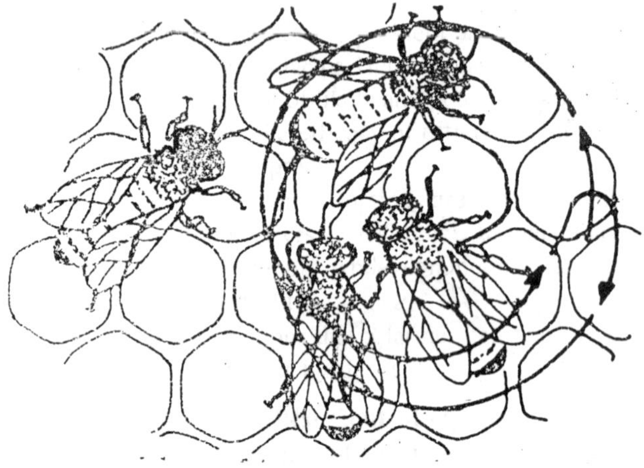

Fig. 3.15 : The round dance of *Apis mellifera*. The upper worker is dancing in the pattern Indicated and is followed and antennated by other workers.

The Tail-Wagging Dance

If the source of food discovered by forager is more than about 30 metres away, it performs a different dance on its return. This is the tail-wagging dance. This dance consists of a short straight run, with the bee turning, to one side and returning by a semicircle to the starting point followed by repetition with the bee ordinarily making the second semicircle on the opposite side of the straight run. Thus, it is a roughly circular dance consisting of two halves. The straight run is emphasised by vigorous shaking of the abdomen from side to side (about 15 wags per second) and usually by a buzzing sound made by the flight musculature and skeleton but without noticeable wing beating. It was found that the further away the source of food, the more slowly the dance was performed, but more wags of the tail were made on the straight part of the run. The tail wagging dance also indicates the direction of the food supply. The signal depends upon the bee's use of the sun as a compass, and of polarized light, if the sun is obscured. If the wagging run is carried out in an upward direction the feeding place is situated towards the sun, while if it is carried out downwards the feeding place is situated away from the sun.

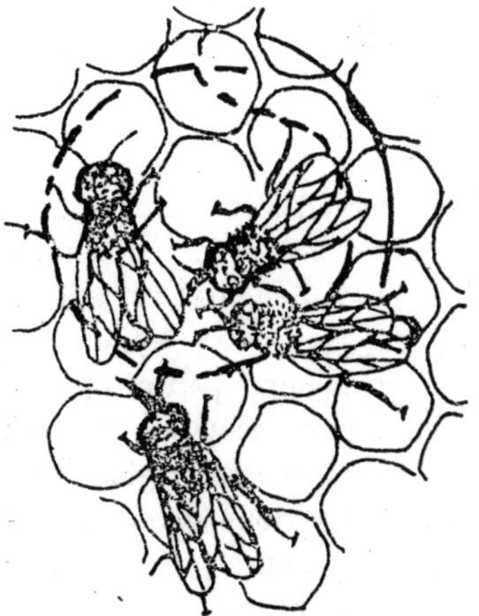

Fig. 3.16 : The tail-wagging dance.

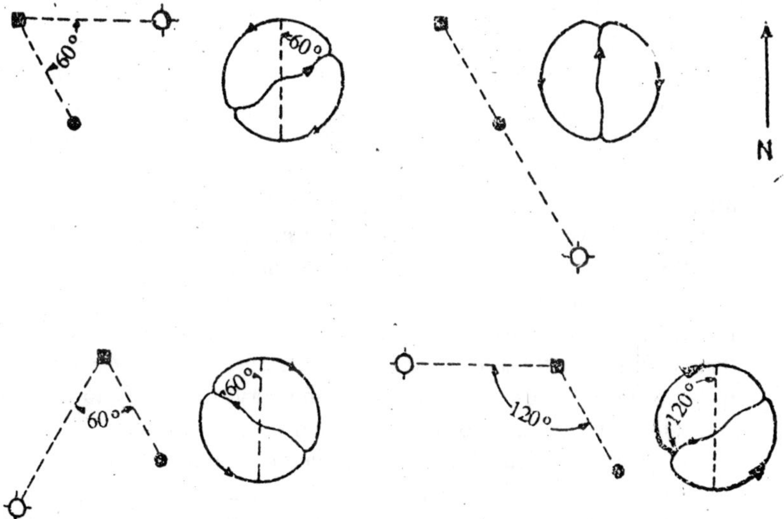

Fig. 3.17 : Diagrams showing the varying orientation of waggle dances of *Apis* workers on a vertical surface at different relative positions of hive (*square*), food sources (*dot*) and sun. The special diagrams can be looked at as viewed from above, with the sun rising in the east (*upper left*), moving across the south (*upper right, lower left*) and setting in the west (*lower right*). In this context, the dances shown illustrate changing dance orientation during one day at a single hive with one food source.

In addition to foraging dances, bee communication behaviour shows many other dances, which are poorly understood. Some of important dances are as follows:

DVAV Dance (Joy dance)

This type of dance is observed under the optimum conditions present in the hive. The DVAV dance is a short form of dorso-ventral abdominal vibration and *Milum* (1955) proposed this terminology. Originally it was named as 'Joy dance' by *Haydak* (1945). While performing this dance, the bee places her front legs on some part of the body of another bee and then five or six shaking movements are made, up and down with the abdomen, with a simultaneous slight swinging forward and backward. This bee then walks further at various locations touching another bee and- performs same dance. Sometimes such type of dance can be seen on a sealed queen cell. After the emergence of virgin queen from queenless colony, many bees perform such type of dance. The DVAV dance is mainly performed by workers doing field duties. Many times DVAV dance, may be intermixed with wag-tail of crescent dances.

Cleaning Dances

This type of dance is stimulated by the presence of particles of dust, hairs or other foreign materials. The cleaning dance consists of a rapid stamping of the legs and a rhythmic swinging of the body to the sides and at the same time bee rapidly raises and lowers the body to cleans around the bases of the wings with the middle pair of legs. Cleaning dance is observed during any time of year including winter. Usually the bee closer to the dancer touches it with its antennae and begins to clean the dancer. The cleaner with it widely spread mandibles touches the thorax of the dancer just under the base of the wings. Dancer bee stops dancing after the touch of cleaner. Then dancer slowly spreads out the wings of one side bends the abdomen and curves the body to the side and somewhat upward so as to accommodate the cleaner. Then the cleaner works energetically to clean around the bases of wings by shear like motions of mandibles. During cleaning cleaner stops from time to time. The cleaner also stands on the last two pairs of legs and the first pair of legs being held in air and works with mandibles as if chewing something. The cleaner continues cleaning with the mandibles over the scutum from the rear to the front, sometimes over the head and in the grooves of thorax. Sometimes the cleaner climbs on the dancer, crawls to the other side and cleans under the opposite pair of wings and then ceases activity. After this cleaning the dancer may clean her tongue, antennae and the body in general. Generally there are ten cleaner bees present at one time on a single comb.

Massage Dance

The bee who requires massaging bends its head in a peculiar way with mandibles wide open and upper of tongue is protruded and dry so as to indicate the other neighbouring bees. One or two neighbouring bees become excited and investigate bee requiring massaging. These bees then climb over and under her, pull the joints of hind, middle and fore legs, mostly touch her sides from below with their antennae,, mandibles; and front legs, cleaning their antennae periodically. The bee need the help turns her head toward the examining bee and the examining bee unfolds the entire tongue, extends the second pair of legs as if sitting on the third pair and constantly cleans the tongue with her, front legs, stroking it from above downward. Sometimes the tongue becomes completely distorted. At the same time messenger bee massage continuously and pulling the sick bee by mandibles or tongue sometimes licking the sick bee. After several minutes the cleaning activity subside and then the sick bee begins to clean normally its antennae, eyes, wings, hind legs and then moves away.

Alarm Dances

Generally the alarm dances are performed by bees when the condition in the colony is critical. Different substances are responsible for arousing alarm dances. Also various disturbances happening to bees as a result of different substances cause arousing of alarm dances. While performing alarm dances, the bees ran in spirals or irregular zig-zags with vigorously shaking their abdomen sidewise. The flight activity of bees is stopped completely and neighbouring bees began to respond to dancers. Schneider (1949) observed such type of alarm dances by keeping sugar solution contamined with dinitrocresol in hive box. After few minutes bees become excited and dances. It was also observed that dancing activity of bees increased with the spread of poison. After one to two hours highest number of bees died. After three to four hours the colony returned to normal and flight activity was resumed again.

Temperature and Humidity Control of Hive

Maintenance of temperature of hive for brood rearing and regulation of hive temperature during summer is essential for normal functioning of colony. This specific function of cooling of hive is assigned to worker bees known as fanners. Fanning of hive is important activity of fanners to control the temperature and humidity. It is also possible that fanning activity of bees may help in controlling the concentration of carbon dioxide. Bees during respiratory activity expire CO_2 and may suffocate themselves if it is not removed efficiently from the hive. During summer bees fan fresh air into the hive at one place and out at another. This ventilation of hive is important as fresh air is circulated in hive and regulates normal hive temperature which is 94°F. In the area where ambient temperature rises upto 115°F or more, under such adverse situation some of the forager bees switch over their duties/and start collecting water from field. These bees collect drop of water and other bees place the drop of water all over the inside of the nest. Fanners fan the air causing water to evaporate, until the hive temperature drops to the normal temperature. In this way in bees efficient biological mechanism operates to regulate the temperature of the hive.

(B)

EXTERNAL CHARACTERS OF BEE

The body of the bee like all insects is divided in to 3 main parts; the *head*, the *thorax* and the *abdomen* and black or brown in colour. All these three body regions are covered by dense felt of hairs which gives them fluffy appearance as most of the hairs are feather like.

Structure of Worker Bee

Head : It is a wide triangular structure with the apex pointed below. It bears dorso-laterally a pair of large *compound eyes* and group of three *ocelli* in the middle of the top. A pair of short but many jointed *antennae* are borne on the middle of the face and probably serve as the tactile and gustatory organs. On the back of the head is present a central opening, the neck foramen through which the cephalic cavity communicates with that of the body. This provides a passage for oesophagus, nerves, blood vessels, tracheae and salivary duct. From the bottom of the head project the specialized mouth-parts.

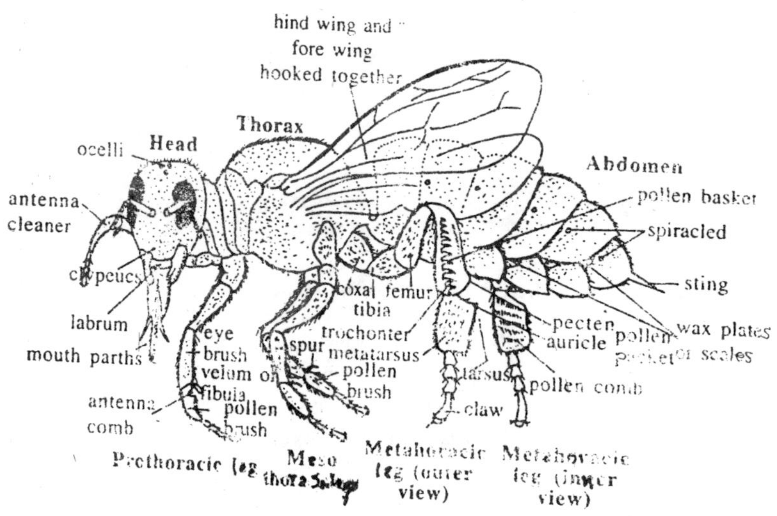

Fig. 3.18 : Lateral view of worker honey-bee.

Mouth-parts

The mouth-part of honey-bee are of rasping and lapping (biting and suçking) adapted for taking nectar from flowers and moulding the wax. Spoon shaped and smooth edged *mandibles, lying* at the sides of *labrum work sidewise* and are used for moulding thc wax and adhering pollen. *Maxillae* lack a lacinia, which the *maxillary pulp* is vestigial and the galeae elongated and blade like. The glossae of the *labium* are greatly elongated forming the *tongue* or *ligula,* which terminates into a *honey-spoon* or *labellum* and carries a midventral groove or *food channel,* overlapped by special hairs so as to form practically a close tube. The labial palps are well developed. The galea, labium and labial palps form a sucking tube.

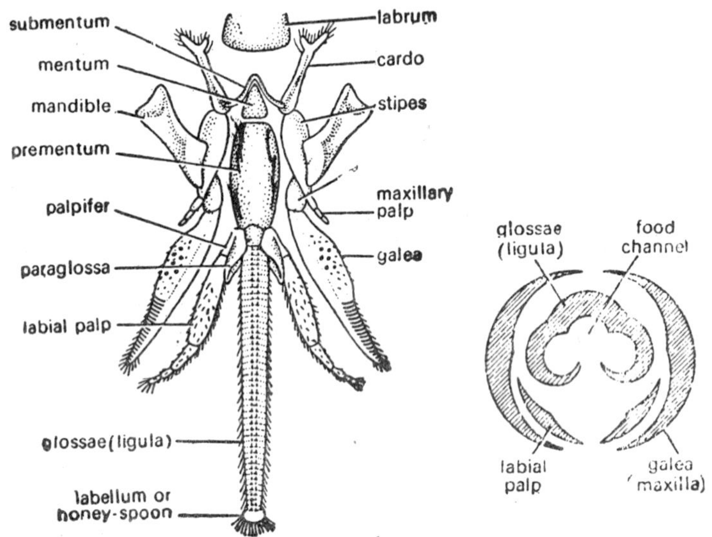

Fig. 3.19 : (a) Honey bee, Mouthparts. (b) Honey bee T.S. Mouthparts.

While feeding, the bee extends its ligula or tongue in to the nectary of the flower. The elongated labial palps and galeae make a protective sheath around the tongue. The nectar rises up through the ventral groove, or food channel by capillary action. Moreover, the shortening of the tongue and sucking action of the pharynx also help in sucking up the liquid food.

Thorax

It is divided into the usual three sagments anterior *prothorax, middle mesothorax* and posterior *metathorax*—each bearing a pair of legs. The mesothorax and metathorax also carry each a pair of wings.

***(a)* Legs:** The three pairs of legs are densely covered with hairs which aid in gathering pollen and are variously adapted. Each leg consists of five parts—an oblong coxa, a short trochanter, femur, tibia and a five—jointed tarsus, terminating in a pulvillus and a pair of claws. The sticky secretion of pulvillus makes it possible for the bee to cling to smooth objects. Tactile hairs are also present.

(*i*) ***Prothoracic legs:*** In each prothoracic or foreleg, the inner distal part of tibia bears bristles forming an eye brush for removing pollen and other particles from the surface of the compound eyes. The distal posterior end of tibia also gives out a plate-like process or velum, fitting over a semi-circular notch in the upper end of the first tarsal segment or metatarsus. The notch bears stiff bristles forming an antenna comb. The velum and the antenna comb together make the antenna cleaner, which serves to clean the antennae drawn in between them. On its posterior face the metatarsal segment also bears bristles forming a pollen brush.

(*ii*) ***Mesothoracic legs:*** The inner distal end of tibia of each of the middle or mesothoracic legs bears a long spine like pollen spur. The spurs are used to remove pollen from pollen baskets and to dislodge wax from wax pockets on the ventral surface of the abdomen and transferring it for comb-building.

(*iii*) ***Metathoracic legs:*** In each metathoracic or hind leg, the outer surface of tibia carries a depression, called the pollen basket, which is partially covered by rows of long curved bristles arising from its margin. Distally the tibia terminates into a row of stiff spines forming a pecten. The proximal upper end of the metatarsus bears a concave plate or lip, the auricle. The auricle and the pecten, working together, form a pollen packer. The inner surface of the metatarsus bears a series of transverse rows of hard bristles forming the pollen combs. While flying from flower to flower, the worker bee gathers pollen on its body hairs which are combed by the pollen—combs. The pectens scrape the pollen from the combs, while the auricles push up the pollen mass into the pollen baskets of the opposite legs. When the baskets are full the worker bee returns to the hive, where pollen is removed by the spurs of the metathoracic legs to be stored in the hive cells. The pollen combs also remove wax scales from ventral abdominal gland cells and transfer them to the mandibles.

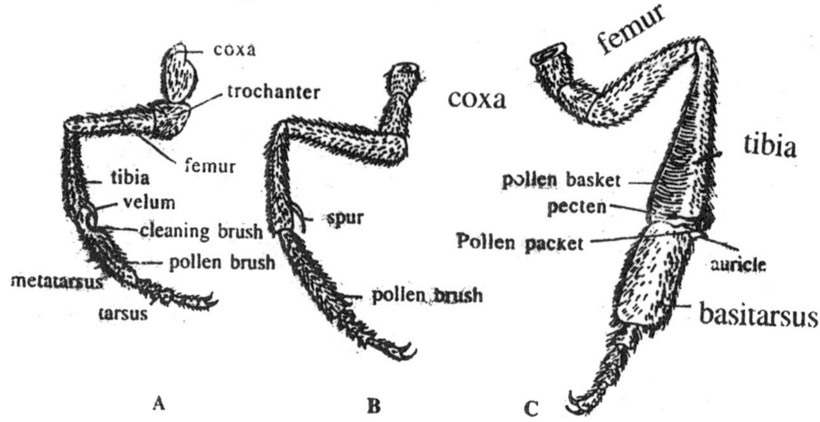

Fig. 3.20 : Legs of honeybee. A-prothoracic legs, B-Mesothoracic leg, C- Metathoracic leg.

(*b*) **Wings:** The two pairs of wings are narrow, membranous and transparent and moved by strong muscles of the thoracic wall. On each

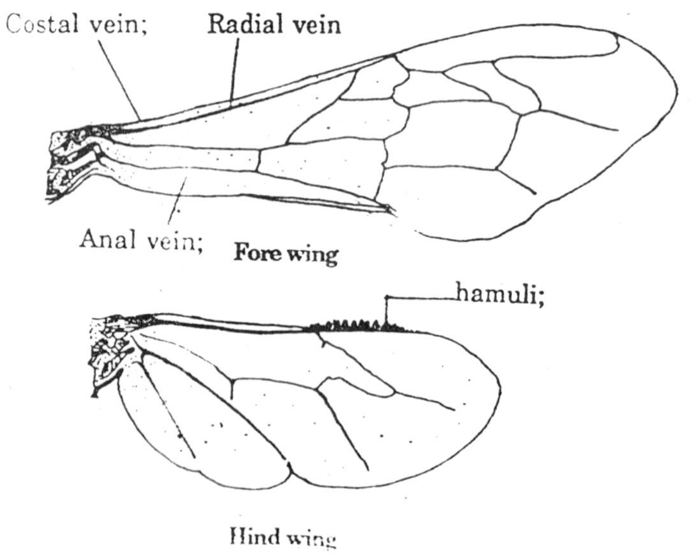

Fig. 3.21 : Wings of worker bee.

side the two wings are hinged together by a series of minute hooks, called *hamuli*, so that they work together during flight. The hamuli can be detached and reattached instantaneously. They were nature's forerunners of modern man's zippers. Wing venation is reduced. The wings vibrate over 400 limes per second during flight.

Abdomen: It consists of six visible segments and bears the wax-glands and the sting.

(*a*) ***Wax-glands.*** These are modified cells on the ventral surface of last four visible segments of abdomen. The wax is secreted through minute pores, forming scales. The wax is masticated by the mandibles before its use for building the cells of the honey comb.

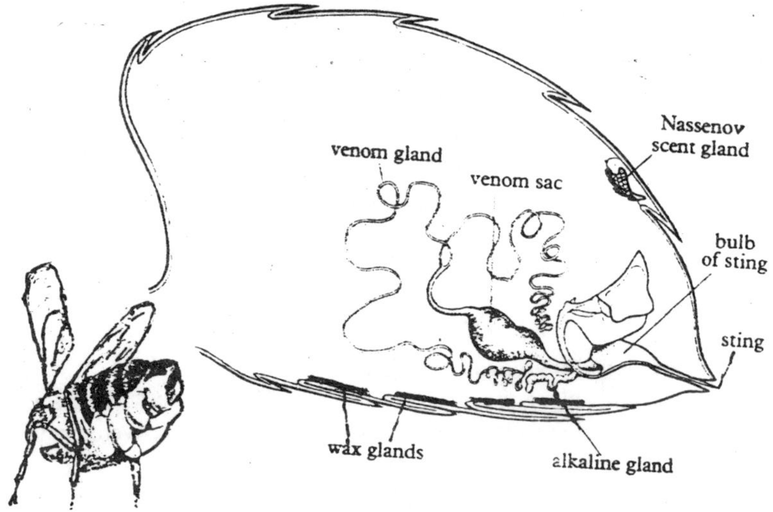

Fig. 3.22 : Worker bee secreting wax (left) from glands on the underside of the abdomen. These and other abdominal glands are shown above.

(*b*) ***Sting:*** The sting is the modified ovipositor and therefore is found only in the queen bee and the genetically female worker bees. This explains why a sting is absent in the drones. The ovipositor has lost its original functions to become modified into a poison injecting apparatus. Bees can be credited with developing one of the first hypodermic needles. The actual sting or terebra encloses a central poison canal bounded by three pieces, a dorsal stylet sheath and two ventral darts, stylets or lancets. The stylets

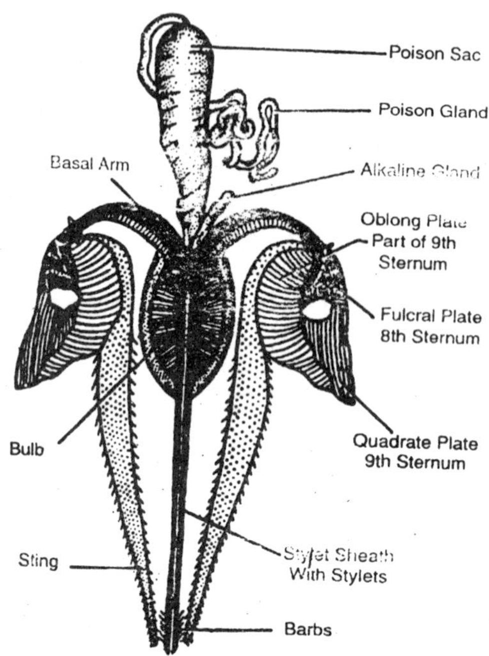

Fig. 3.23 : Sting of honeybee.

are dorsally grooved throughout their length, so as to fit into and slide along two ventral guide rails of the sheath. The distal free tapering ends of the stylets and the sheath are covered with many anteriorly pointed spines or barbs, similar to those on fish-hooks by which the sting is held firmly in the flesh of the victim. The muscles associated with the sting help in the operation of the stylets. Basally the sheath expands to form the bulb, while the sting is continued into a pair of lateral diverging arms. The sting is provided with a set of three plates on either side, serving as levers. The innermost or the oblong plate proximally articulates with the basal arm of the sting, and distally carries a palp-like appendage. A small triangular or fulcral plate is attached anteriorly with the basal sting-arm ventrally with oblong plate and postero-dorsally with a large outer or quadrate plate. Two glands are found associated with the sting. A filiform poison gland discharges its acidic secretion into a large poison-sac, which opens into the anterior end of the bulb. A small alkaline gland also opens separately into the bulb. While stinging, the sheath makes the wound, the barbs hold the sting in the puncture, while alternating stabs by the stylets bury the sting deeper into the flesh. Poison is pumped out of the poison-sac and through the hollow sting into the wound. After stinging the barbs prevent the removal of the sting from the wound so that it remains in the wound along with the glands, sac and the torn posterior end of the abdomen, causing the death of the worker bee. The sting of the queen bee is longer and less barbed. It is used only in stinging rival queens and can be withdrawn from the victim without injury to her own body. It can be used many times. On the other hand the worker bees use their sting only as a last resort; because his always results in their death.

Queen

1. ***Structure of queen:*** The queen is the only fertile female in a hive, having immensely developed ovaries. She is elongated, 15-20 mm long, and is easily distinguished by her longer tapering abdomen and short legs and wings. She has pointed mandibles and shorter mouthparts and sting. She can sting repeatedly as there are no barbs on her sting. She is unable to produce wax and honey or gather pollen or nectar. She is a specialized and degenerate individual with a small brain and without silvery glands.

Drones

2. ***Structure of Drones:*** Flying idly near the hive in the sunshine can be seen the sexual male bees or drones. There are usually 200 drones in a typical colony, depending upon the season of the year. They are intermediate in size, 15-17 mm long but considerably stouter and broader. They possess very large eyes, small pointed mandibles and lack wax producing glands, pollen—collecting apparatus and a sting.

ANATOMY OF BEE—DIGESTIVE SYSTEM

Digestive system consists of a well developed alimentary canal and its associated glands. The alimentary canal is long coiled tube which starts from a terminal mouth and ends postriorly by anus.

Mouth

It is present in the lower wall of the head and opens into the cavity of the *sucking pump* standing vertically in the head.

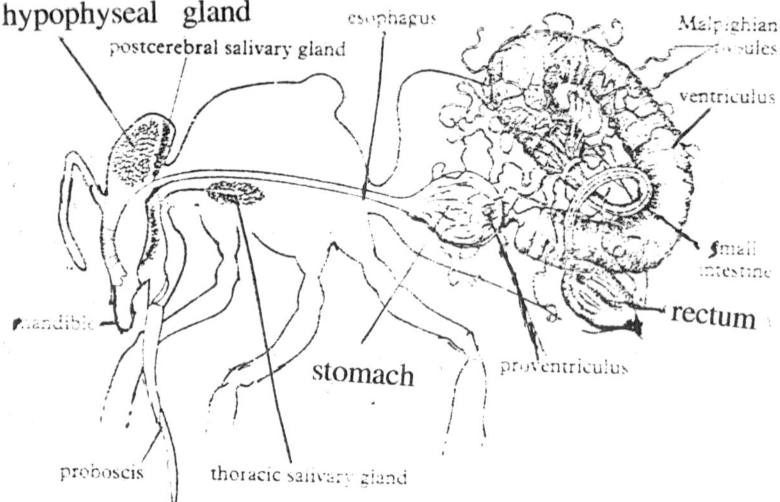

Fig. 3.24 : The alimentary canal and associated glands of the worker.

Sucking Pump

It is large sac with its muscular wall lying in the head. It extends from the mouth to the neck formen, where its narrowed upper end becomes continuous with the gullet. Each lateral wall of the pump is traversed obliquely by a slender rod, extended upward from a plate on the floor of the mouth. The preoral cavity before hypopharynx is the *cibarium. The* sucking pump in honey bee is a combination of the preoral cibarium and the post oral pharynx. The oral plate, the bib like fold hanging from the pharynx represents the hypopharynx and the infolding ending at the opening of the salivary duct into the so called *salivary syring.* The mouth of the bee is the opening into the cibarium between the labrum and the hypopharynx. Five pairs of thick bundles of dilator muscle fibres arising from the clypeus are being attached to the anterior wall of cibarium which is the operative part of the pump. Strong compressor fibres running obliquely cross wise on the cibarial wall are present between these muscles. The liquid food like nectar and honey are sucked up from the canal of the proboscis by the action of the dilator muscles. Construction of the compressor muscles closes the mouth and drives the liquid into the muscular pharynx, from which it is driven into the narrow gullet. Nectar regurgitation and processing are important function of the feeding pump of bee.

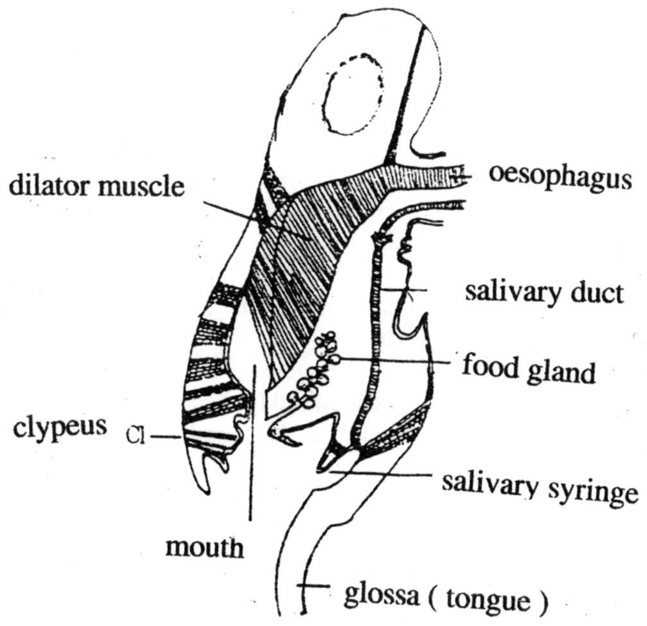

Fig. 3.25 : Sucking pump of honey bee.

The sucking pump narrows at its upper end to form the slender, tubular *gullet or oesophagus,* which has muscular wall and passes the food along its length by successive waves of contractions.

Oesophagus traverses through the neck, thorax and then in the anterior portion of the abdomen where it enlarges into a thin walled sac called *honey stomach,* which corresponds to crop of other insects. It serves for carrying nectar and also as a storage place for ingested food. The inner wall of honey stomach is thrown into several folds and there by making it greatly distensible. Honey stomach is followed by a short narrow part of the digestive canal the *proventriculus*. It regulates the entrance of food in to the stomach. Its anterior end projects like a thick plug into the honey stomach and contains 'X' shaped opening between four thick, bristly, triangular, muscular lips. This structuure forms a stomach mouth and by its action nectar or honey can be retained in honey stomach but pollen is delivered to the next part *i.e., ventriculus.* The posterior end of the proventriculus extends in a long funnel like fold into the anterior end of ventriculus and mostly acts as a valve to prevent regurgitation from stomach.

Next to proventriculus is a long thick, cylindrical sac which is looped crosswise in the abdomen which is *true stomach* or *ventriculus.* It has its inner wall in the from of thick epithelial cellular layer thrown into several crosswise folds, which increase digestive and absorptive surface area and also allow its expansion outside the epithelial layer is a layer of circular muscle fibres and outer to it is a sheath of longitudinal muscle fibres. The epithelial layer secretes digestive juices and enzymes.

The intestine is distinctly divided into two regains, the anterior is a narrow part of intestine and the posterior large pear shaped *rectum.* The anterior intestine is looped or coiled in various ways according to the position of the stomach. The rectum opens by its tapering extremity through the anus into the cavity containing sting. A long thread like tubes called malpighian tubules open at the junction of intestine and the ventriculus, which are the organs of excretion which separate nitrogenous waste substances and salts from haemolymph in which they float and finally discharge them into the intestinal lumen.

ASSOCIATED GLANDS

Salivary Glands

Saliva in honey bees is secreted by two pairs of glands and discharge their secretion finally into the common median duct. First pair of salivary glands lie in the back of the head, while the other pair in the ventral part of the thorax. The pair of glands located in the head are referred as head glands (cephalic glands). These glands are in the form of flat masses of small pear-shaped bodies spread over the posterior head wall. Their ducts unite within the head with the common duct of the thoracic glands.

Thoracic glands : The second pair of salivary glands are thoracic glands which develop from the silk glands of the larva correspond with the usual salivary glands of the other insects. The thoracic glands consists of masses of elongate or tubular saccules and the end of branching duct that lead into a pair of reservoir sacs. A pair of ducts run forward from the reservoirs and unite just behind the head in the common median duct. The common duct then enters the neck foramen of the head and empties into the salivary syringe. A deep depression present between the root of the tongue and distal end of prementum are mostly concealed by the overlapping paraglossae. At the bottom of this depression is an opening that leads into a small pocket of prementum. The wall of the pocket is provided with dilator and compressor muscles and in its inner end opens the common duct of salivary glands. It is a sort of pump for the ejection of the saliva, called salivary syringe.

The saliva is ejected by salivary syringe into the cavity on the labium at the root of the tongue, it is here confined by the overlapping paraglossae. Within the latter is conveyed around the base of the tongue into the channel like groove on the posterior or undersurface of the tongue. The saliva is conducted probably through this channel to the tip of the tongue where it flows out over the smooth undersurface of the flagellum to mix with the nectar or honey being drawn into the proboscis, or it is used as a solvent when the bee is feeding on sugar.

Brood-Food Glands

The brood-food glands are present in the cephalic region of the worker bee. These glands are two long strings of small saccules closely packed in many loops and coils at the sides of the head. Their ducts called axial ducts open separately by two small pores on the lateral angles of the oral plate on the floor of the mouth. As this plate belongs to the hypopharynx, the food glands are called *hypopharyngeal glands*. These glands produce a secretion which is a rich food material called 'royal jelly'. The royal jelly is fed to the queen, drones and worker larvae. The rod like arms of the oral plate give attachment to muscles from the head wall and probably have some function in discharging the royal jelly from the mouth. The food runs down the bib like flap that hangs from the edge of the oral plate and accumulate in the food channel on the base of the proboscis. This open food channel function as a feeding through for other adult bees. They obtain the royal jelly by thrusting the end of the proboscis over the base of the tongue of the feeder bee, the proboscis of the latter (feeder) being turned back, the mandibles opened and the labrum raised. When the nurse bees feed the larvae, the royal jelly is said to be discharged from the partly opened mandibles Fig. 3.24.

Abdominal Glands

It will be seen in Fig. 3.22, four pairs of wax glands are situated upon the underside of the worker's abdomen on the anterior part of the last five sagments each gland being covered by the overlapping part of the segment ahead. Wax is secreted into these packets as a fluid which rapidly solidifies to a small translucent white cake, probably by chemical action rather than by evaporation.

On the upperside of the abdomen, on the front of the last visible segment (segment 7) is a gland called the *Nassenov gland.* This gland produces a scent which, when the gland is exposed and air is fanned over it by the wings, spreads out from the bee as a rallying "call" to other bees. It is used to help collect stragglers when there is a disturbance in the colony, and also at times to mark forage, mainly where a scent is absent in the forage itself. The scent is not peculiar to the colony but is the same for all colonies.

Finally, there are the two glands associated with the sting in abdomen, the long thin, bifurcated, acid or venom gland produces the venom which it empties into the venom sac where it is stored until required, and the short, stout alkaline gland is usually considered to produce a lubricant for the sting mechanism.

Respiratory System

In bees the gaseous exchange for releasing the energy consists of two organs (a) *Spiracles* (b) *Trachae;* which constitute the spiraculo-tracheal system. In adult honey bee there are 3 pairs of thoracic and 7 pairs of abdominal spiracles are present. The first pair of spiracles are largest in size and lie between the prothorax and mesothorax beneath the lateral lobe of the pronotum. They are further covered by a dense fringe of long hairs. The second pair of spiracles are minute and lie between the upper angles of the pleural plates of the mesothorax and metathorax and are normally concealed between pleural plates. While the third pair is fully exposed on the side of the propodeum. The six pairs of abdominal spiracles are situated in the lower parts of the first six tergal plates of the abdomen. The seventh spiracles are not visible externally as it is in the so called spiracular plate associated with the base of the sting. All the spiracles, except second spiracle, have a regulatory apparatus for regulating the spiracular orifice during respiratory activity.

In honey bee, the tracheal tubes are more or less rigid due to spiral intima that serve to keep them open. Hence they are not affected to the pressure changes around them. Consequently in actively flying insect like bees tracheae are dilated to form thin walled balloon like structures, air sacs which expand and collapse in the manner of lungs. Honey bee has large air sacs occupying much of the space in the sides of the fore part of abdomen and extend in the metathorax. Smaller air sacs are also present in the head and in the legs. From air sacs trachae continue as branching tubes that ramify in all parts of the body appendages and internal organs. The ultimate branches of trachae terminate in groups of minute fine tubes called tracheoles and invade tissue cells. The distal parts of the tracheoles are filled with a liquid which absorbs oxygen from the inspired air. The carbon dioxide produced during cellular metabolism is mostly discharged directly into the surrounding haemolymph from which it is carried off by the tracheae and finally expelled through spiracles to the exterior. (Fig. 3.26)

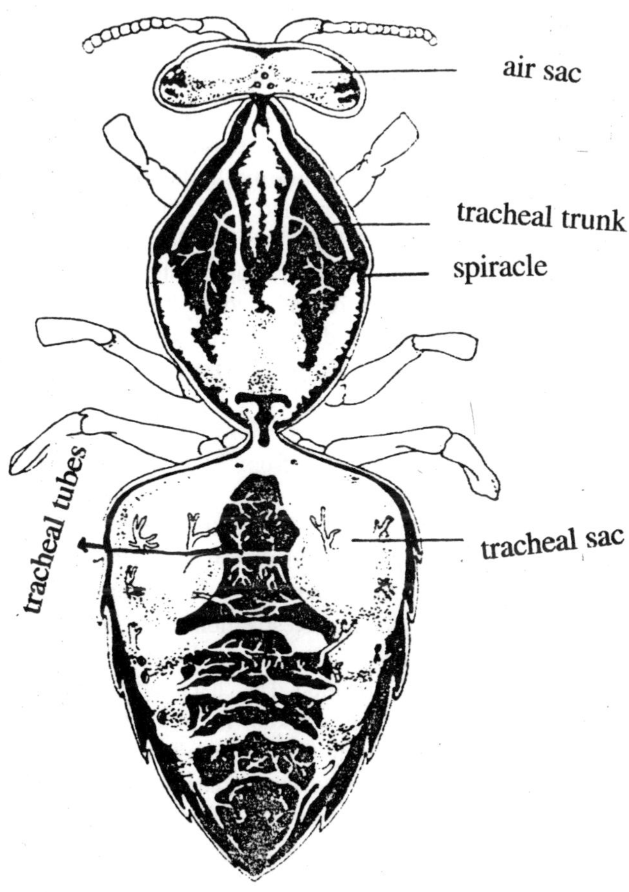

Fig. 3.26 : Spiraculo-tracheal system.

Respiration in the honey bee is effected partly by dorsoventral and lengthwise contractions and expansions of the abdomen and are initiated by the opposing sets of abdominal muscles. Inhalation of air is effected by the expansion of air sacs. According to new concept the spiraculo-tracheal system helps in thermo-regulation particularly in actively flying insects. The air circulation through the tracheal network of various heat generating body compartment brings about thermal homeostasis.

CIRCULATORY SYSTEM

Haemolymph

Like other insects, bees also have open circulatory system in which the blood called haemolymph flows through the spaces in the body. The haemolymph of bee is of a pale amber colour. It contains about 90% of water by volume. Dissolved in it are salts of sodium potassium, calcium and magnesium. The anions are mostly chlorides, some carbonates and phosphates. Haemolymph also contains proteins and their concentration are 2.79 per 100 ml. in nurse bee, 2.349 in drones and 1.3 to 1.89 in foragers. The carbohydrate content in haemolymph of worker bee is about 2.19 per 100 ml. and in drones 8.69/100 ml. The estimations of cations like sodium and potassium concentration in haemolymph have shown to be 7.5 m.eq/l and 46.49 m.eq/l respectively. The ratio of these cations is 0.16. It has been shown that these characteristics changed in queenless colony and the protein contents of haemolymph of foragers increased from 1.89 to 2.39 per 100 ml and that of drone reduced from 2.099 to 1.69. The carbohydrates present in the haemolymph are mainly glucose and trehalose. The carbohydrate contents in the haemolymph of the forager of queenless colony increased three fold and reduced by half in the drones. The cations are also reduced by half in the queenless colony. Investigations have indicated that there may be an androgenic low molecular weight proteins present in the drones but yet it needs confirmation. The characteristics of haemolymph of *A. dorsata* and *A. florea* are not known. Similarly nothing is known about the hormone transporting proteins present in the haemolymph. Studies on immunological comparisons of haemolymph proteins of four species revealed interesting interspecific differences in the haemolymph protein spectra and antigenic properties. Trehalose is known to be the major circulating carbohydrate in haemolymph and is supposed to be under hormonal control.

The haemolysin contains numerous floating blood cells called haemocytes. There are several kinds of haemocytes which are concerned with phagocytosis of foreign micro-organisms as well as coagulation of haemolymph. The haemolymph lack the presence of any oxygen transporting protein, however, it carries little oxygen. The major functions of haemolymph are : distribution of absorbed food material, reception of waste products of metabolism and the transport of carbon dioxide to be eliminated through respiratory-system.

Course of Circulation

The haemolymph is circulated through haemocoelomic cavity by organs of circulation. The circulation of haemolymph is brought about by a pulsatile heart situated in the middorsal line of abdomen from segment VI to II. Anteriorly the heart continues in the thorax as aorta and into the head as long slender tube opening beneath the brain. The lateral sides of the heart are perforated by five pairs of slits called ostia. The haemolymph enters the heart through ostia into the heart tube. The lips of ostia project forward into the heart cavity and function as valves to prevent the backward flow of haemolymph. The aorta has no ostia and is thrown into numerous small loops where it enters the thorax and ends openly below the brain. By rhythmic pulsations of muscular walls of heart, it propels the blood forward to aorta and finally in the cephalic region. The body cavity or haemocoelomic cavity is divided into three sinuses by two partitions called diaphragms. The dorsal diaphragm is in the form of thin membrane supporting the heart. The dorsal diaphragm is streched across the upper part of the abdominal cavity in segments III to VII. The membrane contains five pairs fan-shaped bundles of fine muscle fibers attached laterally on the antenor margins of the tergal plates and spreading toward the heart where the fibers break up into numerous branching fibrils. Thus, the dorsal diaphragm shuts off above in to pericardial cavity containing heart. The lateral margins of diaphragm are free between the muscle attachments thus leaving openings by which blood can enter the pericardial cavity from the general body cavity the perivisceral sinus. A rhythmic contraction of the diaphragm muscles causes the dorsal diaphragm to pulsate in a forward direction. Similarly the ventral diaphragm present above the nerve cord in ventral part of the body extends from the metathorax upto segment seventh of the abdomen. The ventral diaphragm is highly muscular than dorsal diaphragm and beats in a backward direction.

During circulation, the haemolymph is discharged into the head from aorta by contraction of heart and bathes the organs present in head and flows backward through the thorax and the abdomen by fairly well defined channels and also circulating through the antennae, wings and the legs. The pulsations of ventral diaphragm assist the backward flow of blood in the abdomen. The haemolymph from the lower part of the abdomen goes upward into the pericardial cavity where it is driven forward along the sides of the heart mostly by the vibrations of dorsal diaphragm and ultimately again enters the heart through the ostia. The small contractile membranes are present in the head both at the internal bases and in the upper part of the thorax. The neighbouring muscles of these membranes appear to produce the movements of these membranes. (Fig. 3.27).

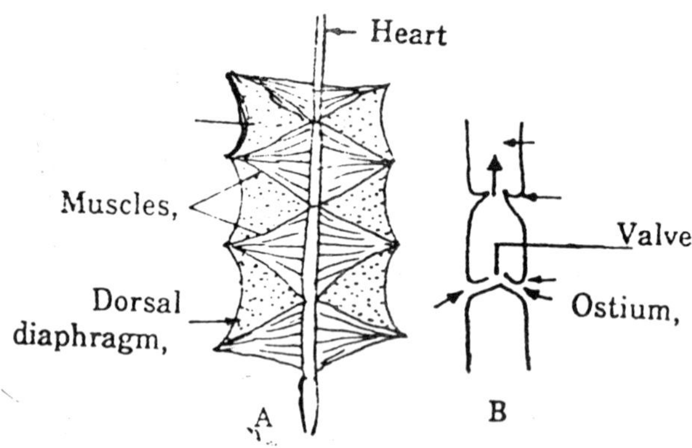

Fig. 3.27 : (A) Heart and diaphragm (B) The course of blood flow.

The body cavity of bees, especially abdomen contains irregular masses of soft, white tissue made up to large, loosely united cells collectively called as fat body. Such fat body cells contain small oily fat droplets in the cytoplasm. The fat tissue is abundant in the larva. The fat tissue of larva .contains large amount of fat, glycogen and protein granules towards the end of larval period. Thus the fat body is a storage tissue for conservation of elaborated food products not immediately required by the individual. Intermingled with fat cells are larger sized pale yellowish oenocytes and nephrocytes present in the form of strands of special cells lying along the sides of the heart. The exact function of these cells is not known.

NERVOUS SYSTEM

It is one of the important body system which is well organised for co-ordination of various life activities in insects. The honey bee has well developed nervous system. The central nervous system of bee consists of a brain or supraoesophageal ganglia, the sub-oesophageal ganglion and the nerve cord. The brain is a major sensory centre which receives the nerves from eyes and antennae. The brain is bilobed and each lobe is called as supraoesophageal ganglion. Both the supraoesophageal ganglia lie side by side and are connected by paired connectives to the suboesophageal ganglion, forming a nerve ring. The suboesphageal ganglion is really formed by fusion of first three ganglion of ventral nerve cord. The brain receives sensory nerves from the eyes and antennae and transmit the nervous impulses from the sensory organs to the motor centres of the ventral nerve cord. Similarly the suboesophageal ganglion supplying nerves to the various components of mouth parts.

The ventral ganglionated nerve cord shows two prominent ganglia in the thoracic region. The first thoracic ganglion is situated in the prothorax and the second, largest in size is situated in the posterior part of the metathorax. This second thoracic ganglion is formed by the fusion of four primary ganglia belonging to mesothorax, metathorax, the propodeum and the first abdominal segment. This ganglion supplies nerves to the muscles, wings, legs, spiracles and other body structures associated with these segments.

In abdominal region, the ventral nerve cord shows five distinct ganglia, out of which first two are displaced forward so that each of it innervates the segment behind it. While the third abdominal ganglion lies in its own segment and gives out its nerves to this segment. The fourth abdominal ganglion lies in sixth segment and supplies nerves to sixth and seventh segment. The last ganglion *i.e.,* the fifth lies in seventh abdominal segment and supplies nerves to eighths, nineth and tenth abdominal segments. (Fig. 3.28).

In drone, the brain is largest as compared to queen and worker. The innervation of 2nd to 6th segments is similar in all the three castes.

In drone nervous system there are six ganglia and one composite ganglion, while in worker there are seven ganglia. Physiologically the nervous system appears to be cholinergic.

The sympathetic and peripheral nervous system are closely associated with central nervous system.

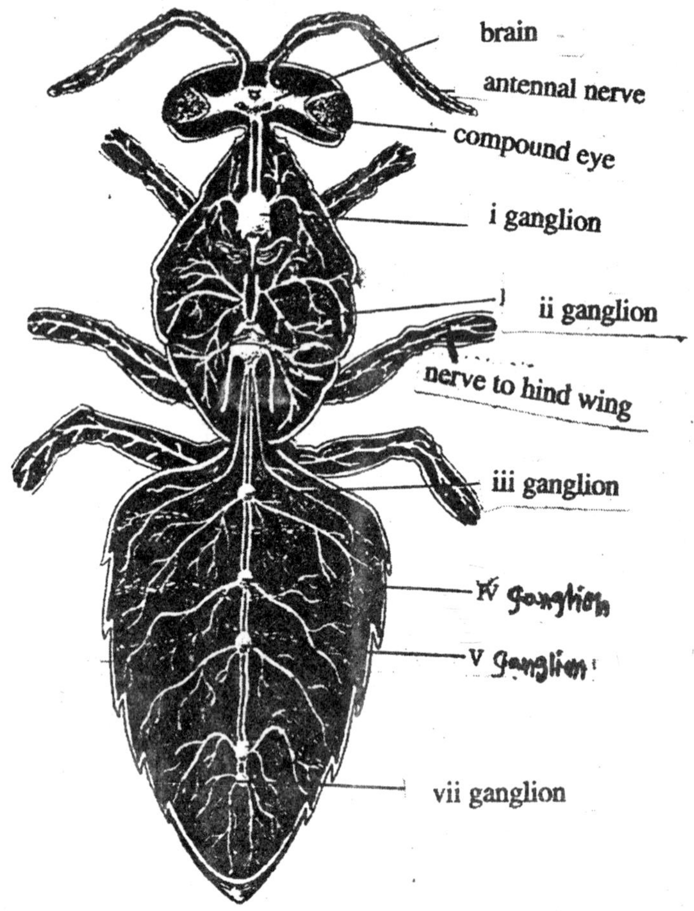

Fig. 3.28 : Nervous system of worker bee.

SENSE ORGANS

The honey bee has well developed sense organs. Reaction to touch or external pressure is probably the most primitive of all the senses. The adult bee has general body surface relatively little sensitive to pressure and touch due to the hardness of its exoskeleton. Hence, most of the sensory nerves of the integument end in cells at the bases of hairs. These delicate hairs are easily moved by contact with objects or air currents. Thus, innervated hair and its associated sense cell form tactile sense organs. The innervated hairs occur on different body parts like body proper, appendages, antennae. There is no definite information on the hearing powers of bees and bee has no known auditory senses.

Bee has thin walled hairs or hair like structures reduced to small pegs and are sensitive to minute particles of matter in air or liquids. Such sensory structures are referred as organs of smell or taste. These organs are innervated by sense cells which send nerve fibers into the hair or peg. The organs of taste or smell are located on antennae and on mouth parts. In addition to this, antennae also have minute oval disks or plates on their surface. Such structures are known as plate organ. Each plate organ has a groove around the margin and covers a large group of sense cells. The plate organs are the main organs of smell in the bee. Honey bee also has sensory organs for stress, strain and bending of integument. These organs are in the form of minute bells or inverted cups sunken in the body wall with nerve ending inside. Such bells or cups are distributed in groups on the various body parts and appendages and are described as 'Olfactory pores'. These sensory organs give the insect in formation concerning its own actions.

Honey bee has photoreceptors in the form of three simple eyes or ocelli located on the top of the head and a pair of large compound eyes. The ocelli or simple eye has one lens for entire retina, while the compound eyes possess many small lenses and the retina divided into parts corresponding with the lenses. The compound eyes form a mosaic image of the object in which each spot image correspond to an ommatidium. Honey bees perceive differences of colour, shape and positron of object. The ocelli keep the insect continually in a state of stimulation to light and thus make the compound eyes more quickly responsive.

REPRODUCTIVE SYSTEM

In a colony of honey bees, there are three castes: viz. queen, drone and worker. The queen being the mother of the colony which has a well developed functional reproductive organs. Similarly, the drones are the functional males with well developed reproductive organs and capable of producing spermatozoa which are being transferred to genital tract of queen during the act of copulation occurring during nuptial flight. The workers are the sterile females posses reduced reproductive Organs. Only under certain circumstances, they become functional and are induced to lay eggs.

Reproductive Organs of Queen Bee

The reproductive system of queen bee consists of paired ovaries, lateral oviducts, common oviduct, vagina and spermatheca (Fig. 3.29). Each ovary is massive, pear shaped structure made up of a slender closely packed tubules, called ovarioles. Each ovariole is narrow at the proximal end containing undifferentiated germ cells and the much enlarged distal end containing matured eggs. Each ovariole has terminal filament in the form of elongated thread. At the posterior end, of each ovary, the ovarioles come together as a lateral oviduct. Both the lateral oviducts run medially and posteriorly and unite to form a short common oviduct.

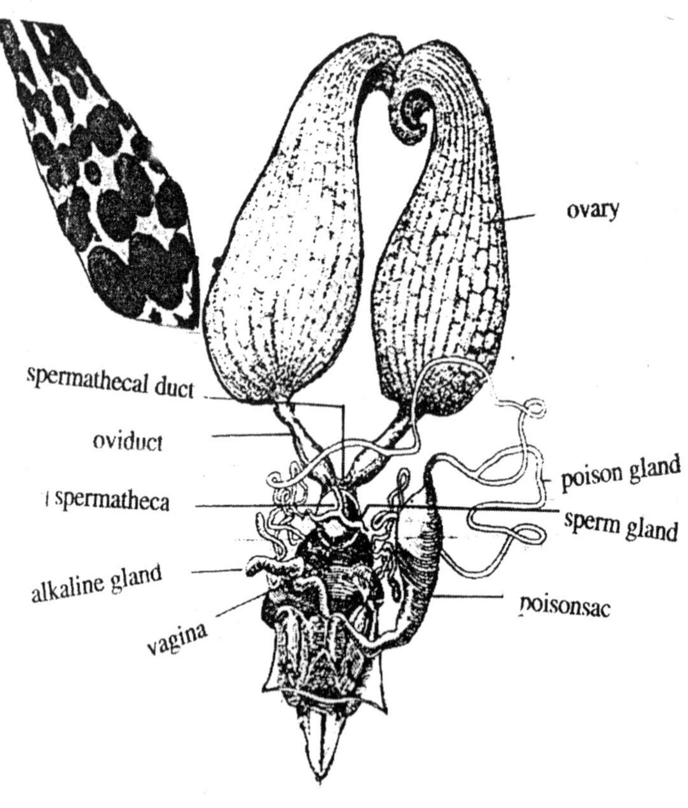

Fig. 3.29 : Reproductive system of queen bee.

The common oviduct is continuous with a wide terminal sac-referred as vagina. It opens to the exterior by a median orifice (vaginal opening) in a depression of the body wall at the base of the sting. At the sides of the genital opening are present two other openings which are the mouths of two large pouches embracing the sides of the vagina. Lying in the dorsal wall of the vagina is a spherical body which is referred as spermatheca or female receptacle for the spermatozoa. The spermatheca has short duct which opens into the dorsal wall of vagina. The paired tubular spermathecal glands open into the distal part of the spermathecal duct.

The reproductive organs of worker bee include paired but reduced ovaries, lateral oviducts, common oviduct, vagina, small sausage shaped spermatheca and terminal sting apparatus. (Fig. 3.30). The ovipositor of the worker bee has lost its original function of egg-laying and has become modified as a weapon of defence, the sting. (Fig. 3.29).

Reproductive Organs of Drone Bee

The reproductive organs consists of paired testes, paired vas differentia, paired seminal vesicles, pair of mucous glands, ejaculatory duct and a large complex penis (Fig. 3.30). The testes are small flattened bodies lying in the sides of the abdomen and they contain primary reproductive cells which latter develop into spermatozoa. From each testis proceeds posteriorly a reproductive duct, the vas deferens. Each vas deferens is coiled at its proximal end and then enlarges into a long slender sac, the seminal vesicle. The spermatozoa formed in testes pass down the vas deferentia into the seminal vesicles where they are temporarily stored with their heads lodged in the soft cellular walls of the vesicles. The posterior end of the two vesicles are narrow and enter the lower ends of a pair of huge mucous glands. Both the mucous glands lie side by side and they open together into a single long outlet called the ejaculatory duct. The ejaculatory duct opens into the anterior end of the large complex structure of penis. At the time of mating, it is everted to discharge the spermatozoa.

The penis is in the form of large complex structure which is differentiated into three parts. The inner half part of the penis is in the form of large pear shaped swelling called bulb. The bulb receives the ejaculatory duct at its anterior end. A pair of dark plates are present in the thick dorsal wall of the bulb. The bulb passes into a narrowed, usually twisted neck or cervix. The cervix has a series of dark cresent-shaped thickenings along its lower side and a fringed lobe projecting from its dorsal wall. The cervix ends in a large thin walled sac called bursa. A pair of crumpled hornlike pouches, the bursal cornua project from its dorsal wall. The bursa opens to the exterior by a wide opening situated beneath the anus and between a pair of small valve like plates. These plates and a still smaller pair at their outer angles are the only representatives of complex, external copulatory structures in the drone. During mating season, the spermatozoa are sent down through the ejaculatory duct in a mucous secretion from the mucous gland which fill the bulb of the penis. (Fig. 3.30).

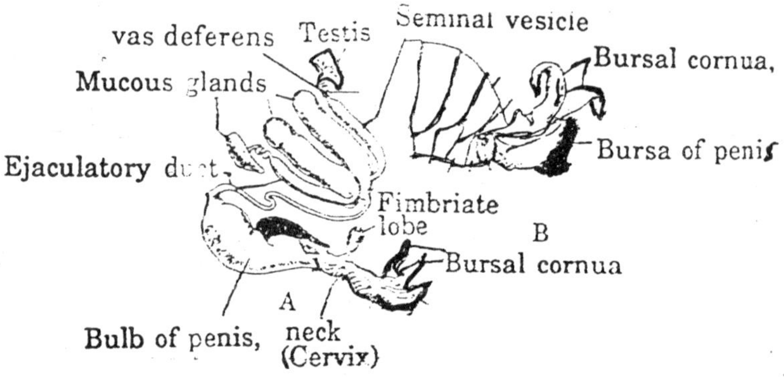

Fig. 3.30 : (A) reproductive system of drone bee (B) Caudal end of abdomen of a drone showing penis.

During mating with the queen it has been supposed that male organ is first anchored by the eversion of its lateral cornua into the lateral genital pouches of the queen. There are many ideas about the eversion of penis. But it is clear that when mating queen finally separates herself from the drone, only the penis bulb remains in her genital tract by tearing at is weakest point between the bulb and the penis cervix. The discharged spermatozoa by drone are first stored in the distended lateral oviducts. After removing the male organ and then mucous from vagina, muscular contraction of female forces the sperms into the vagina. Here in vagina, they are stopped by valve like fold of vaginal wall and cause them to be directed in to the spermathecal duct and then finally into the spermatheca. The vitality of sperms is retained within the spermatheca throughout the reproductive period of the queen bee. The queen mates with several drones in succession during nuptial flight only once in her life time.

(C)

BEE PRODUCTS

The honey bee collects and stores different substances-like pollen, nectar, propolis, water, as well as itself producing wax. Bees utilise these things either in hive making or nutrition, and provide mankind with different types of bee products.

Pollen : Bees may collect nectar, pollen or both. To collect pollen a bee dives down into the flower and covers herself in it. She then hovers while scraping it into the pollen baskets on her back legs, setting up an involuntary fertilisation (pollination) chain when she flies to the next plant. The bee then takes it back into the hive and scrapes it off her legs into a pollen storage cell. Pollen grains are situated round the brood. It is recently known that pollen is necessary for the development of exoskeleton (the outer skin of the bees) of the larvae and adult bees. It also acts as reserve of protein, which is utilised by bees when they cannot find fresh supplies in nature. The chemical composition of pollen varies according to kinds of flowers. The analysis of pollen shows that amount of protein contents varies from 7 to 30%, additionally it also with approximately 5% by weight of pollen consists of fats, oils and waxes and about 20% potassium, and water, enzymes, salts of potassium, phosphorus, calcium, magnesium and iron. A number of vitamins have been found to occur in various proportions in different samples of pollen, but very little is known about the part played by them in honey bee nutrition.

Pollen is widely available in health food shops, usually packed as dried pollen granules. When taken by humans, pollen collected from hives is variously claimed to act as a stimulant for those recovering from illness or surgery; to half impotence and alcoholism; and as a help for anaemia, colitis, and gastritis.

Pollen has been found useful in pernicious anaemia, lowering blood pressure, and increasing hoemoglobin and erythrocyte content of the blood. The discovery of gonadotrophic hormones in the pollen of date palm conferms its use by the bedouins for treating sterlity. Pollen has a beneficial effect on the prostate gland. A pollen preparation (zernilton) is sold in Sweden as a prophylactic for complaints of the prostate and adenomas. Pollen is also used in cosmetic preparations with claims of rejuvenating and nourishing effects for skin. Pollen is biological stimulator and the rejuvenating properties attributed to honey are due to the presence of pollen in it.

Physical Characteristics and Composition of Pollen

Pollen vary greatly in colour, shape and size. Pollen grains range from 6 to 200 microns in diameter. Pollen has very hard outer shell, which cannot be broken or digested. However its germination and exit of most substances takes place through the pores of outer shell. Composition of pollen also varies from species to species. Highest protein content have been reported to be above 40 per cent, but generally ranged from 7 to 30 per cent.

Table 3.5 : Characteristics of pollen.

Water (air dried pollen)	7%
Crude protein	21%
ash	3%
Fat (Ether extract	5%
Carbohydrates:	
Reducing sugars.	26%
Non-reducing sugars.	3%

Besides these, other minor contents include vitamins, organic acids, flavonoids, carotenoids, enzymes, growth factors etc.

Honey : Honey is a sweet sticky fluid made by bees from nectar. This is a typical dictionary definition of honey, and it will probably go on to be equally flattering about the various honey word derivatives; *honeyed*-flattering, coating; *honeymoon*-period of untroubled happiness. The word itself is derived from the Arabic 'han' (that's what they call it), which the Germans changed to 'honig' and English to 'hunig'. Honey is defined in as natural sweet substance, produced by the honey bees from the nectar of blossoms or from the secretions of other parts of the plants, which honey bees collect, convert into honey and store in wax comb to ripen.

Honey has a delightful variety of tastes which are dependent on the nectar sources and the soils of the growing area. The honey produced by a colony at any given time is usually derived from the nectar of only one or few plant species. Bees select the plant species which are with more than 20% sugar content. To change the nectar to honey, the bees convert a complex sugar sucrose to simple sugars fructose and glucose using enzyme produced by special salivary glands.

$$\text{Sucrose} \xrightarrow[\text{Invertase}]{} \text{Levulose (Fructose) and Dextrose (Glucose)}$$

Water is removed from the nectar by evaporation until the sugar content is rich.

Production of Honey

Production of honey is a prime objective of beekeeping industry. Production of honey was almost negligible 40 years ago but has shown good progress during last 40 years.

Table 3.6 : Development of Beekeeping in India

Year	No of been-keepers	No of bee colonies	Honey M.T.	Value of Honey Rs. in lac	Average honey production per colony in kg
1953-54	232	800	128	0.02	1.50
1963-64	57,98	1,64,597	713	37.63	4.33
1973-74	1,50.000	5,22.714	2,435	365.25	4.65
1984-85	2,00.000	8,68.000	5,500	950.00	6.33
1990-91	2,40.000	10,61.000	9,288	2,322.00	8.75
1993-94	2,3.000	6,78.000	5,529	1,382.25	8.15

Table 3.7 : Honey production from different bee species

Apis cerana	–	8-10 kg per colony
Apis mellifera	–	10-15 kg per colony
Apis dorsata	–	10-25 kg per colony
Apis florea	–	500 g-2 kg per colony

Physical Characteristics of Honey

Honey is a super saturated solution of sugars. Most of the physical properties vary as per water content in honey.

Viscosity

Resistance to flow is termed as viscosity. Viscosity depends mainly on the moisture content of honey. Viscosity of honey as high as 600 poise at 11.5 to 13°C reduces to 2.6 at 71.1°C.

Refractive Index

Refractive index of honey greatly depends on moisture content of honey and the refractive indices and specific gravity at different moisture levels are given in the Table 3.8.

Table 3.8 : Specific gravity and refractive indices on honey at different moisture levels (after Wedmore, 1955)

Moisture content %	Refractive index	Specific Gravity	Moisture content %	Refractive index	Specific Gravity
15	1.4992	1.4385	21	1.4840	1.4008
16	1.4966	1.4353	22	1.4815	1.3941
17	1.4940	1.4321	23	1.4789	1.3871
18	1.4915	1.4315	24	1.4763	1.3803
19	1.4890	1.4143	25	1.4717	1.3735
20	1.4865	1.4075			

Density and Specific Gravity

Specific gravity is also termed as relative dersity. This also depends on temperature and moisture like other physical characteristics. Relationship of specific gravity with water content is similar to that with refractive index.

Hygroscopicity

This property means the ability to absorb moisture from the surrounding atmosphere. It increases the moisture content in honey, which may lead to fermentation. Among all the sugars, levulose is most hygroscopic which varies in different honeys. Thus water content increases with increase in relative humidity.

Granulation of Crystallisation

Among the sugars present in honey, dextrose (glucose) content in honey is more liable to crystallise. The crystallization depends on the dextrose content, moisture content, and storage temperature of honey. Raw honey contains crystal producing nuclei including dextrose crystals, wax particles, pollen, dirt and other foreign materials. Granulation is retarted or delayed by liquefying dextrose crystals at 40°C and by eliminating insoluble matter by filtration.

Biological Properties

Honey is biologically active substance and these properties characteristically include fermentation due to presence of yeast and inhibition of bacteria called antibacterial activity.

Yeasts and Fermentation

The sugar tolerant (osmophilic) yeast cells are present in honey. These cells cause fermentation. Fermentation is seen by the presence of air bubbles or foam of CO_2 on the surface. The yeasts reacts with sugar (levulose and dextrose) producing alcohol and CO_2.

Honey fermentation : Alcohol + $CO_2 \rightarrow$ Acetic acid + CO_2. The rate of fermentation increases with increase in moisture content, storage, temperature, and time. Generally all raw honeys should be considered to contain yeasts. Storage below 10-11°C will prevent fermentation. Heating at 63°C for 35 minutes and 67°C for 7.5 minutes destroy feasts completely in Indian honey.

Antibacterial Activity

Honey is known to possess antibacterial property, due to its acidity, hyper-osmotic property, and evolution of H_2O_2 (hydrogen peroxide) by glucose oxidase system in diluted honey. The growth of bacteria is inhibited by these factors and is denoted by inhibition number. Antibacterial activity of honey decreases or is lost on heating and long storage at high temperature. Various pharmaceuticals and food preservative applications are attributed to honey due to its antibacterial property.

THE COMPOSITION OF HONEY

Moisture Content

Generally dandies honey contains more than 20% moisture. The keeping quality, particularly granulation and fermentation of honey, mainly depend upon the moisture content. Hence great care is necessary to measure the moisture and decide the mode of processing for quality control of honey.

Sugars of Honey

Total dissolved solids in honey are as high as 85% and almost all are carbohydrates or sugars. The simple monosaccharides are the building blocks of more complex type sugars. Dextrose (Glucose) and levulose (Fructose) are monosaccharides ($C_6 H_{12}O_6$) and their proportions vary in different honeys. Fructose is generally predominant and tastes sweeter than glucose. Sucrose is most common disaccharide ($C_{12}H_{22}O_{11}$) and is 2-5% in honey. High sucrose is considered to be due to adultration. Additionally, other carbohydrates called "honey dextrins", analogous to starch are also present with the use of new instrumental analytical methods for separation and identification revealed several sugars in honey, including maltose, nigerose, kojibiose, ketose, raffinose etc. Presence of over 19 minor sugars in honey have been confirmed by using high performance liquid chromatography (HPCL).

Acids

Honey is known to be acidic due to the presence of formic-, acetic-, butyric-, malic- and succinic-acids. Most predominant is gluconic acid which is derived from dextrose. pH of Indian honeys ranged from 3.25 to 4.50. Acidity of honey contributes to the taste, but is not evident to great sweetness of honey. Honey gives sour or acidic taste due to increased acidity on long storage or on fermentation.

Minerals

Minerals in honey vary widely and are grouped as ash. Generally dark honeys contain more minerals than light honeys. These minerals include K, Na, Ca, mg, fe, cu, Zn, phosphorus etc. Honey contains much less sodium than potassium, hence it is suggested to use the Na/k ratio to detect honey adulteration with high fructose corn syrup.

Enzymes

Enzymes are most important part of honey as these are added during the inversion of nectar into honey by the bees. The most important enzymes reported in Indian honeys are is follows.

Invertase : Nector sucrose is converted into glucose and fructose by invertase, added by the bee and the conversion continues even after extraction.

Diastase : It is also called amylase which destroys starch when added by the bee during ripening. Diastase varies greatly in honeys and is found to be naturally low in the Indian *A. Cerana* honeys.

Glucose Oxidase : During ripening of nector, bees add glucose-oxidase which oxidises small amounts of glucose to glucone, lactone and to gluconic acid. Gluconic acid is the principal acid, and the acidity thus formed contributes to the stability of the ripening nectar against fermentation. In the reaction one molecule of hydrogen-peroxide (H_2O_2)

this produced for each molecule of glucose oxidised. Ripening of nector is also stabilized by this H_2O_2 and helps in preservation of honey. It is shown that most of the antibacterial activity in honey is due to the production of H_2O_2.

Catalase : It decomposes the H_2O_2 in honey. Other enzymes in honey included acid phosphatase which removes phosphate from organic phosphates. Alkaline phosphate in honey was also reported.

Proteins and amino acids

Proteins are found in small quantities in different honeys. Proteins originate mostly from bees and from the forage plants. Among Indian honeys contained an 1-20% proteins. Amino acids in honey comprise of large part of non-protein nitrogen. White (1975) have reported 18 amino acids in honey.

Vitamins

Very low amounts of vitamins have hardly any nutritional significance in honey. Various vitamins found in honey are riboflavin, pentothenic acid, Niacin, Thiamine, Pyridoxin, Ascorbic acid.

Volatile Constituents

Attributes to the colour, flavour and taste. The most important content is hydroxy methyl furfural (HMF). Colour varies very light to dark, depends upon mineral contents. Colour is also related to flavour and taste. Many light honeys are of mild flavour and dark honeys have more pronounced taste. Aroma or flavour depends upon its plant source. Some of the honeys are known for its fine flavour like litchi, plectranthus, etc. some have characteristic aroma like jamun, sunflower, mustard etc. The taste of honey is attributed to the sugars, gluconic acid and proline, besides volatile constituents.

Hydroxy-Methyl-Furfural (HMF)

Hmf has been considered as an indicator of the heat exposure to honey and possible adulteration with artificial sugar syrup that invariably contain Hmf. In honey, fructose in presence of acids produce Hmf. This reaction starts right from the ripening of nectar in the hive. The rate of production of Hmf increases with temperature. Fresh honey from the hive contains no or very low Hmf and 30 year old honey contained 200 mg/kg. In general tropical honeys like India contain high content of Hmf.

Patterns of Honey Production

Most of the beekeepers in India are illiterate and untrained. They maintain very small number of bee colonies for honey production. However, the trend is changing and large number of commercial beekeepers have come up, especially, with *A. mellifera*. Bee colonies of *Apis mellifera* are a real boost to the honey production in Punjab. Beekeeping with *Apis mellifera* has also gained good popularity in Bihar, West Bengal, Uttar Pradesh and Jammu and Kashmir. Other features of honey production patterns are as follows:

- Honey is collected by traditional honey hunters or tribals. Generally bees are killed by fire. Comb is cut and squeezed along with pollen, brood, wax etc. Such honey is highly contaminated and gets spoiled due-to fermentation.
- Indigenous *Apis cerana* bees are kept m suitable boxes (hives). Honey is collected by extraction with the help of centrifugal machine. This can be none repeatedly without disturbing the colonies.
- *Apis florea*, small wild species is of little use for commercial production, since it is very low yielder. Some may collect honey, though small quantity for its medicinal value as per ancient literature. However, recent reports show that as much as 300-400 tons of honey is collected from *Apis florea* in Gujarat by the State forest Corporation.
- Honeys are classified mainly as per their floral origin. If collected from single type, unifloral such as litchi, *Juman,* mustard etc. or mixed floral-multifloral.
- Honeys colleted during particular season of the year are named accordingly such as spring—basant, Jamun blend or Greeshama or even winter or *kartika* honey.
- Honey is named according to the place where honey is produced such as Mahabaleshwar honey in Maharaahtra, Coorg honey of Karnataka, Kashmir honey of J & K, English honey of United Kingdom and so on.
- Honey collected from the combs by centrifugal machine is generally termed as "apiary honey" or "extracted honey". Honey obtained by pressing or 'squeezing' of combs is called "pressed" or "squeezed' honey which is generally applied to wild or forest honey of *Apis dorsata* or *Apis florea.*

- Comb honey may be available in pieces of the original comb, cut and wrapped (cut comb) or pieces of comb in glass jar of liquid honey (chunk honey). Very small frames filled with honey by the bees or sealed honey in comb is termed as "section comb". In India, however, this type of technique is still not prevalent.

Thus different patterns of honey production in India needs considerable improvement, commercialization and mechanisation at different levels of operation.

Constituents

An average sample of honey would contain 38% fructose (also known as levulose), 17% water, 31% glucose (also known as dextrose) and 2% sucrose plus minute amounts of other sugars, oils and enzymes (diastase, invertase, saccharase, catalase, peroxidase, and lipase). It is predominantly energy-giving carbohydrate food. The amount of protein is very small, and if the honey has been filtered to give a clear appearance the protein may be reduced to nil. Natural honey also contains small amounts of vitamins—thiamin, ascorbic acid, riboflavin, pantygiothenic acid, rydoxine and niacin, filtering may remove them also, and overheating the honey for any reason will destroy them. Other constituents of honey are salts of Ca, Na, K, Mg, Fe, Cl_2, PO_4, S and I, additionally also contains organic acids (Malic acid, citric, tartaric, oxalic, formic, gluconic, lactic and succinic acid). Many aldehydes, alcohols, and esters contribute to the colour of honey.

Uses of Honey

Honey has a high calorific value; 1 kg of honey contains about 3150 - 3350 calories. It is used as a food for newborn kids, the aged. Dextrose present in honey is quickly absorbed in the blood stream, so acts as immediate source of energy, so it is liked by athletes after hard exercise. Honey in combination with other foodstuffs like curd or butter acts as a best delicious dish. It can be used on fruits, cereals, in salad dressings, canning and preserving and for flavouring meats and vegetables.

Honey is used as a important ingredient in Ayurvedic medicines, which acts as a laxative, blood purifier, preventing against cold, cough, fever (honitus, madhuvaani etc.); and a curative against sore eyes (Netranjan), for ulcers of tongue, sore throat and burns. Its regular use is recommended in severe cases of malnutrition with heart attack, impaired digestion. It is also useful for diabetic and allergic patients.

There are endless recipes which involve the use of honey which is used as a more distinctive sweetener than sugar. Honey is used in manufacturing different kinds of meal. Mead is one of the oldest drinks known to man. Mead contains about 12-14% alcohol and should have at least 2 years to mature; in fact the longer the better. Mead is basically made from yeast, water and honey, while interesting variations include Melomels (fermented honey and fruit juices) and Metheglin (fermented honey with herbs and spices).

Honey is also used in preparation of chocolates, confectioneries, biscuits, cakes etc.

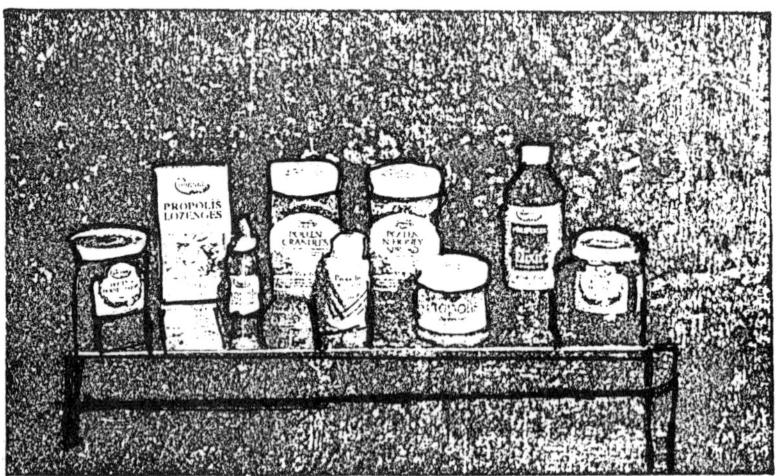

Fig. 3.31 : Some Comvita honeybee health products.

Consumption of Honey

All over the world, except India and developing countries honey is consumed as food (table honey) or in food products. But in India honey is being consumed as medicine only. Although honey is produced in almost every-country in the world, its consumption is restricted only to the affluent societies. Such countries even import honey to fulfil their demand (mainly in Europe and U.S.A.). In recent years, Japan and Saudi-Arabia have increased their consumption and import of honey to a great extent. Almost 80-90 per cent honey is consumed directly as table honey in the neighbouring countries like Pakistan, Napal, Bhutan, Bangladesh, China.

In India, wild honey is mainly consumed in rural and semi-urban areas in unpacked form. However, 75-80 per cent of apiary honey is consumed in urban areas. Out of total estimated 10 per cent consumption of honey in industrial sector, the Ayurvedic and other pharmaceutical preparations consume large quantity of honey. It is estimated that the rate of consumption of honey in India was only 8.40g as against 200g per capita of the world consumption. Table 3.9 given below summarises the consumption of honey in different countries as compared to India which is negligible. Thus, there is a great scope to increase our per capita consumption which will not only take off malnutrition but also of other health problems.

Table 3.9 : Use of honey in different countries.

Item	India	China	USA	Europe	USSR (former	Japan
(a) Domestic consumption % a As food	—	38	55	60	65	35
(b) As household Medicine	90	47	—	—	—	—
Industrial consumption (%)	10*	5	45#	40	35	65@
Per capita consumption (g)	8.4	120	540-1200	640	340	—

Bee Wax

Besides honey, bees produce large quantity of wax. Bees wax is a true wax, secreted by the four pairs of wax glands on the ventral side of the abdomen of worker bees of about two weeks of age, and it is synthesized from reducing sugars of alimentary origin. It is generally secreted in the form of wax scales, and it has been estimated that about 1.5 million wax scales are required to produce one kg. of wax. Amount of honey consumed in producing a kg of wax varies somewhat among calories, the average being 8-40kg. The pure wax, as found in the scales that bees secrete, is invariably white, irrespective of whether the bees had fed on sugar syrup or dark honey. Yellow hues in the wax combs are caused by the fat-soluble carotenoid pigments that originate from pollen.

Extraction of Wax

Wax taken from the hive is full of all sorts of bits and pieces which have to be removed before you have pure wax. The easiest way to do this is with a solar extractor. Patent designs can be brought in bee shops, or you can fairly easily make your own.

Extraction

A solar extractor is basically an insulated container with a double glazed lid. The old cappings, foundation, scrapings, old frames, etc. can be piled in on top of a metal grid over a dark coloured tray. The extractor is put out on a sunny day facing directly at the suns and with lid closed the inside temperature soon soars past the 145°/147°F degree stage at which wax melts. The melted wax run down the tray, where it is filtered by fine mesh and drips into a container leaving the residue behind. When you have finished, the wax block should be stored for future use. Wax can become brittle in extreme cold, so a moderately warm room is best.

In a crude way, wax can be extracted by scrapping the comb and dipped in water for hours and then boiled for few minutes. Wax melts and molten wax is seen at the top. In third method the comb is scrapped out, taken in a simple cloth bag and then it is put in water and boiled for few minutes. The melted wax sips through cloth bag and floats on water. After cooling it forms flakes and then are collected.

Physico-Chemical Characters

Beewax is yellowish to grey brown, solid. It has an agreeable odour. It is brittle, insoluble in water, but completely soluble in ether, chloroform. Melting point is 63.5°C to 65°C. Specific, gravity is 1. It has about 15 separate chemical components, and contains 70.4 to 74.7% complex esters of monoatomic alcohols (melissyl or mericyl alcohol, ceryl alcohol etc.) and fatty acids, 13.5 to 15% free acids (cerotinic, melissic, the oleric group etc.) and 12.5 to 15.5% saturated hydrocarbons. In addition it contains colouring matter and fragrant substances giving it colour and also pleasant smell.

Various plant growth substances such as myricil alcohol, N-triacontanol, gibberlin $4A_3$ and rape oil steroid have been detected in and isolated from bee wax.

Table 3.10 : Composition of *A. mellifera wax.*

Component	% fraction	No. of compounds	
	Major	Minor	
Hydrocarbon	14	10	66
Monoesters	35	10	10
Diesters	14	6	24
Triesters	3	5	20
Hydroxy-monoester	4	6	20
Hydroxy-polyester	8	5	20
Acid esters	1	7	20
Acid polyesters	2	5	20
Free acids	12	8	10
Free alcohols	1	5	—
Unidentified	6	7	—

Wax Products and Uses

Beewax is utilised in many industries to prepare polishes to shoes, furnitures, waterproofing, cosmetics like cold creams, cleansing creams etc. It is also useful in engineering industries, railway transport, pharmaceutical and confectionary industries.

Wax is also useful in medicines as adhesive, in preparation of soaps, plasters and ointments. Beewax is rich in vitamins and vitamined sweets can be prepared that retain their value for several months. It is also used in horticulture for grafting. Beewax candles are the finest candles. Not only do they have a heavenly smell, but they also burn a long time. They are however very expensive to buy, which is why the beekeeper with excess wax can do nothing better than make his own. Recent industrial application of beewax include dentistry, electronic, mechanical moulds etc.

Propolis (Bee Glue)

Propolis is the sticky substance which the bees used to glue up their hive. Given the time, they will stick their boxes together and frames to the boxes, and that is why you need the wedge end of a hive tool to open them all up again.

The bees make propolis by collecting sticky resinous globules from the bark and leaf buds of certain trees. The name is derived from Greek-

pro meaning "in front of" and *polis* meaning 'city'. Quite simply it is a substance which the bees use to seal up and protect their city.

The first thing bees do when they occupy a new hive, or any hollow nesting place in the wild, is to coat the inside walls of their home with a very thin layer of propolis, so thin that it cannot be seen with the human eye. They also coat the inside of every wax cell before use for laying or filling with honey in a similar fashion. The reason is thought to be that propolis contains natural antibiotics which protect the colony from brood diseases including bacterial and fungal growths. They use propolis to glue everything together, and will mix it with wax to fill up any undesirable holes.

Composition

On an average propolis contains 55% resin and balsam, about 10% essential oils, upto 30% wax and 5% pollen. The colour and aroma of propolis varies according to geographical distribution.

1. Resins Flavonoids Phenolic acids and esters	45-55%
2. Waxes and Fatty acids originated from bees wax, (but many are of plant origin).	25-30%
3. Essential oils (Volatiles)	10%
4. Pollen, Proteins and Aminosacids	5%
5. Other organics and minerals	5%

Propolis as an Alternative Medicine

Propolis is known to have wound healing powers. In the past propolis was widely used as an surgical antiseptic; it lost its popularity with the advent of modem medicine. 'Comvita' is a major New Zealand producer with an attractively packaged range of propolis based products that include Propolis Herbal Elixir, Propolis Tincture, Propolis Ointment, Propolis Toothpaste and Propolis Capsules.

These products are described as "all natural infection fighters" which are variously said to help throat infections and common cold, mouth and gum disorders, gum decay, some skin conditions, resistance to general illness, pain from stomach ulcers, minor infections, and fungal skin complaints.

Numerous investigators have demonstrated that propolis has anti-microbial properties against various bacteria and fungi. Propolis was reported to be effective in healing wounds. It was popular in folk medicine for removing corns. A piece of propolis was softened by heating and a thin layer smeared on the corn, which was then lightly bandaged. The corn and its root came out completely after a few days. In the Soviet Union, propolis is used in veterinary practice, in ointments, for treating cuts, abscesses and wounds of animals. A propolis ointment prepared with Vaseline and sunflower and henbane oils in proportions of 1:1 and 1.5:1 was more effective in necrobacillosis in cattle than other remedies even without removing the infected part of the surface. It is obviously a mild irritant and probably promotes normal tropics.

Propolis also acts as a local anaesthetic. The anaesthetizing strength of its 25 per cent solution was greater than that of cocaine and procaine. Propolis is a good anaesthetic in dental medicine. Propolis is also believed to cure burns, external ulcers, and eczema in humans.

Propolis ointment prevents radiation skin reaction in patients who have to undergo radiation treatment. Inhalation of propolis in diseases of the upper respiratory tract and the lungs (*e.g.*, bronchitis and tuberculosis) has given good results. The treatment is simple and can be employed at home as well as in out-patient departments. For an inhalation 60 g. of propolis and 40 g. of bees wax are put into an aluminium or enamel vessel (300-400 ml), which is stood in a large metal bowl of boiling water. The mixture is inhaled for 10 to 15 minutes in the morning and evening over a period of two months.

Royal Jelly

The brood-food glands or hypopharyngeal glands are present in the cephalic region of the worker bee. These glands are two strings of small saccules closely packed in many loops and coils at the sides of the head. Their ducts called axial ducts open separately by two small pores on the lateral angles of the oral plate on the floor of the mouth. As this plate belongs to the hypopharynx, the food glands are called *hypopharyngeal glands*. These glands secrete a secretion which is with rich food material called 'royal jelly'. The Royal jelly is fed to the queen, and drones, worker larvae for the first two or three days. From then on it is reserved exclusively for future queens.

Royal jelly is a milky liquid contains about 18% protein, 10 to 17% sugar, about 5.5% fat, 1% minerals and vitamin B complex with little amount of vitamin A, C, D and E. It is demonstrated that it also contains a gonadotrophic hormone. It is acidic in nature. Royal jelly supplied to the queen larva differs significantly from that given to the workers and drones. From the 4th day onwards the drone and worker larvae are fed by mixture of honey and predigested pollen. This difference in nutrition is responsible for determining whether a given larva grows up into a queen or worker. It is proved by the experiment of transferring either the eggs or newly emerged larva from worker to queen cells or the reverse. So this experiment indicates that the royal jelly contains some essential accessory food substance, which is not present in honey or pollen, and acts as a determining factor of sex (female queen).

Table 3.11 : Composition of royal jelly.

	(%)
Water	57-70
Proteins (%); Nx6.75)	17-45
Sugars (%)	18-52
Lipids (%)	3.5-19
Minerals (%)	2-3
Vitamins	Rich in vitamins

Commercial Production of Royal jelly

On a commercial scale, royal jelly can be produced by modifying the standard techniques of mass queen rearing. By this system, strong queen right cell building colonies are used, in which the queen is confined into the chamber below the queen excluder, and the upper chamber is amply supplied with combs of pollen and honey. Into the middle of the upper chamber are placed frames of unsealed brood, to attract the young nurse bees from the lower chamber. Young worker larvae are then transferred into artificial queen cell cups that have been primed with a drop of royal jelly, on wooden bars that fit into a special frame that is placed adjacent to combs of unsealed brood. These colonies are fed constantly with sugar syrup and pollen substitute or supplement. Ideally a strong colony can adequately care for 45 newly grafted cells per day. Once every 5-7 days, frames of unsealed larvae from the lower brood chamber are raised above the queen excluder adjacent to the queen cells. On 3rd or 4th day, depending upon the age of the grafted larvae each queen cell will contain maximum amount of royal jelly (250-300 mg) in *A. mellifera*. The cells are then removed, and cut down to about the level of the royal jelly by means of a sharp, hot, thin bladed knife. The larvae are then removed with forceps and discarded. Then the royal jelly is removed either by a royal jelly spoon or vacuum apparatus. It requires an average of 1000, 3-day-old cells to produce 500gm. of jelly. Royal jelly should be strained through 100-miesh nylon cloth to eliminate bits of debris and then immediately stored under refrigeration (2°C) until required for use.

Royal jelly can be collected by cutting off queen cells, removing the larvae, and scooping out the contents which can be stored long term in the freeze. Specialist techniques need to be used to produce large numbers of excess queen cells to do this, and commercial royal Jelly production relies on queen rearing on a massive scale. Approximate productivity is likely to be 28.35 gms of royal jelly collected from 100 or more queen cells-no wonder it's so expensive.

Uses of Royal Jelly

Royal jelly has a reputation as a panacea, approdisiac and rejuvenator. It is exported in large quantities from the People's republic of China, Japan and several other countries. Much research has been done to test royal jelly and use it to treat disorders of the cardio-vascular system and gastrointestinal tract. Royal jelly normalizes metabolism, has a diuretic effect, can be used to prevent obesity and emaciation, builds up resistance to infections, regulates the functioning of the endocrine glands, and is good for arteriosclerosis and coronary deficiency. Royal jelly is a tonic restoring energy, getting rid of the feeling of indisposition and improver appetite.

In China, 1000 tons of royal jelly is produced every year. It is used to make medicine and nutritional supplements. Nutritive products of royal jelly are ginseng, royal jelly tablets, liquid, crystals, even wine, chocolate and cosmetics.

Royal jelly is a claimed cure all which is normally mixed with honey and sometimes propolis as well. It is said to improve body resistance against influenza, illness or infection, and some believe that it stops you growing older. Like other bee products it is safe to say that this splendid sounding product does you no harm and probably does you some good.

Bee Venom

The seventh segment of a bee is its sting. The sting is barbed, and it catches so effectively in human skin that it disembowels the bee which will then die. The sting consists of acid and alkali glands, the secretions of both glands together form the venom.

Bee venom is transparent, with bitter burning taste, acidic and contains formic acid, HCl, histamine, tryptophan, sulphur, volatile oils, magnesium sulphate, traces of copper, calcium and other substances. It also contains some peptide groups called melittins, which are responsible for haemolysis of blood. The enzymes hyaluronidase, phospholipase and other substances are also present. At ordinary room temperatures venom dries up, it is soluble in water and acid, activity is reduced by oxidising agents like potassium permanganate, and when kept dry toxicity retains for several years.

Table 3.12 : Composition of the bee venom.

Content	% Dry weight
Water	88%
Enzyme	
Phospolipase	10-12
Hyaluronosidase	1-3
Proteins and Peptides	
Melittin	50
Apamin	1-3
Mast cell degranulating	
Peptides (MCD)	1-2
Secapin	0.5 to 2.0
Procapine	1-2
Other peptides	13-15
Amines	
Histamine	0.5-2.0
Dopamine	0.2-1.0
Amino acids	0.5-1.0
Sugars	2
Phospholipids	5
Volatile compounds	4-8

Bee venom is obtained by *ether present* in a glass bottle which irritates the bees which release the venom but it is not a reliable method. A improved method is by passing a weak electric current through a special device fitted at the bee entrance of the hive. As the bees pass through entrance, they receive a mild shock and release their venom which fall into a piece of glass placed there for the purpose, the venom rapidly dries and forms the crystals which could be picked up.

Recent investigations shows that bee venom acts as antibiotic, and used in different diseases. Bee venom causes a lowering of blood pressure, haemolysis and contraction of muscles and block nerve muscle and ganglia synapses. Clinical observations have shown that bee venom is an active remedy in cases of acute rheumatic carditis. Bee venom is also useful in neural disorders affecting the sciatic, femoral and other nerves. In folk medicine bee venom has long been used to treat certain eye diseases as iritis (inflammation of iris). Bee venom is also used to treat hypertension. Some bee-keepers, and even medical workers, consider that all illness can be treated with the bee venom and used it in gynaecological and children's diseases. Bee venom must be used with care and only under medical supervision, especially when treating children and elderly people, who are very sensitive to it.

Many examples could be cited in which patients with hypertensive conditions improved soon after they began to work in an apiary where they were stung by bees. Their headaches disappeared, their fitness for work improved and their blood pressure dropped almost to normal. Use of venom is increasing and application methods of bee venom include natural bee stings, subcutaneous injections, electrophoresis, ointment, inhalation and tablets.

(D)

BEE KEEPING EQUIPMENTS

Modern bee-keeping is based on a scientific knowledge of the structure, life-history, habits and habitats of honey bees and it began with the invention of the *artificial hive* in 1789. The first centre of apiculture based on modern techniques with some sophisticated equipments was started at Lyallpur (All India Research Station) in Punjab in 1945. In countries like UK and USA the bee-keeping is done with sophisticated equipments, but elsewhere they do not take such luxuries for granted. Now-a-days, India have number of centres which are with modern techniques to handle the artificial hives. An apiary bee-keeper requires following equipments for handling the bees successfully.

The Hive

The *artificial hive* is quite simply the bee's home designed for accommodating bee-colonies. There are many different designs of hive around the world, but the two you are most likely to see in the UK are the single walled box (National) hives and double walled (WBC). The WBC is the traditional style of country hive which you expect to see in the back gardens of straw cottages. WBC stands for *William Broughton Carr*, the famous bee-keeper who invented it at the turn of the century, and though the prettiest ones are painted white, it is now more usual to see them in unpainted hard wood which will last a life time with very little upkeep.

The National works on the same principle as the WBC, but lack its characteristics outer walls which are known as '*lifts*',—lift them off and you will find a similar floor, brood box and supers. However, be warned that brood box and supers are not interchangeable from a National to WBC or vice versa; so once you have opted for one type of hive you should stick with it. The frames are the same size, but while the WBC brood box and supers have 10 frames the National's have 11 or give the hive slightly more capacity. The *Langstroth, Dadant* and *Smith* are other types of popular hives, using different size components, though the principles and working them remains the same. As *Langstroth* hive is world's best selling hive including India, the detailed construction is given as follows:

1. Bricks or Hive Stand

The hive must be mounted on bricks or a hive stand to keep it clear of the ground. The length of stand may be kept about 15.24 cm to 22.86 cm. The main purpose of stand is to support the box and prevent direct entry of predators.

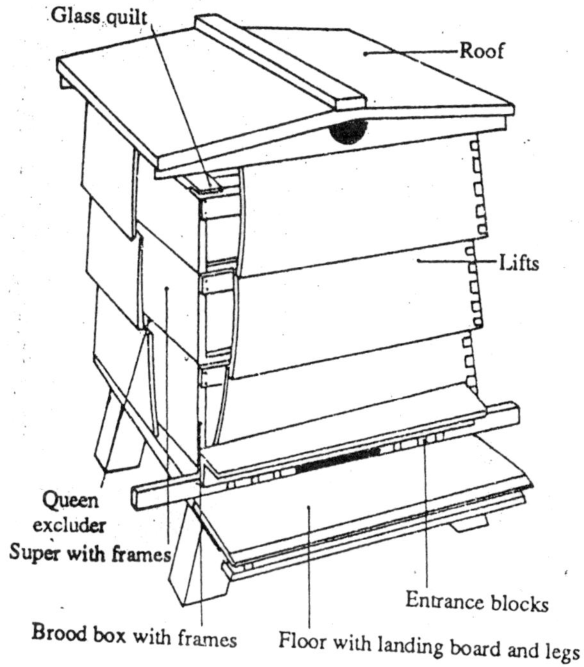

Fig. 3.32 : Cutway view of a WBC hive with broodbox and honey supers.

2. Entrance Block

The entrance block on a modern WBC is a two-part slider which can be adjusted from fully open to a few "bee space" holes, or completely closed. A bee space is 4.7-6.3 mm square to allow one worker to pass in and out—if the hole is any larger the bees will follow their natural instinct and block it with propolis.

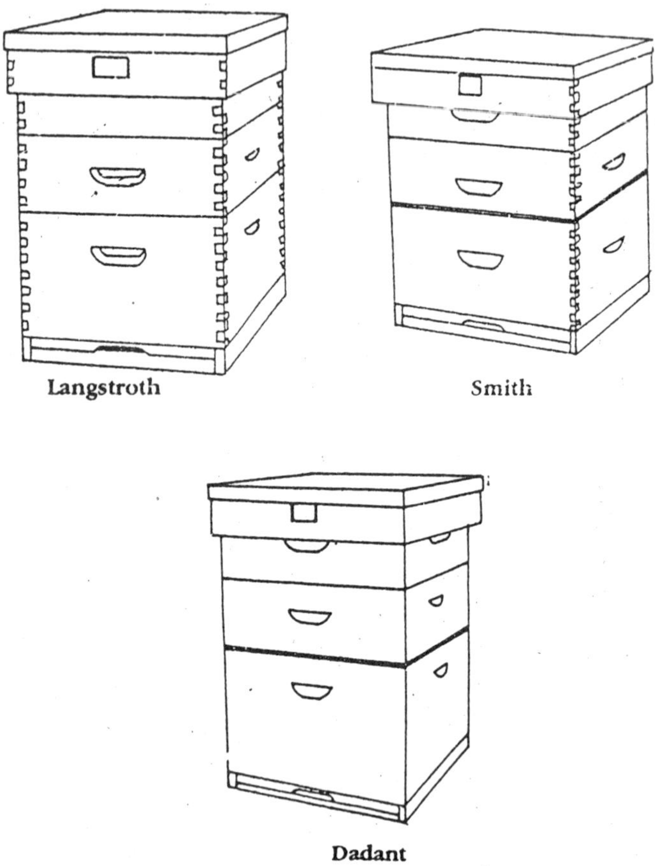

Fig. 3.33 : Langstroth, Dadant and Smith are all single wall hives similar in design and working on the same principles as the National. All these single wall hives are more practical to work than the double wall WBC with its outer 'lifts' and are also less costly. They simply consist of floor, brood box, supers and roof, and can easily be made by the carpenter. The Langstroth is very popular in the USA and elsewhere since Its large capacity can give very good returns. However full supers and brood box are heavier to lift than a smaller National. Smith or WBC which are therefore to be preferred by those who doubt their weight lifting abilities, particularly when dealing with angry bees'.

The entrance block on a National is a more basic square ended piece of wood. One side has no cut-out to completely close the entrance; the other has a large cut-out to allow the bees a limited amount of access.

The entrance block can be prepared either by taking a piece of wood 55.88 cm long, 40.62 cm broad and 2.3 cm thick. You should also reduce the entrance to a minimum when feeding your colony in preparation for the winter, but when winter comes the entrance should be completely opened to keep hive well ventilated. The layer is then likely to come from mice looking for a home; to keep them out the entrance is covered with a 'mouse guard'—a zinc strip perforated with 9.5 mm round holes which a mouse cannot force its way through.

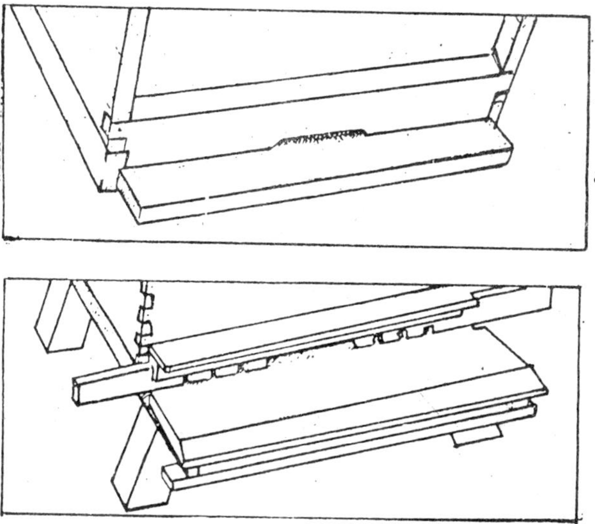

Fig. 3.34 : Entrance boxes.

In simple terms a hive is composed of a floor, boxes for the bees to live in and a roof on top. There are two types of boxes—brood boxes and supers. They work on same principles and are constructed in the same manner; the difference is that brood boxes are deeper and have a different role.

Each hive has a single brood box, and is the bees' home, accommodating the frames on which the queen lays her eggs. The depth of a brood box is the same on a National or WBC (22.5 cm), but while the former will take 11 frames the latter is one frame shorter and will only take 10 which is why they are not compatible. Both are 'bottom space' boxes, meaning that there is (9.5 mm) space allowing bees to run to and fro beneath the frames and up the sides; on a 'top bee space' box they would run across the top of and down them.

Brood chamber is made of 2.3 cm thick wood. Its outer length is 50.80 cm and on inside is 45.5 cm. While breadth on the outside is 40.62 cm and on the inside 36.1 cm, and its height is 23.3 cm.

Frames for the Brood Box and Foundation Sheets

Frames are four sided wooden supports for the wax comb which the bees live and work on. Wax foundation is commercially made from pure bees wax to fit a brood box or super. It is available with worker (small sized) or larger, drone size cells.

The foundation sheet is generally wired to give it some structural support. The frames are held the correct (9.5 mm) distance apart by removable metal or plastic ends. You can buy frames with specially shaped 'self-spacing' side bars, but it is more difficult to get a good tight fit. Metal ends are easier to handle. They allow easy access to each frame, or you can pack them in tightly (useful for transport) by opening out the flaps on each end frame. The self-spacing frame has dimensions 48.26 cm long, 2.54 cm wide and 2.3 cm thick.

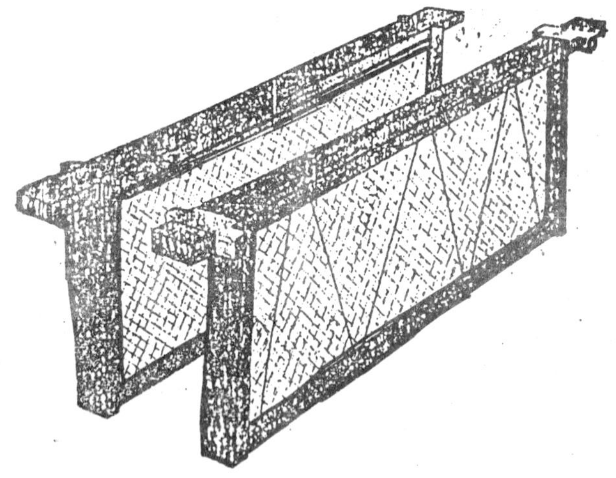

Fig. 3.35 : Removable metal ends give the correct spacing for the frames. Self-spacing frames (left) are an alternative with shaped side bars

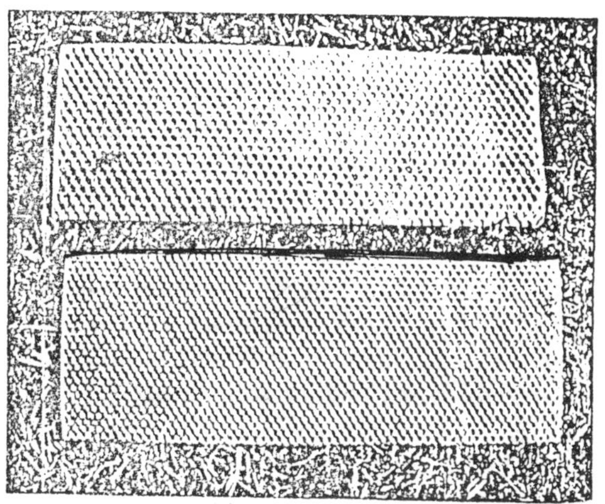

Fig. 3.36 : Drone foundation (top) has larger cells than worker (bottom). It can be used In honey supers, but this is not recommended for the novice.

Super

'Super' in Latin for on top of, and the supers are an exact fit on top of your brood box. National and WBC supers are 15 cm deep, taking 11 and 10 frames respectively. The dimensions of the super and frames

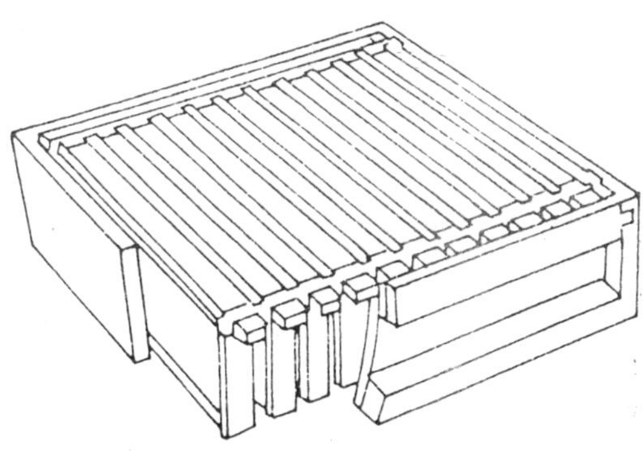

Fig. 3.37 : A National super is the same dimensions as the brood box but it shallower with shallower frames. If buying home-made/ equipment check that brood box and supers fit together with no gaps for the bees to escape.

should be same as those of brood box and the brood box frames respectively.

A super takes the frames which act as the bee's food and honey store. In the first year of bee-keeping you may well only need one super complete with frames to go on top of your brood box. By the second year you are likely to need a total of atleast three supers.

Glass Quilt/Crown Board/Clearing board

More confusion with three versions of a very similar item. You need a cover for the top of the brood box or super beneath the roof of the hive, and for most of the season we would strongly recommend you use a glass quilt. At other times you will need a crown broad/clearing board which is effectively one and the same thing.

The glass quilt is a glass cover set in a wooden frame, with a single opening for feeding the bees in the central bar which can be closed with a Porter bee escape. The glass quilt was invented some years ago by Arthur Chitty of West Sessex, and it keeps the bees in and lets you look at them with impunity.

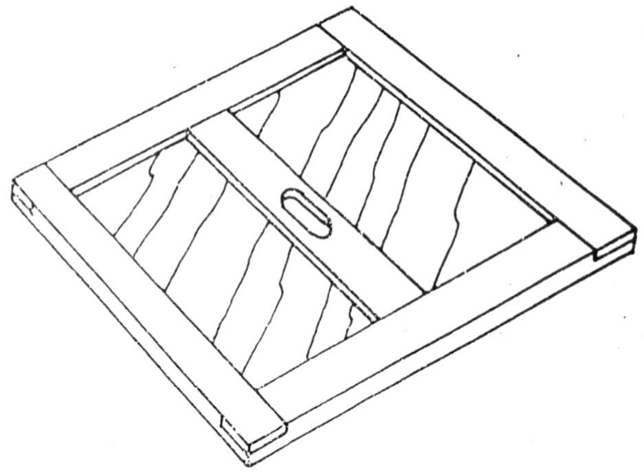

Fig. 3.38 : The glass quilt allows you to lift the roof and watch your bees. The feeding hole can be closed.

A crown board is the glass quilt's predecessor—the same thing in wood, but with two openings for bee escapes. Its disadvantage is that you can't watch the bees through it; its advantage is that it absorbs moisture and doesn't cause condensation as the glass quilt does in late autumn or winter. To overcome this condensation some bee-keepers insulate their glass quilts with polystyrene; the trouble is that the bees tend to eat it, so we would recommend changing to a crown board for the colder weather when there is not much to see anyway.

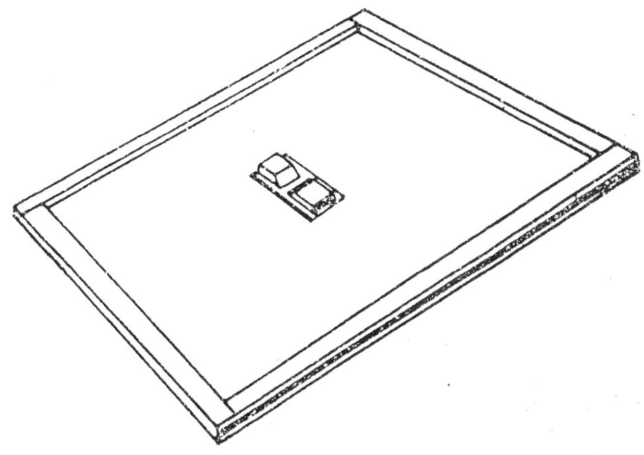

Fig. 3.39 : The underside of a crown board with a single bee escape In position. Most have two feed holes.

A clearing board can be a glass quilt or a crown board, whichever is spare or handy. It is used to clear the bees down from the honey super which you wish to remove, and this is achieved by means of blocking the opening (or openings) with a Porter bee escape. This ingenious device is fitted with a spring which leaves enough space for two workers at a time to get down into the next box, but they can't get back up.

All the workers will want to go back down to the queen once they have deposited their supplies in the honey super, and after 24 hours of this downward migration the top super should be empty of bees, allowing you to steal it away with relative impunity.

Queen Excluder

The purpose of queen excluder is to prevent entrance of queen bee into the super chamber. For obtaining pure honey free from extraneous matter, it is quite necessary to separate the brood nest from the surplus honey stores of the super chamber. This separation between super and brood chamber is done with the help of queen excluders. Queen excluder is made up of zinc sheet (metal rod) or round wires assembled together, 3.8 mm apart. The workers having thorax varying from 2.3 to 3.5 mm can easily pass through these perforations but not the queens as its thorax dimensions vary from 4.3 to 4.5 mm with the use of queen excluder, queen is unable to reach the super and the brood nest becomes limited to lower hive chamber.

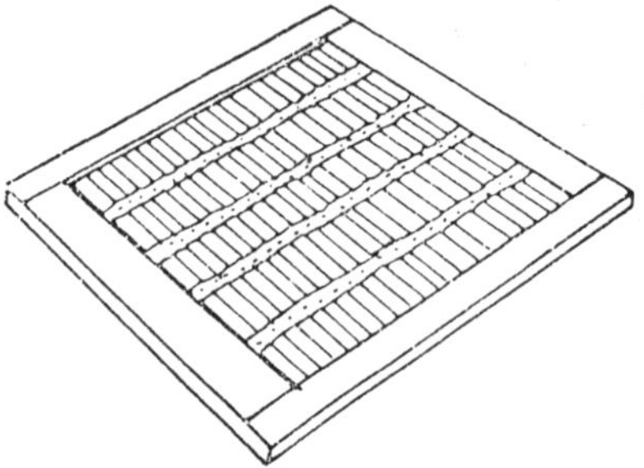

Fig. 3.40 : The modern version of queen excluder with metal rods is much better, though the bees dislike its cold feel.

The Roof

The roof is the sloping top cover which helps for the rain water to shed off from the hive, as well as leaves some necessary space for ventilation over the top box of the hive.

Most roofs are made of wood, covered by thin alloy for extra protection against the elements. The problem is that water clings to the alloy, and bees that settle on the wet roof can drown on it. To overcome this you should roughen the surface of the alloy—use a scattering of sand on a wet coat of varnish, or you can buy proprietary non-slip deck finishes designed for yachts. The National roof is available in two depths. The 15 cm roof is likely to blow off in exposed situations, but

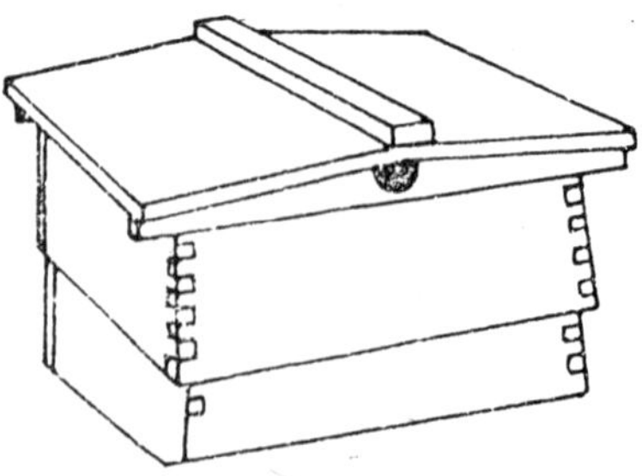

Fig. 3.41 : The WBC has a cottage style roof and outer lifts.

is somewhat clumsy to handle; the 10 cm version is fine for town or suburban use.

Rustic and Feral Hives used for *Apis Cerana*

Beekeeping with *A. Cerana* has been practised in India since the times immemorial. The various types of shelters had been used for rearing honeybees which are as follows:

Log Hives

In India log hives are still used at higher altitudes from about 2000m above sea level for *A. Cerana indica.* A log hive consists of a tree trunk,

Fig. 3.41(a) : A large comb of *Apis cerana* in a log hive with movable frames in Kerala.

hollowed from inside, closed at both the ends having a small hole of suitable size along its length. The horizontal log hive have dimensions of 60 × 75 cm long, with entrance hole of 6 mm in the middle of the log. The two open ends are closed with a piece of wood, tin or mixture of cowdung and clay in the horizontal log hives, strips of old combs at two ends are fixed with molten water candle's droppings. The space between old combs from centre to centre is 3 cm which provides enough space for the movement of bees. Each end of the log hive is provided with 4 to 5 such old comb strips and the central ones are left empty of the bees to build new combs themselves. In such a horizontal log hive the bees build combs parallel to the ends of the log, only the combs at the two ends of the log are removed for honey extraction and the central combs with brood and honey are kept intact.

The vertical log hive is similar to the horizontal except that it is kept in an upright vertical position instead of a horizontal position. The honey production from these hives is normally below 5 kg as compared to movable frame hives.

Though the log bar hives are still based in some areas for *A. cerana* but they have not become popular with the beekeepers as they are more interested in modern movable frame hives.

Fixed wall hives

Beekeepers in hilly areas have been keeping bees in wall hives. In kashmir valley, Doda areas of jammu and Himachal Pradesh, hollows are provided in the mud walls. An entrance is kept to the outside. The hollow cavity is covered from the inside of the room and with bamboo matting or wooden plank by plastering with mud. The honey is collected from the end of the hive inside the house and only combs with honey are removed. Fixed wall hives are circular, rectangular or square in shape and of different dimensions. A small triangular round or rectangular hole either at the bottom, centre or top on the outer side of the hive facing east or south is provided for the entry of bees. Each recess or cavity in the wall is 40 to 60 cm long and 25 to 30 cm. deep. The side walls of the wall hive are plastered with mud or cowdung, where as flour or roof are provided with wooden planks. Bees generally build the combs on wooden room parallel to the entrance of the hive. After the honey harvesting, the rear cover is replaced into its original position. It has been found that *A. cerana* colonies in wall hives abscond and migrate less frequently than those maintained in modern movable frame hives. This is possibly due to the reason that they are subject to less human disturbances. Furthermore the height of wall hives is generally 6 to 8 feet above the ground and are less prone to attack of enemies.

Fig. 3.41(b) : Wall hives near Badarwah, J & K. The hive on the right had a strong colony that swarmed, leaving the combs to wax moth infestation.

Miscellaneous hives

All sorts of boxes; clay, pitchers, tree barks, and mud receptacles of variable dimensions are used by beekeepers for rearing bees. These are generally suspended in open verandah or open carridor of a house immediately below the roof to attract the swarms of bees.

Modern Bee Hives

In India, beekeeping with movable frame hives started in the later part of the 19th century (1881-84) in South India. Father Newton, designed a small bee hive with movable frames for *A. cerans,* which is commonly used all over India (1930). But it was suitable for southern India, as compared to northern states. In view of these constraints, Mutlo (1956) designed a new hive for larger sized *A. cerana* as found in higher altitudes and he named it as "jeolikote villager's hive". This hive was widely adopted not only at higher altitudes but in plains as well; and near about 15 different types of hives with different dimentions come into existence for beekeeping with *A. cerana* through India and were called as modified Newton's or villager's hive.

In view of this multiplicity of hive designs at that time, a need was felt to evolve different, definite standards of bee hives suiting to different sub-species/geographic ecotypes of *A. cerana* found in different climatic zones and beekeeping areas of the country. Therefore in 1956 an "Apiry Industry Sectional committee" was constituted by Indian Standard Institutions (ISI) to lay down ISI standards for bee keeping equipments and honeybee products. Consequently following standards for hives were laid down: Type A—modified Newton type

Type B—modified Joelikote villager type,

when the above standards were adopted only arbitrary classification of plains and hill variety of *A. cerana* was available. But recently three sub-species of *A. cerana* have been identified which include:

1. *Apis cerana cerana* : J and K and Himachal Pradesh area.
2. *A. cerana himalaya* : North eastern area of India including Manipur and Assam.
3. *A. cerana indica* : Southern India including Tamil Nadu, Kerala, Karnataka and other areas.

Appliances for Handling Bee Colonies

Smoker

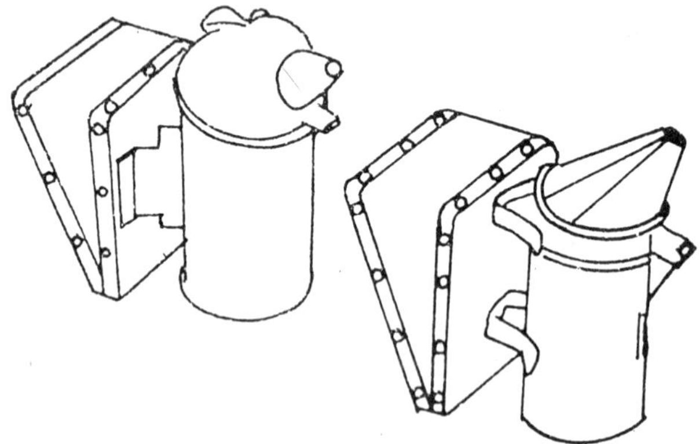

Fig. 3.42 : All smokers Work on the same principle, with a bellows attached to the main fuel canister. The design is unfortunately very antiquated. A modern smoker which keeps going whenever you need it is sorely needed by the bee-keeping community.

Table 3.13 : Dimensions of various parts of *Apis cerana* type A beehives with ten frames (in mm)

Description	Distance between centres of two adjacent frames = 30 mm (bee space = 7 mm)			Distance between centres of two adjacent frames = 31 mm (bee space = 8 mm)			Distance between centres of two adjacent frames = 32 mm (bee space = 9 mm)		
(1)	Length(L) (2)	Breadth(B) (3)	Height(H) (4)	Length(L) (5)	Breadth(B) (6)	Height(H) (7)	Length(L) (8)	Breadth(B) (9)	Height(H) (10)
Floor board	361±2	356±2	50±	361±2	366±2	50±2	361±2	376±2	50±2
Brood frame:									
Outside	230	–	165	230	–	165	230	–	165
Inside	210±2	–	145±2	210±2	–	145±2	210±2	–	–
145±2									
Brood chamber:									
Outside	286±2	356±2	172	286±2	366±2	173	286±2	376±2	174
Inside	240	310	172	240	320	173	240	330	174
Super frame:									
Outside	230	–	85	230	–	85	230	–	85
Inside	210±2	–	65±2	210±2	–	65±2	210±2	–	65±2
Super chamber:									
Outside	286±2	356±2	92	286±2	366±2	93	286±2	376±2	94
Inside	240	310	92	240	320	93	240	330	94

Inner cover (crown board)	286±2	356±2	22	286±2	366±2	23	286±2	376±2	24
Roof (top):									
Outside	328±2	398±2	100±2	328±2	408±2	100±2	328±2	418±2	100±2
Dummy board	330±2	–	165±2	230±2	–	165±2	230±2	–	165±2
Division board	236 –	–	One-end 182 Other end-194	236	–	One-end 183 Other end-195	236	–	One-end 184 Other end-196

Source: Indian Standards Institution 1981.

Table 3.14 : Dimensions of various parts of *Apis cerana* type B beehives with ten frames (in mm)

Description	Distance between centres of two adjacent frames = 31 mm (bee space = 8 mm)			Distance between centres of two adjacent frames = 32 mm (bee space = 9 mm)		
(1)	Length(L) (2)	Breadth(B) (3)	Height(H) (4)	Length(L) (5)	Breadth(B) (6)	Height (H) (7)
Floorboard	431±2	366±2	50±2	431±2	376±2	50±2
Brood frame:						
Outside	300	–	195	300	–	195
Inside	280±2	–	175±2	280±2	–	175±2
Brood chamber.						
Outside	356±2	366±2	203	356±2	376±2	204
Inside	310	320	203	310	330	204
Super frame:						
Outside	300	–	105	300	–	105
Inside	280±2	–	85±2	280±2	–	85±2
Super chamber						
Outside	356±2	366±2	113	356±2	376±2	114
Inside	310	320	113	310	330	114
Inner cover (crown board) Roof:	356±2	366±2	23	356±2	376±2	24
Outside	398±2	408±2	100±2	398±2	418±2	100±2
Dummy board	300±2	–	195±2	300±2	–	195±2
Division board	306	–	One-end 213 Other-end 225	306	–	One-end 214 Other-end 226

Source: Indian Standards Institution 1981.

A smoker consists of container to hold a slow burning fuel; a bellows to make it burn faster and produce smoke when required; and a nozzle to direct the smoke at the bees. The classic style of smoker is copper bodied and will last a life time. You can buy the same type in tin plate which is cheaper, but will corrode after a few years. Larger, commercial size smokers using the same principle are also available (only in tin plate), designed for working on up to 20 hives at a time!

Another type of smoker has a clock-work driven fan in place of a bellows. You wind up the clock-work, and simply press a release button to activate the fan when you want smoke. This has the advantage of being a one-handed operation, and makes it considerably easier to keep the smoker alight. The same principle could no doubt be used with an electric motor.

Lighting a Smoker

If you take care with lighting a smoker it will stay alight and puff sufficient smoke for at least 30 minutes, which is more than enough time to inspect two hives. Even if you are unlikely to need smoke it is always a wise precaution to have your smoker ready to hand and lit when doing any work on a hive that interferes with the bees.

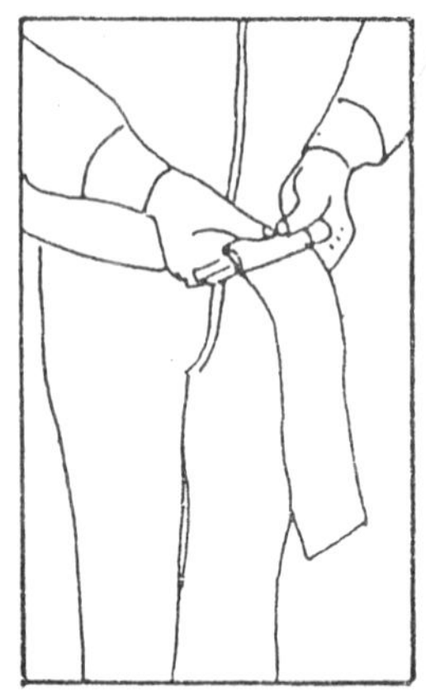

Roll up some cardboard. If it has been treated with a fire retardant, you should soak your cardboard and allow it to dry.

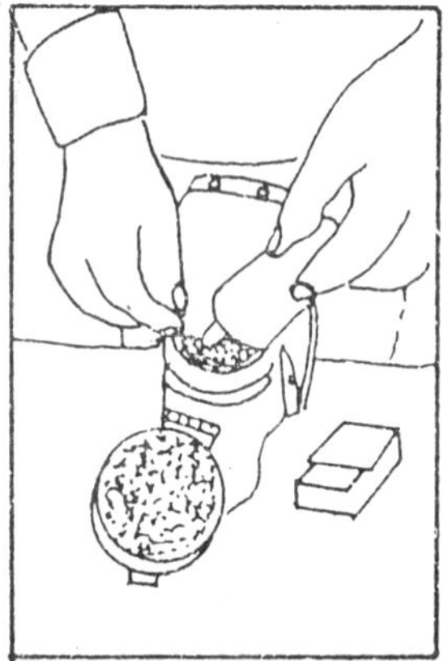

Pull out one end to make a wick. Light it in a draught free place. Your smoker should already contain some fuel.

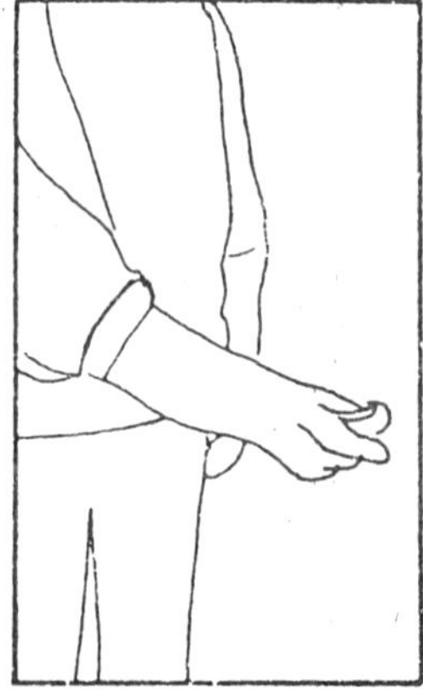

Wave it back and forth until it is well alight, with flames shooting out of one end. A glove may be useful.

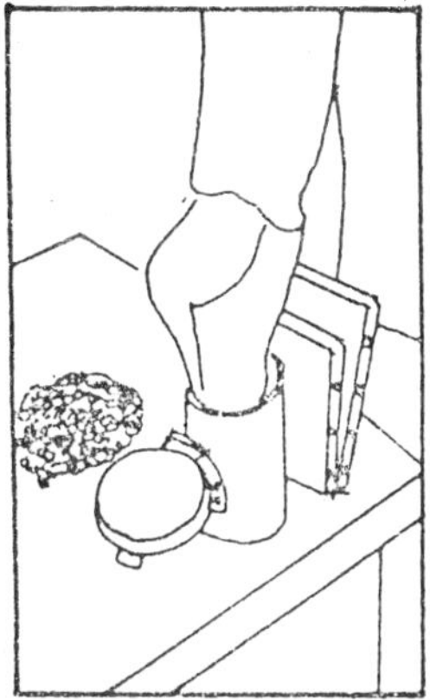

Carefully push the cardboard down into the smoker with the lighted end downwards. Have more wood shavings (or similar fuel) ready,

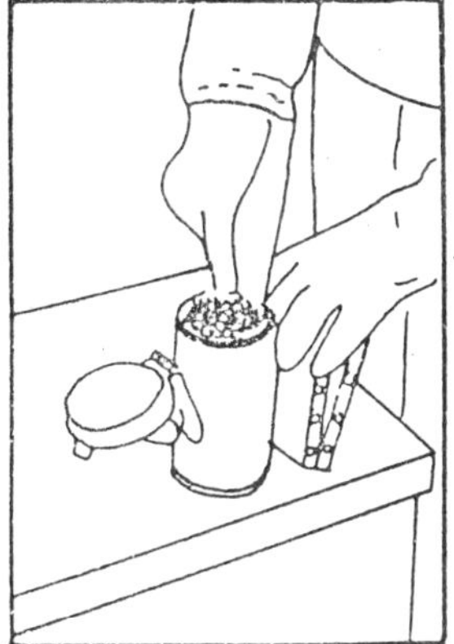

Cover the top of the cardboard with this fuel. Close the top of the smoker, and give a few pumps on the bellows to make sure it is going.

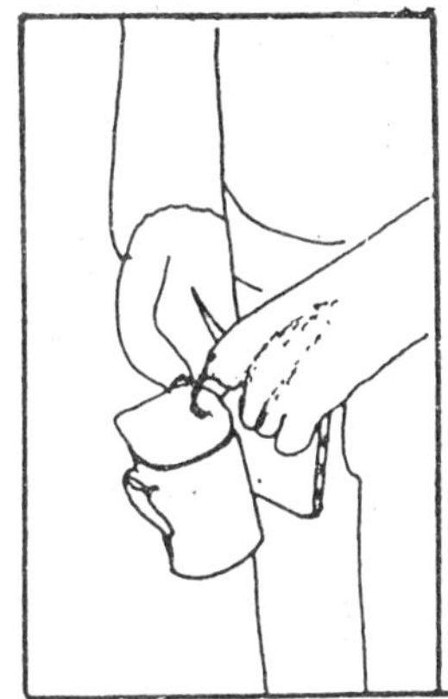

The smoker should stay alight for at least 30 minutes and can be refueled as necessary. Block the nozzle with grass to put It out.

Fig. 3.43 : Lighting a Smoker.

Ideal fuel to get a smoker going is corrugated cardboard paper. This can be ripped off any old cardboard box, but in many cases will have been saturated with a fire retardant to prevent it burning. To get rid of this you have to soak it in water, and then dry it. Then, when you are ready to go to the hive proceed as follows as shown in figures.

It is worth emphasising what the smoker is and is not for. It is used to sedate the hive, but if used too enthusiastically may well make the bees become extra aggressive. The basic technique is to first give a gentle puff of smoke it the entrance to make them head for the honey and suck it up ready to abandon home-but don't smoke them too hard or you will drive them out of their minds.

You then wait for three minutes before taking the roof of the hive off. By then the bees will have been sedated by their large intake of honey. As you remove the glass quilt (or clearing board) which is covering the super or brood box give a few gentle puffs round the edges to clear your way, but no more or you will begin to antagonise them. If you oversmoke them you could drive out the whole colony.

Hive Tools

A purpose-designed hive tool is invaluable for the following reason. The bees like to glue everything together with 'propolis'. Thus the super is glued hard to the top of brood box; the glass quilt is glued hard to the top of the super, and the frames are glued hard to one another. You could look on it as a sort of advanced form of draught proofing.

There are two basic types of steel hive tool available with slightly different characteristics. Unfortunately a single tool which combines the best of both has not been invented.

The Yellow Hive Tool

The yellow painted hive tool is wider and heavier, and is generally favoured by commercial bee-keepers. One end is used to lever open thing held together by propolis and does so without too much damage to the surrounding wood; the other acts as an efficient scraper of unwanted propolis and other gunge. The lever end is the width of a frame, and if the box isn't full can be twisted between two frames to make the right space to insert a new frame.

The Red Hive Tool

The red hive tool is slimmer, with one end incorporating a notch designed to lever each frame clear of its box and then be lifted out., The

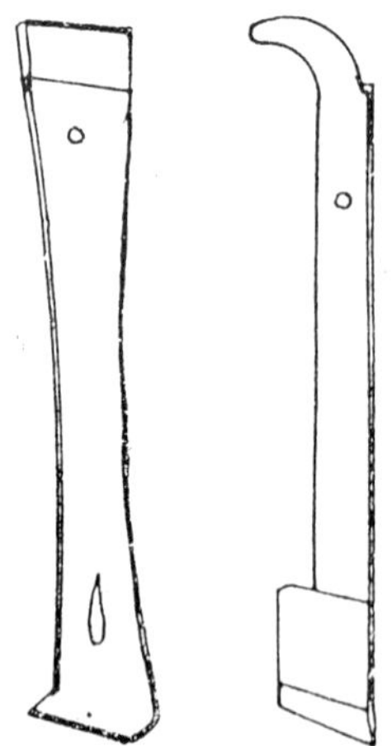

Two types of hive tool for breaking propolis bound supers and frames, usually yellow (left) and red (right)

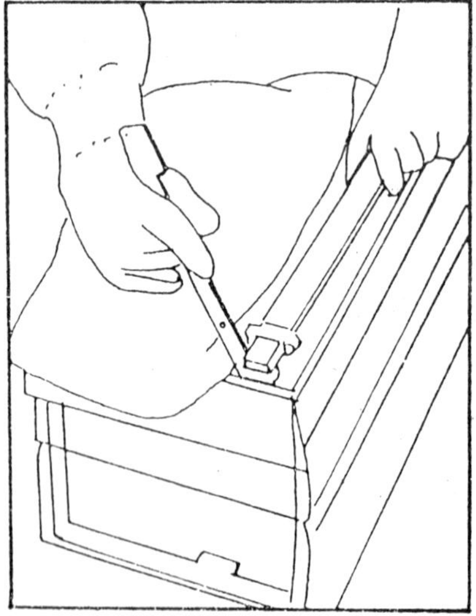

Fig. 3.44 : Hive Tools.

The truth is that you are best off with both kinds of hive tool. Like the smoker the hive tool is an aid, but remember to use it in a calm, methodical fashion so as not to antagonise your bees. They dislike sudden movement or bangs, which are liable to result if you blunder ahead without carefully thinking out what you are going to do before starting to tackle a hive.

Protective Clothing

Number one priority is to protect your face and head. Due to the pollen collecting hooks on their legs been often get hopelessly entangled if they land on a head of hair or in a beard, and when you panic and try to get rid of them they sense your fear, panic and sting you in return. Your face contains some highly sensitive areas which don't take kindly to stinging—a sting on the mouth or eyelid can result in severe and possibly dangerous swelling.

The simplest protection for this area is round hat with a veil that hangs well away from your kin. Bees have a natural tendency to go up, so take care that they cannot get up inside the veil.

This can only happen if you don't adjust it properly, but if a bee does get inside :

1. Remain calm.
2. Move well away from the hive.
3. Take the hat off to remove the bee, or if necessary squash it between gloved fingers.

Never attempt to remove the hat while standing by the hive. The bee may sting you as you do so, immediately attracting other bees to sting you in the same area.

The next priority must be gloves. Washing-up gloves are sting proof, but they may expose the wrists and are sweaty. We would recommend buying good quality goatskin bee-keeping gloves with gauntlets which offer complete protection.

With the hat, veil and gloves you can make do with ordinary clothing, so long as you remember to cover all exposed areas and take note of the fact that bees go up. A bee flying up a loose trouser leg may appear amusing to others, but if it's your trouser leg you will wish that you had tucked it into boots or thick socks, or some make do with bicycle clips.

Remember also that the bee sting will penetrate tight jeans, thin socks, etc. Loose fitting, relatively thick or tough clothing is necessary to keep that little barb away from your skin.

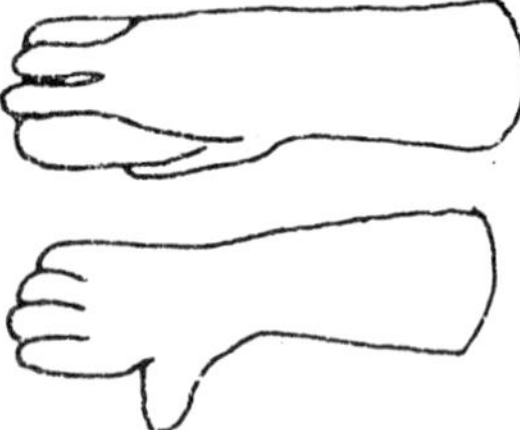

Fig. 3.45 : Bee gauntlets make for safe handling of bees.

Fig. 3.46 : Two kinds of veil. The type which can be attached (usually by zipper) to a bee suit is preferable

The Well Dressed Beekeeper

The best kind of bee-keeping suit is an all-in-one overall incorporating a hood which can be unzipped and flipped back off your head. Made in polycotton we would definitely recommend this type of peace of mind, comfort, and a good selection of pockets for storing hive tools and other small items.

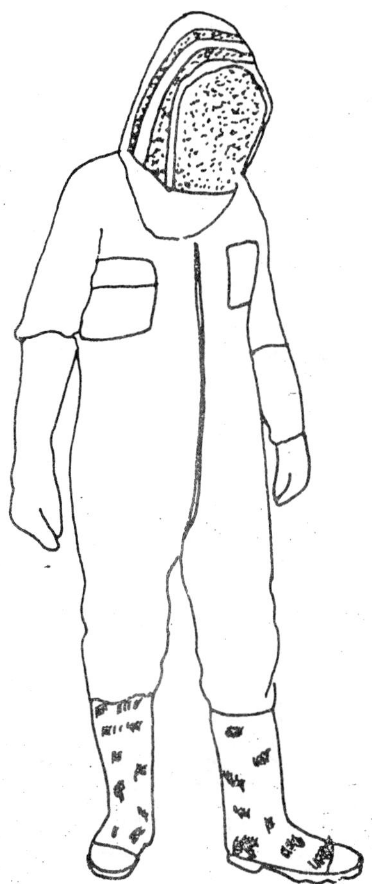

Fig. 3.47 : Overall.

Bee brush or Whisk Broom

It is always used to brush off bees from a honey-comb before it is taken away for extraction. A goose or turkey's pinion feather with attached primaries makes good brush for this purpose. The brush should be gently used to avoid inducing aggressiveness.

Feeders

Bee-keepers use various types of feeders for feeding sugar syrup to bees particularly in dearth period. The object of feeding is purely to get the bees to draw out the wax. The feed stimulates the wax glands of wax making workers, and like little robots they go to work.

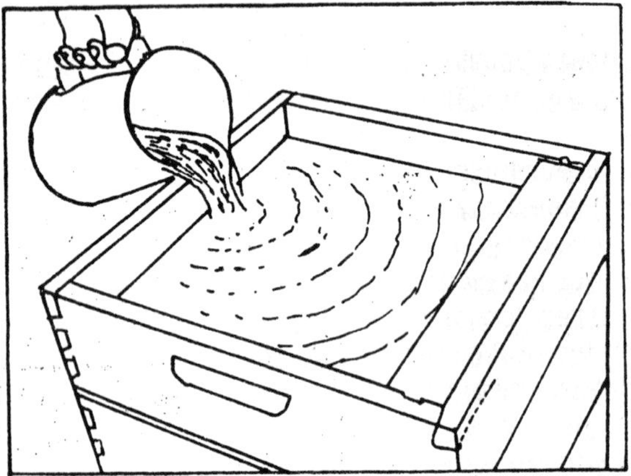

Fast feeder in action. This is the Miller design, which takes approximately I gallon of sugar syrup at a time. With exactly the same dimensions as a super it will fit neatly on top of the hive

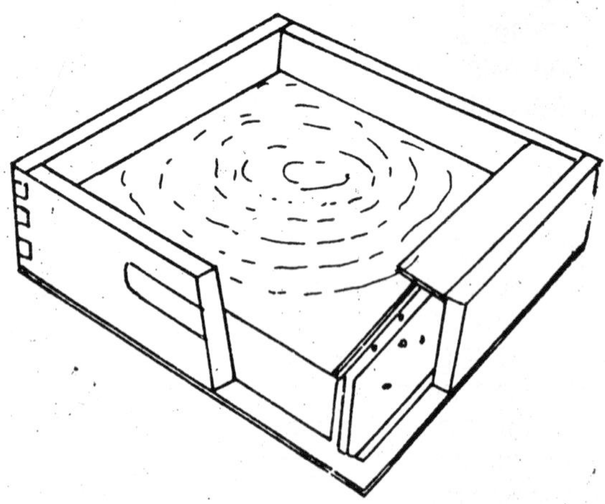

Fig. 3.48 : Being fast workers, the bees can empty the syrup in 24 hours. You then refill the feeder until they have been supplied with the necessary amount for winter. A fast feeder is a very worthwhile Investment

Very often, the division board feeder made up of wooden trough of the regular Langstroth frame dimensions with shoulders may hang in the hive like any other frame and with a wooden strip to serve as a float is used. A another type, a lever-lid in empty tea tin with holes in the lid and an inverted glass jar with holes in the lid are also good types of feeders.

Honey Extractor

This is most important instrument for separating pure honey from comb. It is available in varieties of forms with the frames arranged tangentially or radially and is operated with manual belt, chain or gear drive or electric power. It works on principle or whirling of a decapped honey combs in a cage enclosed by a drum and honey is thrown out under the centrifugal force. During the process the honey accumulated at the bottom drains out through the spout and is collected in bottles.

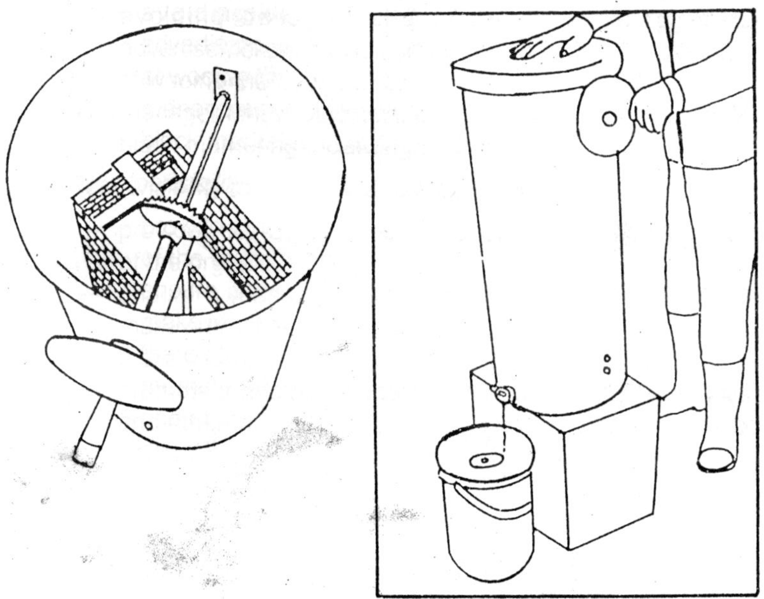

Fig. 3.49 : A typical manual extractor will take four uncapped frames dropped into the sides of a cage with metal ends removed. Start spinning slowly; work up to full speed; reverse the frames after a couple of minutes and repeat.

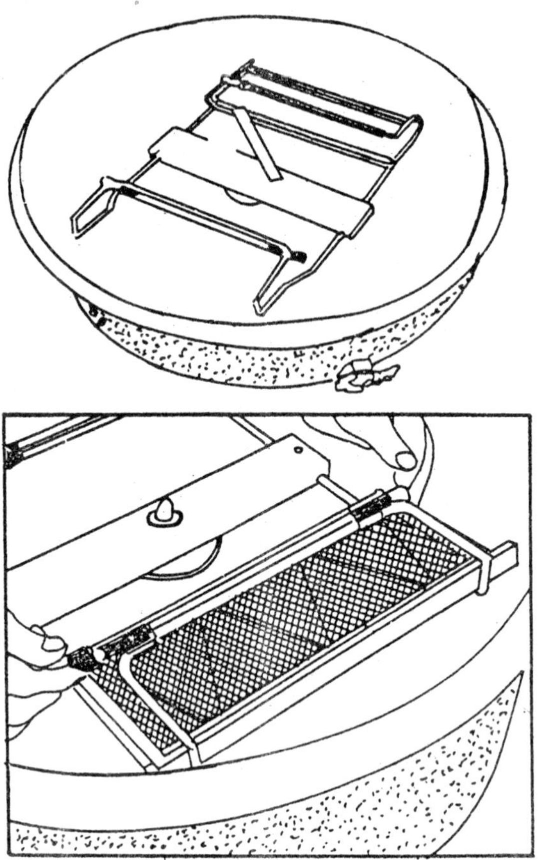

Fig. 3.50 : An electric extractor which resembles a huge plastic saucer. Two uncapped wired frames are locked in place at both ends of the rotor arm. With the machine plugged into the mains the speed of the revolutions can be adjusted; start off slowly and then build up speed. Otherwise you risk spinning the wax completely out of the frames with too much centrifugal force. The main disadvantage of this machine is that it will only spin two frames.

Uncapping Knife

Bees after depositing the honey in storage cells, they seal the cells by capping with wax. Such honey combs before placing them in honey extractor for extraction of honey need the removal of wax cappings of cells. These cappings are removed with the help of a special-knife called as uncapping knife. It helps in emptying honey from storage cells during honey extraction process.

A sharp knife is needed to evenly slice off the honey's wax cappings

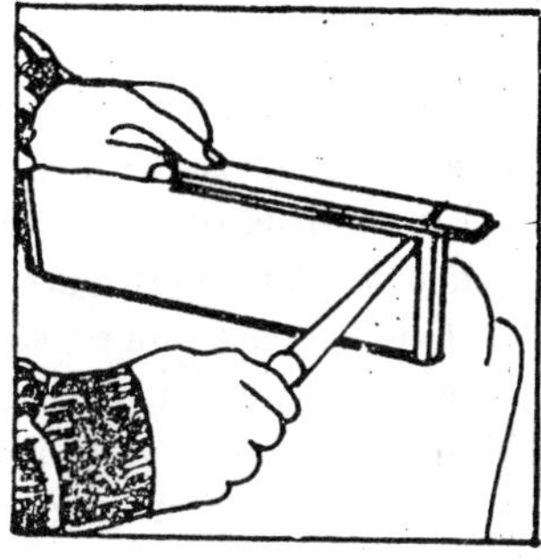

If the wax is wired, you can spin it off If not, cut out the whole comb

The unwired comb can then be cut into five pieces stored in cartons.

Fig. 3.51

Catching a Swarm and Equipments

A first sign of a swarm in action is a lot of bees flying around the hive in a seemingly hapazard manner. This can happen upto 24 hours before the actual swarm. The bees who are leaving then issue forth in a great cloud, usually around midday. The first swarm is called the 'prime swarm'—approximately 25,000 bees or half the hive following the old queen. Within 20 minutes the swarm settles on a handly tree or wall, while their scouts go off to search for a new permanent home. This temporary home is usually within 50 yards of the hive. The swarm honey surrounding the queen in a great mass, and will usually stay a couple of hours before heading to the final destination.

Swarm catching is a May/June occupation. The basic law of Swarms is 'finder's keepers', unless the original owner still has it in view. The swarm can be captured with the help of swarming equipments like skip-(a hive made of straw) or swarm catching basket, a white sheet, soft cloth, smoker and two drawn out frames. You may also need a flat board, a long handled pruner, step ladder etc. During catching basket is kept upon the hive gently directing the bees into it. Sometimes the bees are shaken into a box held on the underside of the limb or caught in a swarm catching bag from an inaccessible limb. The swarms can be helped in settling down early by sprinkling water into the hovering swarm (lingering near by) with the help of syringe.

Fig. 3.52 : Swarm catching bag.

Queen Cage

It is important to be able to recognise the queen. Once she is found, a round queen cage is used to isolate her on a comb so that she can be marked. This is essential for introducing queen to queenless colony so that the bees become acquainted with the fact that she is indeed a queen. Today, different types of queen cages are in use. They are Miller queen introducing cage. Smith introducing cage. queen mailing cage and wire gauze cage are common types.

Queen Cell Protector

The bees make queen cells to produce new queens, using larvae less than three days old. These cells are found on the bottom bars of the frames; supercedure queen cells or emergency queen cells. Such cells are protected in a queen cell protector until it is accepted by the bees.

Dividing Board

It is a solid brood shape wooden frame or, partition, used to divide or limit the brood frames. It keeps brood nest warm and well protected from bee-enemies.

Fig. 3.53 : The queen is isolated with a queen cage and marked with a small blob of paint before being freed.

(E)

BEE MANAGEMENT

Almost everything about bee-keeping runs in threes:

There is a three-year system to bee-keeping. Buy a small 'nucleus' of bees in the spring of year I and they will grow to form a nice colony. In their second year the colony will reach maturity, with fully drawn out frames for them to work on and lots of honey for you. In the third year they will want to swarms and you will want to prevent them doing so.

This is an ideal sequence, but they may confuse you in number of ways—Swarming in the first or second year, dying out one winter, or failing to reach maturity. As a bee-keeper you have to spot the problems, and know how to remedy them.

- There are three 'castes' of honey bees-queens, workers, and drones.
- The queen lasts three years as an effective egg-laying machine, though she can live to six years if the colony allows her too—a most unusual occurrence.

- ***Three days for eggs***

The eggs are laid in the frames by the queen, each egg becomes a larva which soaks up the food that the worker bees supply it with and after three days' growth this diet will decide if it will grow into a queen or worker.

- ***Three ways of obtaining bees***

1. Buy a 'nucleus'—a small colony of around 10,000 bees headed by a young queen, clustered on four or more frames.
2. Catch a 'prime swarm' made up of approximately half an established colony which has left its old home headed by the old queen.
3. Buy or be given a secondhand colony. You need to be very careful that you are getting disease-free bees and equipments.

- ***Three minutes wait***

Bee-keepers use a 'smoker' to help control bees if they become unsettled. If the bees sense a whiff (puff) of smoke they think their hive is on fire; they dive down into the combs to fill themselves up with honey which will act as emergency rations if they have to abandon the hive. This gorging on honey makes them soporific and less inclined to be aggressive; after the initial smoking the bee-keeper waits three minutes for them to get this stage.

- ***Three days before feeding a nucleus or swarm***

If you have just 'hived' a nucleus or swarm, it is usually best to wait three days before feeding those bees. The reason is that they are liable to be comparatively weak and too disorganised to protect themselves against 'robbing' by other bees if you feed them straightway. They should have sufficient stores and no more, which will allow them to survive.

- ***Three days honey in a swarm's honey sacs***

Bees which about to swarm suck up enough honey to last them three days before they need to find more stores.

- ***Three feet or three miles if moving a hive***

If you move your hive a couple of miles from its original site, they will continue to fly back to that original site every time, hive or no hive, and will wait there patiently wondering what to do.

The rule of thumb is that you should only move a hive less than three feet, or more than three miles. If it is less than three feet the entrance can be aligned with the old entrance so they can find their way in; if it is more than three miles they will forget the old flight pattern, and establish a new flight pattern for the new site.

Before hiving the bee colonies, it is necessary to select a proper site. When selecting a site for your beehive, you are looking for an area which is :

1. Relatively dry and flat;
2. Not exposed to the full force of the wind;
3. Clear of damp undergrowth;
4. With a fair amount of forage close to hand;
5. With a flight path well clear of humans so that the bees are safe and not a nuisance-particularly important in town;
6. Easily accessible for carrying and storage—important in a big garden;

7. Undisturbed and quiet;
8. Not in the same as disruptive elements such as, horses which like to kick beehives with their hooves.

To start a colony, wild combs with their full Complements—queen, workers and drones may be removed and transferred to the frames of artificial hive. Since locating wild colonies is tedious and difficult for a beginner, natural swarms are trapped during the swarming season. After a month or so the bees are transferred to the artificial hive. It is also possible to start a bee colony with nuclei (queen) obtainable from *apiaries*. The hives should be spaced at least 2 metres apart and should be capable of being manipulated from behind. They should be placed preferably facing east, in a dry sheltered spot and in the vicinity of pasturage.

After hiving the bee colonies, it is necessary to do the routine management like regular colony inspection, feeding—cleaning, watering etc. These activities favour the bees which help in maximum yield of honey.

When bees are working normally, then bright sunny days are safest for colony inspection. During rainy, cold, windy climate or at night hours, bees should not be disturbed. While visiting to the colony the bee-keeper should not be strongly smelling of alcohol or horse-dung and should not be in profusely perspiring state. He should wear an overall and a bee veil. After lighting the smoker, approach the hive from side to avoid interfering with the flight of bees. Lift the inner cover little with the hive tool after giving few puffs of smoke at the entrance, then direct smoke into hive and drop the inner cover in place. Remove the inner cover after a few moments and place it upside down in front of or against the hive. The frames should be pried apart with the hive tool and scan them one by one. The frames should stand in a vertical position preferably over the hive. For examining the other side of the comb, it should be manipulated as shown in figure. The vigilance on is essential during these manipulations keep back the frame in the hive on which queen is located. The frames should be handled gently and jerking and crushing of the bees must be avoided so as to avoid their stinging. No doubt that the stings are painful, but the odour of the venom irritates the other bees and makes them difficult to be managed.

If bees sting, then take out the sting by finger nail or use sharp edge of the hive tool. It will reduce poisoning. It is wrong to squeeze it out with finger tips. The sting spot should be washed well in water. Rubbing the wound only aggravates irritation. Some persons particularly women if stung by a single bee develop rashes all over the body and have difficulty in breathing. After a prolonged experience the person handling bee colonies eventually become immune.

Cleaning

If the frames and hive box on the bottom and sides show the presence of dirt or accumulated debris, then with hive tool scrape out the dirt and debris from the frames as well as from hive and make it clean. This process of cleaning is highly essential to maintain the boxes and frames clean so as to prevent unhygienic conditions. Poor hygiene will provoke the swarming behaviour. At the same time cleaning gives idea of activity of wax-moth if it is present in the colony. Generally a hive tool which is a piece of flattened iron with hammered down edges is used for scrapping bee glue and superfluous pieces of comb from the various parts of the hive.

After the inspection and cleaning of colony the box should be closed tightly without leaving any chinks and crevices.

Feeding

During death period, at the time of artificial increase, hiving of swarms, packing the bees for winter and too small nuclei artificial feeding becomes essential, otherwise honey is the best food for bees and bee-keeper should leave enough amount of it with them at all times. The syrup made-from white crystalline sugar is the best substitute food for bees. Use of jaggery, molasses etc., are not recommended as these substances undergo fermentation in honey comb cells and cause dysentery.

The sugar syrup consists of two parts of sugar in one part water; and few grains of tartaric acid. This is ideal for bees to be fed before winter. During remaining time of the year, the syrup is prepared from one part of sugar and one part of water. Sugar crystals should be thoroughly dissolved in water and if necessary it should be heated. For preparing fine-grained candy, heat two parts of sugar in one part of water with 10 grains of cream of tartâr per kilogram of sugar to 115.6°C temperature by continuous stirring the mixture. Cool the mixture to 60°C and again start vigorous stirring and continue until the material becomes fluid after an initial stiffness and looks milky white. This syrup is poured in cardboard moulds and candy cakes about one inch of thickness can be prepared.

Generally sugar syrup is given to bees in a feeder. Feeder consists of a wide-mouthed bottle of 1-2 kg capacity with a few holes in its cover and is overturned over the frames. Division board feeder will also serve the purpose well. The sugar syrup can also be fed to bees in small dishes with a few floating straws. The bees suck the syrup by sitting on the straws. Leaking feeders should not be used. In humid and cold weather preferably fine grained candy can be fed by placing about one inch thick slabs of it over the frames.

Watering

During the active periods, bees collect much amount of water required to dilute the winter stores of thick honey except during a very heavy honey-flow period. Particularly bees collect much amount of water during spring. Large number of worker bees congregating at the edge of streams and pools and sucking up the water are commonly observed. They often visit to house drains of towns and a dripping tap. It becomes necessary to provide them water by different methods, small apiary; one jam jar can be inverted over a board with grooves scored in it, so that enough water seeps out to keep the board moist. The water with little sugar added in first supply and placed in a sunny spot is immediately taken by bees. Some bee-keepers put salt in water which is equally attractive. In large apiary, a tub supported on a stand and a tiny hole, bored in at the bottom to allow a continual drip onto a board placed below provides enough water for bees to collect.

Seasonal Management of Bee Colonies

Seasonal fluctuations adversely affect apiaries. Therefore bee-keepers should handle their bee-colonies in each season in such a manner that the colonies are well prepared for coming honey-flow period.

Seasonal Management

Beekeeping in the country is practised in the following four major geographical regions.

1. Southern peninsular region.
2. Indo-Gangetic plains.
3. Northeast region.
4. Northern and northwest region.

The first three regions represent subtropical to tropical zone, while the fourth constitutes sub temperate to temperate zone. In the former; major swarming and flow season is during February to May. June to September is a wet season with floral dearth that often results in heavy wax moth attack, October-January is a minor swarming and flow season. In the fourth region, November-February is an acute dearth period. March-May is a build up and flow season and June and July is again a dearth period. August to October is a minor swarming and a major flow season especially from plectranthus in the hills of H.P. and Jammu-Kashmir. November and December is a dearth period due to severe winter, sometimes, accompanied by snow fall. Management of bee colonies vary according to the climatic conditions and forage availability in each season.

Spring Management : The advent offspring, particularly in the northern parts of the country, marks the beginning of warm weather and blooming of several tree species and cultivated crops. The active season of the bees starts with the first income of pollen and nectar. The rate of egg laying of the queen improves and brood rearing picks up. The winter packing is removed in places like Srinagar, Jammu & Kashmir, where the colonies are packed for winter. The hive is cleaned, paying particular attention to remove wax moth, rat nests and other pests. If any drone combs are present these too are removed. The queen is examined for its vigour and egg laying capacity. If the queen is old and not perfect the colony is marked for subsequent requeening at the earliest. If the liquid food stores are low, stimulative feeding (50 per sugar) is given so as to boost the brood rearing. Well drawnout worker brood combs are given to the colony to help it increase its brood rearing but such additional combs should not be given when not needed, because the bees may not be able to cover this additional space, and the colony will be exposed to unexpected changes in temperature and to pests. New frames should preferably be given with full comb foundation sheet.

The colony is inspected once every 10-15 days to check for requirement of additional frames with foundation, and to replenish sugar feeding if needed. The emphasis should be on building up the colony strength to its maximum. Weak or queenless colonies are united with all age strength colonies to make them strong. Production from one strong colony is always more than that from two weak colonies. The entrances of comparatively weak colonies are not widened to more than 3 cm to enable the bees to effectively guard their nest against predators.

Flow Management : Management of colonies during the major nectar flow is important, because on these depends the colony's performance with respect to honey production.

Preflow preparation : Preparation for the main flow begins in the previous autumn or winter, particularly in places where winter is severe. At the start of winter, the beekeeper ensures that the colony has a young queen, a large number of young worker force, adequate stores for use during winter and for brood rearing during early spring, adequate comb space and adequate protection during winter. Generally a good flow follows the swarming session. During the active brood rearing season, therefore, swarming is kept in check. This can be accomplished by removing capped worker-brood frames from colonies showing congestion, and giving them to comparatively weaker colonics. Frames with comb foundation are given to replace the removed brood combs. It is a good idea to give these sealed brood combs to weak colonies having a new queen. Brood combs with eggs and young larvae of the weak colonies are given to the strong colonies from which sealed brood combs are removed. By this process, the lower egg laying capacity of the queens of the strong colonies is compensated by providing their colonies with brood combs having eggs and young larvae, laid by the young and vigorous new queen of the weak colonies. The eggs and young larvae also demand attention of nurse bees in the strong colonies, thus reducing its swarming impulse. This management ensures that all the colonies in the apiry are populous and are ready for flow.

Usually steps are taken to strengthen colonies at least six to eight weeks in advance of the flow, so that a large worker force of field age is ready at the start of nectar flow. It is essential at this time for the colonies to have ample stores to enable active brood rearing and build up of colony strength.

Strong colonies are provided with supers having, preferably, drawn out empty combs. Alternatively, super frames with full comb foundation sheets may be given sufficiently early in the season to have drawn out combs ready at the time of flow.

Deep supers have to be used in this case. Presence of drone combs or full of stores in the brood nest creates congestion. They restrict the space for egg laying and brood rearing. Honey bound combs should be replaced with drawn out combs or frames with foundation sheets. If any sealed honey stores are present in some portions of brood combs, the sealing is opened, so that bees can remove the honey from the cells and transfer it to the supers. The cells thus cleaned are available for brood rearing.

The start of honey harvest period is indicated by the whitening of the honey cells and increase in colony weight. At this time, drawn out worker brood combs are given on the sides of the brood nest. This helps in maintaining the tempo of brood rearing and in keeping up the supply of field bees. In places like the rubber plantation areas in the southern states of India, where no pollen flow accompanies the nectar flow, it is important to provide suitable pollen supplement that ensures continued brood rearing.

Supers are provided as and when needed. It is important at the same time to remember that unnecessary addition of supers leads to consumption of large quantities of honey by bees to cover the frames and keep the brood nest warm. According to Singh (1943b) the Indian bees need a maximum of 2 deep or 4 to 5 shallow supers in the flow season in any case, it is advisable to take out completely or two-thirds capped honey frames, extract the honey and to give back the empty combs to the colony. Such extracted empty combs stimulate nectar gathering. Extracting honey and replacing supers at intervals during litchi flow gave more honey yield than removing honey once at the end of flow. It is advocated that honey should be extracted only from super chambers and leaving the stores in the brood combs for the colony for use after the flow season. Because of the excessive relative humidity in the tropical climates, the moisture content in the honey is high. Provision of auxiliary ventilation holes to supers and top cover, as well as placing the colonies in open instead of dense shade, enables bees to ripen honey quickly. At high altitudes and in localities where the flow is in autumn, *e.g.,* Himachal Pradesh, precautions are necessary against cold, by providing some insulation to the hive. Extraction of honey is done in a bee-proof room during the day time, or in the open during late evening or night time. Empty combs are returned to the colonies for cleaning. When the extraction is done at the end of the season, it is best to take precautions against robbing.

Post-flow operations : Most of the supers are removed after the end of the flow and preserved properly for use in the next season. A long dearth often follows the main flow season. It is necessary, therefore, to allow adequate stores within the colony. Sugar syrup (65 per cent) is also fed as and when required, depending upon the length of dearth period. During dearth period, feeding should be carefully given to avoid robbing. Short distance migrations are sometimes possible and profitable for the sustenance of colonies. Extra combs are required to be removed with decreasing strength. This is also a period when enemies like bee eater birds are most active at many places. Therefore, efforts should be made of make the colonies comfortable and keep vigil on enemies and diseases.

Monsoon management : In the tropical and subtropical regions of the country, June to September represents the monsoon or wet season.

Bees face several problems of pests, predators, excessive humidity and starvation. Provision of adequate stores during this period is the most critical aspect for the survival of colonies during monsoon. Protection is given to the colonies against rain water, high humidity, ants, wasps and other pests. Weak colonies are united and crevices, cracks or any opening other than the entrance gate in the hive are plugged. The entrance space is restricted to 6 mm in height an 50 mm in length to prevent entry of wasps. It is preferable to fix a strip of queen excluder sheet at the entrance. Removal of uncovered combs before the monsoon starts and ensuring that the colony is well populated with young bees, covering all combs are measures that prevent wax moth infestation and wasp attacks. Use of ant wells and keeping the hive surroundings clean take care of the ant problem. The hives are kept on stands sloping towards entrance in order to drain out water and prevent its accumulation inside the hive.

A major wet spell of the monsoon is generally over by the end of August, after which there are intermittent sunny days. At this time weeds, grasses and many *Kharif* crops flower and bees start foraging for pollen and/or nectar. On a clear sunny day the colonies are quickly inspected and old, black combs uncovered by bees or infested by wax moth are removed. This management prevents post-monsoon desertions and makes the colonies strong and ready for the ensuing flow or for division.

Summering : In most of the northern plains of India bees need protection from sun and heat. In May and July, the westerly winds are .very hot and ambient temperatures often exceed 45°C. Known as "loo" these hot winds are a serious problem to the bees. Colonies are placed under the thick shade of a tree. The thick canopy of trees also reduces the damage due to bee eater birds. Provision of cool and clean water for bees within the hive, wind breaks in the form of a hedge or live fence around the hives, and use of ant wells to prevent ant attack are measures that help bees in successful summering. Two layers of a gunny leaves are arranged on the lop of the hive to provide shade and humidity to the colony. The leaves are replaced after about 20 days. Wet gunny cloth is also often used to cover the colony for the same purpose. The cloth is kept moist by periodically sprinkling water on it. Rice straw mattings can also be used in place of gunny bags. This practice lowers down the hive temperature by 2 to 3°C.

Winter management : In the upper Himalayan region bees experience severe winter from November to March, and as much as 30 per cent of the colonies are lost annually due to poor wintering. Unlike in summer, management in winter is more for maintaining colony strength with young bee population and a young vigorous queen, nutritional status and hygiene. Colonies with adequate stores can overwinter successfully. Sufficient stores should be ensured, may be by feeding concentrated sugar syrup before winter packing in hilly areas. This facilitates bees in makings compact cluster that has liquid stores close by. Winter is a flow period in north Indian plains and colonies expand from October onwards. Therefore management in there areas is different *i.e.,* adding more combs. All cracks and crevices are sealed and protective measures are taken against cold winds and rain. Shah and Shah (1987) observed that winter packing affects ventilation resulting in a dampening of the hive that leads to dilution of honey stores and development of moulds on the combs. Bees develop dysentery after consuming the diluted and infected honey stores. The life of the hive itself is shortened. They do not, therefore, advise winter packing. In the extreme northern region where the temperature drops down to near freezing point or where there is a snow fall, winter packing is necessary. Colonies are packed well with a tar paper, leaving entrance and ventilation holes open. The boxes are kept tilted towards the entrance side to prevent water accumulation within. Beekeepers in Kumaon hills in Uttar Pradesh use double-walled hives that provide protection against severe cold and help colonies remain strong even during winter. When winter packing is provided, the

Fig. 3.53(A)

colonies are unpacked at the end of February or early in March, after commencement of forbidding activity.

Supersedure and Swarming : Young queens produce more brood than one year old queens. The queen's performance declines drastically after 1.5 years. A gradual reduction in the number of fertilized eggs or increase in drone brood, increase in scattered brood and a reduction in pheromone output mark the failing old queen, At this time bees prepare to replace the old queen under the supersedural impulse. A few queen cells are built mostly on the face of the combs or occasionally at their edges. The queen lays fertilized eggs in these few cells. The supersedure cells are few in number, well built and spacious and the queens developing in these cells are fed liberally and turn out to be of superior quality. When the new queen emerges, gets mated and starts laying the old failing queen is killed. Thus the young queen supersedes the old failing queen.

Reproduction of the colony to produce new colonies is termed swarming. In most parts of India, reproductive swarms occur normally during sensor with abundant supply of pollen and nectar. An increased rate of egg laying and brood rearing precede swarming. As a result of the intensified brood rearing, the colony population comes to consist largely of very young bees. The young nurse bees secrete royal jelly, the supply of which now becomes much above the demand within the colony. Increased number of worker bees also leads to congestion within the hive and the supply of queen's pheromones becomes too inadequate to reach all the worker bees. These factors trigger the swarming impulse. The first symptom of this impulse in the colony is the construction of drone cells and drone rearing. When the first batch of drone brood comes to pupal stage, a number of queen cells are constructed on the lower edges of combs. More queen cells are, raised under swarming (9.0) than under supersedure (2.5) impulse. Under both the impulses, queens are raised from the eggs laid in cells constructed earlier.

When a few of the queen cells are sealed at pupal stage, the mother queen and worker bees leave the colony in a swarm and settle in another location to establish a new colony. The queenless colony left behind with ripe queen cells soon gets a new queen that mates 5-6 days after emergence. If the parent colony is very strong, the daughter queen may also leave the colony with some workers as an afterswarm. A second daughter queen then takes over the colony in the same way as the first. Issue of afterswarms, however, depends upon the strength of the colony and availability of virgin queens or mature queens cells. Swarms are normally issued in the forenoons.

The special features of the swarm queen cells and queens reared in them are; (a) the cells are constructed with planning and deliberation, so that, the quality of the cells and the queens reared in them are excellent, (b) the growing queen larvae get ample supplies of royal jelly, because of the abundance of nurse bees in the colony, and (c) swarming takes place when the forage availability is in plenty and the weather is congenial. Swarming takes place before a major flow. This affects honey production in the colony. For beekeepers who rear bees for honey production, swarming results in serious losses. Even for the commercial beekeeper who raises queens for sale or for queen rearing, natural swarming is a time consuming process fill of uncertainties. Swarm control and prevention are thus taken up as a part of routine seasonal management. Management for swarm prevention consists of removing the factors that induce swarming impulse, like removing sealed brood combs to decongest the colony, providing frames with comb foundation for drawing out combs, providing additional hive space by giving supers or by deliberately dividing the colony and removing the queen to a nucleus hive to fulfill the swarming impulse, and uniting the nucleus after the season.

Absconding as a survival strategy : Hunger swams or colony desertions are frequent after a prolonged period of dearth, generally following winter in the north or monsoon in the south. This can also be described as a resource induced absconding, resulting from a scarcity of nectar, pollen or water. Absconding swarms leave their nest and search for areas with better resources. Great predation pressure in tropical habitats is also a reason for absconding. Frequent swarming ensures survival of the species, by offsetting the high colony death rates and escaping from predation pressure.

Queen Rearing

Colonies produce-queens under swarming, supersedure or emergency impulse. The quality of the queens produced under the emergency impulse is poor, because the colony is queenless and disorganized. In the hurry to get a new queen the bees transform a number of cells with aged worker larvae into queens. The resultant queen larvae do not get proper or adequate nourishment. Compared to these, the queens produced under the swarming or supersedure impulse are carefully nurtured from the egg stage in ready-made queen cells. Grafting is one of the popular methods of queen rearing that involves creation of all the three impulses in varying proportions, but fully under controlled conditions.

Grafting technique : Mass queen rearing by grafting method is possible in *A. cerana* and it is basically the same as for *A. mellifera* but essentially the queen cell cup size should be smaller. The weight of the queen is directly related to its performance. Woyke (1971) found in *A. mellifera* that each day's increase in age of the larva grafted for queen rearing decreased the weight of the queen produced, as also the number of ovarioles in the ovary. Even the diameter and the volume of the spermatheca decreased with an increase in the age of the larva grafted. Queens reared from larvae aged 3 days, mated with fewer drones and had less semen in the spermatheca than queens raised from younger larvae. Even in the case of instrumentally inseminated queens, the number of spermatozoa in queens from older larvae was less, despite the fact that there were plenty of spermatozoa in the oviducts. These facts indicate the importance of choosing the larvae of the right age --as young as possible, but never older than 3 days for rearing queens.

The colonies selected for raising the queen cells are called the cell builder colonies. Different starter and finishing colonies can be used and the queen cell frames are transferred from the former to the latter, a day after the grafted larvae are nursed in the starter colonies. This transfer may, however, involve an important gap in feeding tin growing larvae, and impair the quality of the queens produced. In the tropical climates, this also involves exposure to atmospheric factors, that may affect the larvae. It is, therefore, advisable to have only one colony to start and finish the queen cells.

The colonies must be populous with young of bees, well stocked with pollen and honey and must have ample brood. Preparation and conditioning of these colonies is necessary a week before the grafting operation. Unsealed brood from these colonies is removed and given to other colonies, and sealed worker brood is retained. The colonies are further strengthened by giving them sealed worker brood from other colonies, such that they have about 7 days old workers in abundance at the time of grafting, to nurse the growing queen larvae. On a warm evening all the broad frames with larvae up to 3 days of age are removed and given to other colonies and frames with brood of advanced stages and stores are given to the cell starter colony. This creates conditions similar to those under swarming impulse. The queens are also removed and kept in small nuclei with one or two-frames with bees. Artificial queen cell cups are fixed on wooden bars with wax blocks, or wooden blocks coated with bees wax, and frames with these queen cell cups are placed in cell starters. This ensures proper cleaning of the cups by bees and the cups also acquire the characteristic odour of the colonies.

Grafting is a delicate operation. Care should be taken to avoid damage to the young larvae due to exposure to atmospheric factors. Grafting is done preferably in the late mornings. When the weather is clear and warm, but not very' hot. Atmospheric temperature of 25-30°C and relative humidity of 50-60 per cent are okay at the time of grafting. Grafting can most ideally be done in grafting room with controlled conditions. Colonies showing a combination of desired characters are selected as donors of larvae for queen rearing. Larvae of 12 to 24 hours of age are grafted into the wax cups, that are primed earlier with royal jelly. Care is taken to .select larvae of the same age, since queenless bees in the cell raiser colony prefer older larvae to younger ones in their hurry to get a new queen. This may result in rejection of the larvae. Each frame can have 10 to 12 cups fixed on two wooden bars. This number is usually optimum for the Indian bees, and maintains the proper ratic of grafts and nurse bee strength of the colony, so that all the larvae receive maximum quantity of royal jelly and grow into superior queens. The frame is checked after 24 hours to find out the acceptance of the larvae. With a skilled hand, the acceptance of the grafts would be about 80 per cent. Rejected larvae are removed and replaced with a new graft, preferably of the same age and with least disturbance to the other accepted queen cells.

Double grafting is also used, in which a young larva is grafted, but is replaced the next day with another less than 24 h old larva. Since the second larva gets additional royal jelly given to the first grafted larva, the resultant queen is heavier, than that raised by a single graft.

The nuclei, prepared to receive the mature queen cells are of two types. When the queens are intended for requeening the existing colonies, the purpose of the nuclei is to take care of the virgin queen from the time it emerges to the time it starts laying eggs. For this purpose the nuclei can be small. Baby nuclei made with frames one-fourth the size of normal brood combs serve the purpose quite well. For the purpose of preparing the nucleus to receive the queen cell, normal brood combs are carefully cut into pieces of suitable size, and each piece is tied to a baby frame and kept in the nucleus. Each nucleus can have 3 to 4 such baby frames. The nucleus is provided with ample stores in one or two frames and a good supply of young bees. In the alternative case of making a new colony with the new queens, nuclei are made with the usual size brood frames. These nuclei are 4 to 6-framed, and allow the nucleus colony with the new queen to grow into the normal colony.

Mating nuclei are prepared a day prior to queen cell distribution. On that day in the evening, nuclei are prepared with a couple of combs with brood in all stages-eggs, larvae and pupae, and a frame or two of stores. The nuclei are inspected on the next morning, and any queen cell, if started is removed. After this, the mature queen cells are distributed among them. The queen cell is securely placed between two brood frames. Two days after the queen cells are given the nuclei are inspected to check the virgin queens. On the 6th, 7th or 8th day after emergence the queens fly out on their mating flights. Before the queens go out on mating flights, it is desirable to give to the nuclei a fresh comb of brood in all stages so that the bees in the nuclei do not leave the hive with queens on their mating flights as mating swarms. It is important also to provide adequate stores in the mating nuclei. The mating nuclei are arranged in such a way that the outgoing virgin queens have no difficulty in orientation and return to their own nuclei after the mating flight. An important prerequisite is to ensure that the mating apiary is stocked with

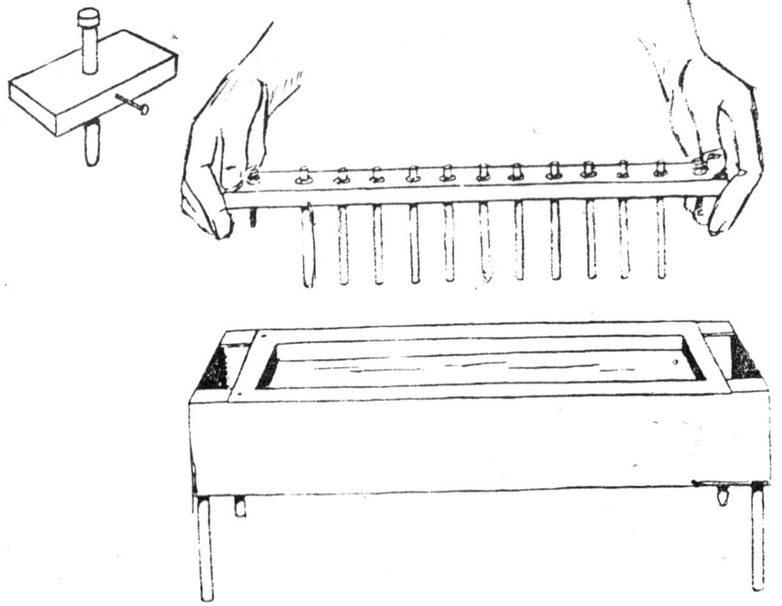

Fig. 3.54 : A former (left) for making in artificial queen cup and a multiple former (below) being dipped into the wax tank.

ample mature drones produced from selected colonies. Normally, on the 10th or 11th day, the first batch of eggs are seen on the central comb indicating successful mating.

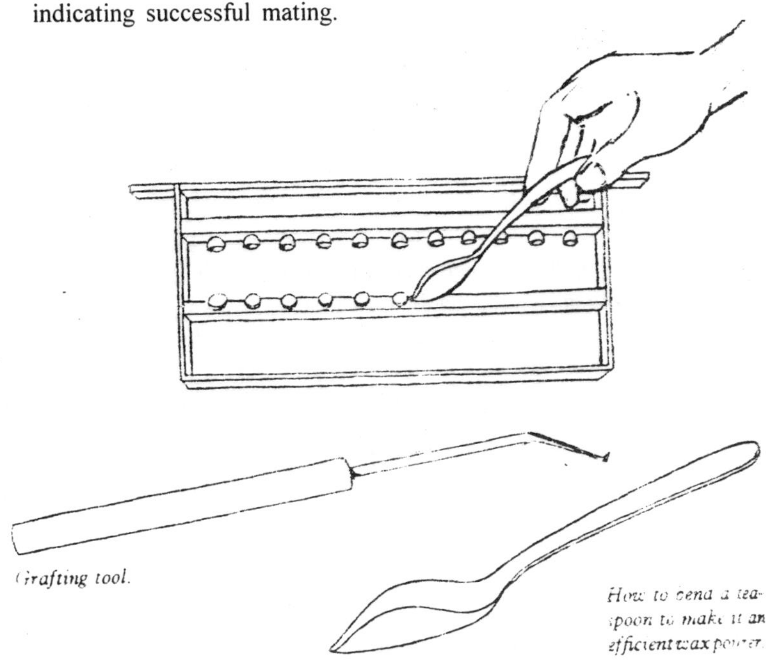

Fig. 3.55 : Artificial queen cups being fastened to a bar with wax. The bar can then be swivelled round so the cups hang vertically.

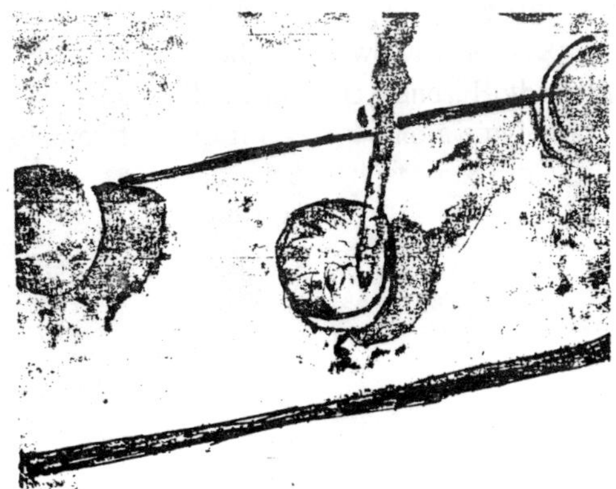

Fig. 3.56 : Grafting a larvae into an artificial queen cup containing diluted royal jelly.

Fig. 3.57 : A grafting frame with cell ready for distribution to mating nuclei in the lower row the fifth cell from the left is too small for use and the end cell has been rejected by the bees.

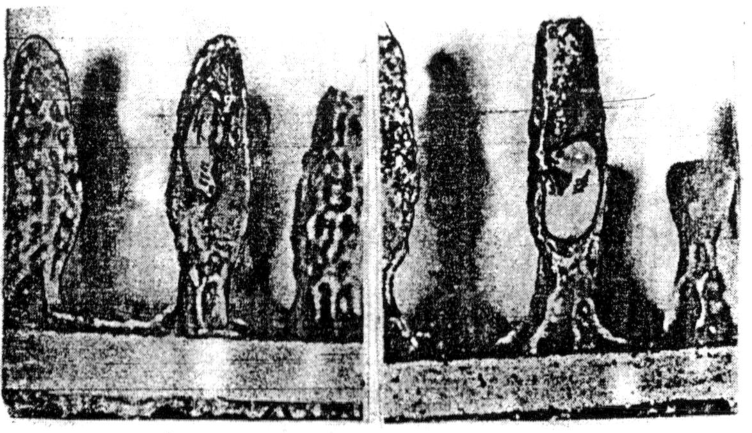

Fig. 3.58 : *Grafted cells opened to show the contents. The larva (left) is almost fully fed, of good size and well supplied with royal jelly. The pupa (right) is well developed and large.*

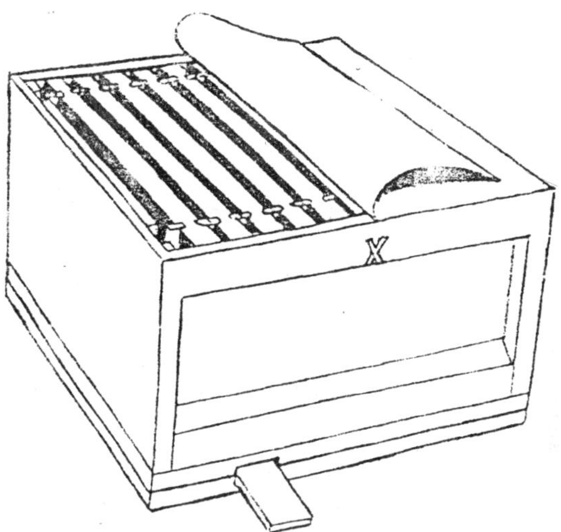

Fig. 3.59 : *A double nucleus box showing the central partition, a mark on the side, a canvas quilt and one of the entrances.*

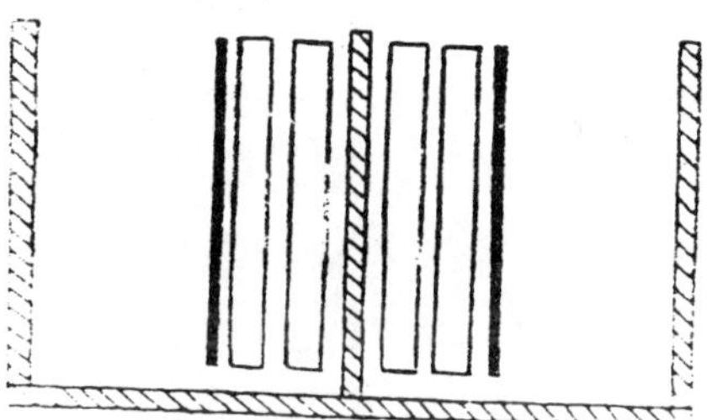

Fig. 3.60 : *A double nucleus box with two frames and a dummy board on each side of the central partition.*

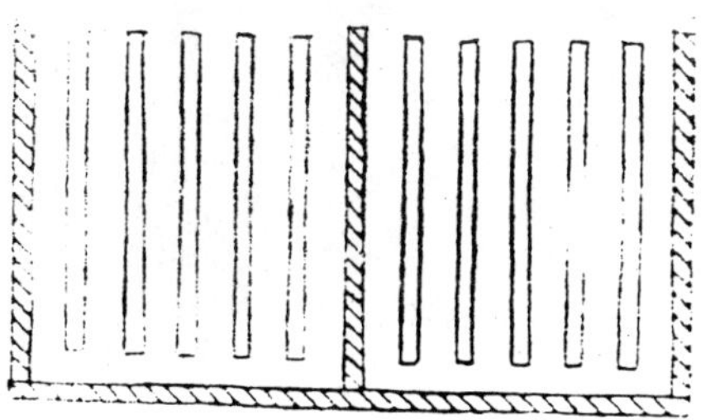

Fig. 3.61 : *A double nucleus box with two five frames on each side of the central division.*

Mating yards

The apiary should have drones produced in the colonies. The quality of the drones produced in natural colonies is not known, and the performance of the queen fertilized with such drones may not be good. It is therefore, advantageous to have separate mating yards for the purpose of getting the queens mated with the desired type of drones. In planned bee breeding programmes, isolated mating yards are usually selected in the farming or orchards areas in the plains where the population of natural colonies is insignificant. In these isolated mating yards the chances of mating between the selected queens and drones from selected colonies are quite high. The mating yards are selected in places that have abundant of bee forage during the period of queen mating and for at least a month after, so that the new colonies have enough time for brood rearing and developing. The colonies are given adequate protection against pests and predators like bee eaters. The latter cause a good number of mating queen losses. To prevent the bee eater problem the mating nuclei are located at a distance from any perching places like electric cable lines.

(F)

BEE POLLINATION

Pollination is the transter of male pollen from the anthers to the female stigma of the same or another flower of the same species. Some plants will self-pollinate; for many pollination by wind (enomophilous); birds or insects (entomophilous); water (hydrophilous) is desirable, or in some cases a necessity. Pollination may be cross involving transfer of pollens between two different flowers. Bees of all kinds are the most effective external agents, due to following some morphological adaptations.

1. Specialised tongue to collect nectar.
2. Body is hairy and the hair are branched; so that pollen grains get entangled very easily.
3. On the metathoracic legs, pollen basket is present which stores pollen. Since these baskets are open, the chances of pollination are increased.
4. The design of the first two pairs of legs are adapted for transferring pollen from the body to the pollen basket. There is also a pollen brush on each mesothoracic leg.
5. Most of the pollinating insects are seasonal, but bees are available throughout the year. This important feature utilises the services of bees effectively during that part of year, when the other natural pollinators would be absent.

The situation today is that intensive farming and the use of insecticides and herbicides has so reduced wild bees and other insects that domestic honeybees worked on a migratory system and becoming more and more important. Since they are the only insects that can be 'worked' by man.

In India, use of cattles, goats on hire basis is a usual practice to increase the fertility of soil. In foreign countries like California, North and South Dacota, Minesota etc. the bee-keeping is practised on hire basis for pollinating the crops naturally. Israel is considered as the country of Honey and milk; this achievement is only due to beekeeping in modern way. Now-a-days in India agricultural scientists are also trying to follow the same path. From the recent research data it is evident that the monitory grains are 10-25 times more, with the pollinating services of bees than that from honey, wax and other bee products.

Agricultural scientists from abroad calculated the increase in production of grains and fruits *per hector* due to use of bee as pollinating agents; strawberry (1.5 times), coffee (1.75 times), brinjal; (1.75 times), mustard (2.25 times); guava, coconut (3 times), sunflower (7 times). They suggests that we can also make use of beekeeping to get maximum yield of fruits, grains and vegetables.

In India, bees can be profitably used for increasing yield. In our country 4 types of bees, the dammer bees *(Trigona and Melipona)*, the rock bee, the hive bee, the little bee are useful as plant pollinators.

Many of our Indian oilseed crops like mustard, niger, sunflowers, safflower, pulses like tur, beans, forage legumes, vegetables like pumpkins, cucumber, tondii, raddish, cabbage, snake gourd, bitter guard (Bhopla), ridge gourd, carrot, cauliflower and onion; fruits like melons, orange, mousambi, lemon, pomegranate, guava, apples, pears and other crops like cotton, coffee, cardamom require cross pollination via honeybees to affect fruit or seed setting and also for improving the quality of seed and fruit.

Now-a-days sunflower cultivation is encouraged for oil. It is a self-sterile plant and therefore needs an efficient natural pollination or moreover in summer this crop is more dependent on bees as other natural pollinators are absent.

As we are aware that bees obtain their food (pollen and nectar) only from the inner part of flower (reproductive part only) as compared to other insects which make use of calyx, corolla etc. for food and destroy floral parts. Thus the bodies of bees come in contact directly with reproductive structures; this fact makes the bees an efficient pollinator.

Experiments have been conducted in actual fields regarding the crop yields by pollination with the help of honey bees alone, or with bees and other natural pollinator alone or with the help of wind as pollinating agent. It has been that honey bees increase the crop yields to a greater extent.

Genetic researches are being conducted to evolve strains of bees which will be crop specific. This area of research therefore holds a tremendous potential for the utilisation of bees for pollination of particular crops. However, yet such studies are only at the experimental stage.

The number of colonies required to pollinate a crop is a variable which depends on the pollination needs of the crop, its size, its flower density, the amount of forage for the bees, the number of competing pollinators, and effectiveness of honeybees on that particular crop. *Two or three colonies* per hectare is a rough and ready average for most crops, with local variations. India has about 50 million hectares of land under bee dependent crops like oilseeds, legumes, vegetables and fruits. Many of these crops require 3 to 9 bee colonies/hectares. We may roughly estimate that about 150 million bee colonies are needed, while we have about 0.5 million colonies presently.

The Central Bee Research Institute, Pune has taken up experiments on the effect of the bee pollination in various crops. The results are as follows-mustard 131%, Sunflower 511%, Niger 112%, Sunflower 675%, Linseed 232%, Onion 178%, Carrot 500%, Radish 700%, Brinjal 125%, fruits 90%, Coffee 83%.

MANAGEMENT OF HONEYBEES FOR POLLINATION

Quantity and quality of crop produced depends upon the proper management of honeybee colonies before and at the time of flowering. Honeybee colonies for pollination should be managed as efficiently as possible to harvest maximum benefit. Following aspects should be taken into consideration while managing honeybees for pollination:

(a) Foraging strength of colonies

(b) Concentration of honeybee colonies needed

(c) Foraging efficiency of colonies and their distribution in the crop.

(d) Moving and conditioning of colonies to crops

(e) Increasing the attractiveness of crops

(f) Increasing the proportion of pollen gatherers

(g) Breeding honeybees for pollination

Foraging Strength of Colonies

Colonies to be used for pollination should be at least of 8-9 frames strength. In areas where brood rearing is diminished or ceases during winter, colonies should be fed with pollen substitutes/supplements in addition to their feeding of sugar syrup. Generally, 8-9 frame colonies are more useful for pollination than 12-15 frame colonies because deterioration in foraging conditions discourage foraging relatively more from large colonies than from small ones. When colonies are hired for pollination, the size is usually assessed as the number of combs occupied by bees or brood or the number of seams between combs occupied by bees or more rarely, the area of comb occupied by brood.

Honeybee colonies must be removed immediately when the flowering of the crop to be pollinated has ceased so as to avoid over exploitation of the wild flora to the determinant of native pollinators.

Concentration of Honeybee Colonies Needed

For the pollination of a given area of a crop, the number of colonies required is based on the experiences and the assumption of the growers and crop specialities rather than on experimental results. The honeybee population needed to meet the pollination requirement of the crop however, depends mainly upon the density of flowers, their attractiveness, competing insects and crops, type of crop whether self or cross-pollinated, the age of orchard/crop and location of the orchard crop. Younger orchard trees will have fewer flowers to be pollinated and will need fewer colonies of bees. Orchards/crops prone to wind or that lack warmth and sunlight due to geographical conditions may have the need of additional colonies to compensate the reduced bee activities during marginal weather.

The number of colonies required for optimum pollination of crops also depends upon their pollination requirements. Orchards having more number of self-fertile cultivars require less number of honeybee colonies per ha as compared to self-unfertile cultivars. When it is known that a crop is difficult to pollinate because the species is relatively unattractive to honeybees then more than 2.5 colonies (often 3 to 6) per hectare are sometimes recommended. In contrast, when a crop is attractive but its flowers are sparse or can be self-pollinated, fewer than 2.5 colonies per hectare are suggested.

Generally colonies for orchard pollination should have minimum of 4 to 6 brood combs. The number of colonies required are determined as below:

(i) Increase the concentration of colonies progressively until it makes little or no difference to the concentration of foragers on the crop. To overcome competition from more attractive crops, attempts are sometimes made to increase the strength of colonies to saturation level. For the pollination of fruit trees, colonies should be provided for maximum rather than for minimum pollination; too great a set can be corrected by pruning and thinning but too small a set is a direct loss.

(ii) Place a honeybee colony in the centre of orchard/crop and record the number of bees per unit area at different distances from the hive and later fruit/seed set at these distances. Determine the appropriate distance at which fruit/seed set is significant.

Considering this distance as a radius, calculate the effective area pollinated by one colony. Finally, total colonies required per unit area can be calculated.

(iii) Relate the number of foraging honeybees per square metre bloom area or per branch or per plant or per given number of flowers needing pollination with seed or fruit set. In both the above methods feral populations of bees should be taken into consideration.

On the basis of various characteristics, colony requirement for optimum pollination of some crops has been determined. The number of honeybee colonies required per hectare are given in Table 3.15.

Table 3.15 : Requirement of honeybees colonies for pollination

Crop	Colonies/ha	Crop	Colonies/ha
Alfalfa	10 to 12	Medicago (Lucerne or alfalfa)	6 to 18
Almond	2 to 4	Melon	2.5
Apple	2 to 3	Onion	8 to 10
Berseem	3 to 4	pears	4 to 5
Buckwheat	3	Pumpkins	0.2
Carrot	6	Rapeseed	4
Cauliflower	2	Red clover	
Cherry	2.5	Late flowering crop	3
Citrus	2.5	Early flowering crop	6 to 8
Cotton	1.0 to 2.5	Sunflower	2.5
Cucumbers	6 to 20	Watermelon	2.5
Hairy vetch	3.5 to 7.5	Whiteclover	0.2 to 10
Kiwi fruit	8		
Ladino clover	2.5 to 3.5		

Foraging Efficiency of Colonies and Their Distribution in the Crop

Depending upon various factors, namely availability of forage, weather conditions and physical features of the area including shelter belts, honeybees forage at a considerable distance from the hive; initially they confine to the vicinity of their hives and then gradually expand.

It is found that when colonies were put in a new location, the foraging range of bees from their colonies extended upto 200 metres on the first day; 300 metres on the second and third day and 800 meters on the fourth day; strong colonies expanded their foraging areas more rapidly. In general, economic distance for nectar collection does not exceed 0.4 km for *A. C. indica* and this is important in the context of use of bees for pollination.

General tendency of the bees to forage as near the hive as possible is more for pollen gatherers than for nectar gatherers. Weather has pronounced effect on foraging. In poor weather, most bees restrict to foraging on nearby flowers or remain in the hive.

The effective pollination of a crop expends upon the distribution of honeybee colonies in it. In a field, colonies should be distributed singly but it is convenient to both growers and beekeepers to put colonies in groups as large as possible separated by appropriate distance. However, it is important to find the maximum size of groups that can be distributed equally throughout the crop concerned so that the overlap between adjacent groups is sufficient to prevent the number of foragers decreasing midway between them. To obtain an even distribution of bees on a crops influence of any topographical gradient and wind direction should also be taken into consideration. Hives should be placed at such a site so that bees can fly out against the prevailing winds and back to hive in the direction of the wind.

Moving and Conditioning of Colonies

Generally colonies are moved at night but when day time moves are necessary and colonies are likely to be exposed to hot dry conditions, colonies and hives should be sprayed with water. Moving colonies for longer distances and for longer period often kill their brood and receive a set back. All sealed honey frames should be extracted before moving the colonies. During favourable conditions in spring, moving colonies nearness to a good source of forage, generally overcome the damage done to them during their move and good honey crop produced by the moved colonies being a reflection of their larger size. However, under poor foraging conditions, moving colonies has a detrimental effect produced because of the size difference and poor honey crop.

Following points must be kept in mind while using bees for pollination:

(i) colonies should not be taken to a crop needing pollination until it is sufficiently flowing to be predominant species in the locality.

(ii) colonies should not be taken to crops until flowering has begun or until there are enough flowers for the bees to work because if the foragers have previously been visiting other flower species in the locality they will not readily forsak them.

(iii) when the crop needing pollination is less attractive to bees or has a short flowering period, a delay in taking colonies until flowering has begun always increases pollination.

(iv) colonies should be replaced by others as soon as the bees start working in areas or on species other than those intended. In orchards, colonies could advantageously be exchanged between orchards, several miles apart every third or fourth day and could, if necessary be returned to their original sites 7-10 days later.

(v) in orchard crops like apple and kiwi fruit, it has been observed that dense growth of white clover reduces attractiveness of honeybees to main crops. Therefore, white clover growth should be mown at regular intervals to reduce competition.

Increasing the Attractiveness of Crops

The bees can be directed to the crops in various ways:

(a) by feeding sugar syrup containing the 'scent' of target crop to the colonies. This is done by immersing flowers of the crop in sugar syrup for some hours, strain off the flowers and feed the syrup to the colonies.

(b) the crop can be sprayed with sugar syrup to which bees are encouraged to visit but the effect of spraying is very local and temporary. Bees will generally forage for the sugar syrup drops and their numbers may not actually increase on the flowers after the sugar syrup droplets have exhausted.

(c) by applying the attractive odours present in the pollen (*e.g.*, C_{18} straight chair trienoic acid) to a crop would increase the population of foraging bees but they might be diverted from the flowers and spend much time in scaching the leaves/stems for pollen odour.

(d) by increasing the attractiveness of the crop itself to bees by long term selective breeding of certain crop species and simultaneously for the factors responsible for this attractiveness.

Increasing the Proportion of Pollen-gatherers

Pollen gatherers are more efficient pollinators than nectar gatherers. Therefore, for effective pollination, it is beneficial to increase their number and it can be achieved in the following ways:

(i) Feeding sugar syrup—it should be given only when pollen from the target crop is available.

(ii) Remove stored pollen from colonies and give them extra combs.

(iii) Increase brood rearing of colonies by feeding them with pollen of pollen substitute. However, pollen substitute should be as attractive as the pollen available in the field otherwise it will diminish pollen gathering.

(iv) Pollen foraging can also be induced through some manipulations in the hive. It can be done by directly opening the hive entrance on the brood area or by extending the hive entrance as a narrow tunnel to the brood combs when they are at right angles to it or by using narrow strips of wood immediately beneath the brood comb when they are parallel to it.

Breeding Honeybees for Pollination

Pollen collection in a colony is directly related to its colony needs. Successful selection of pollen hoarders, could result in colonies that collect more pollen irrespective of their needs and so would be better pollinators than available colonies. High line colonies had 2-15 times as much hoarded pollen as low line colonies. Such selection could be especially helpful in increasing honey bee pollination of some species but continued selection could be counter-productive if brood rearing space became limited by the presence of stored pollen.

It has been found that when colonies were put beside a *Medicago sativa field* some collected a much greater pollen than others. Later they found that colonies headed by sister queens that had been inseminated from their brothers were more similar in the proportion of *M. sativa* pollen they collected than those colonies headed by unrelated queens and consequently selected lines showing high and low preference for *M. sativa.* There was continuous increase in the percentage of pollen collectors in the high preference line and decrease in the low preference line upto 6th generation of selection. Selection of a seventh generation in the high and low preference lines produced little change in the percentages of alfalfa pollen collected over the fifth and sixth generations and they concluded that a plateau had reached in both lines from which further improvement was unlikely. This work showed that tendency to collect *M. saliva* pollen was heritable and remarkably few lines were needed to achieve the separation into high and low preference lines.

RECENT TRENDS IN POLLINATION WITH BEES AND FUTURE THRUSTS

Now a days there is a great emphasis on the use of 'pollen dispensers' or 'pollen inserts' to disperse pollen from selected cultivars/varieties through honeybees. This is most popular in orchard crops. The device is fitted to the hive entrance and the outgoing foragers are forced to walk through pollen of required compatible cultivar and the foragers carry some with them when they go for foraging. However, there is a need to find out whether the use of honeybee colonies with pollen dispense in an orchard without pollinizers would give more cross-pollination than the use of honeybee colonies alone in an orchard adequately provided with pollinizers. The device can also be usefully employed in the hybrid seed production of many crops.

Pollen loads collected by bees in their corbiculae can be successfully used for pollination of crops after removing the inhibitory effect of honey or honeybee's hypopharyngeal glands secretion on pollen germination. It has been reported that the use of 50 per cent sucrose solution first to wash the loads and then in a sequentially lower sugar solutions, and finally to dry the pollen in cold dry air. It has been found 40-60 per cent pollen germination after 1-2 years of freeze preservation. The method for removing the inhibitory effect on pollen loads needs to be improved for better pollen germination. Further, the method by which bees of particular colonies can be directed to collect pollen from specific cultivars needs to be standardized.

Recently use of synthetic pheromones particularly the synthetic Nasanov pheromone has been advocated to direct the bees to visit the crops which are marginally less attractive. Encouraging results have been obtained on onion and apple. Similarly, synthetic alarm pheromone can be used to repel bees from unrewarding food sources or from pesticide treated crops. The use of these synthetic pheromones on commercial scale, however, needs to be tested.

To increase the cross-pollination of crops by way of quick pollen transfer through honeybees, Free *et. al.* (1991) have developed a pollination enhancer. Inserting the rows of soft nylon bristles at the entrance so that foragers while entering and leaving rub against them, increases both the number of pollen grains carried by the departing foragers. This technique, however, needs testing on large number of crops particularly those which are used for hybrid seed production.

There are various ways to increase the attractiveness of crops to honeybees. Recently the application of attractive odours present in the pollen (*e.g.*, C_{18} straight chain trienoic acid) has been advocated to increase the population of foraging honeybees on a crop. However, there is a need to devise a method to apply this odour only on the floral parts and not on the leaves/stems so that bees do not waste much of their time in searching all these parts without pollinating the crops.

Pollination in Green Houses/Cages

Honeybees, bumble bees, blow flies and others have been used to pollinate flowers in the greenhouses, glasshouses or cages. Attempts have been made to increase the population of solitary bees by inducing them to occupy artificial nests. Many crops like *Brassica, Cinchorium, Raphanus* etc. have been reared by providing artificial nesting sites and managed for pollination in various parts of the world. Use of honeybees to pollinate fruit, vegetable and ornamental flowers in greenhouses has increased considerably. They have been used to pollinate strawberry, brussels sprout, muskmelans, onion, runner bean, tomatoes etc. in greenhouses or in large pollination cages.

Pollination by honeybees inside the greenhouses depend upon type of material used for constructing green houses. A cage influences the light intensity, temperature, humidity and wind speed to which plants inside are subjected. However, the extent to which it does so varies with different weather and climatic conditions and with different types of cages. Honeybees seen unable to forage in greenhouses made of ultraviolet opaque polymethyle methacylate sheet and fibre glass but foraged well in air inflated polythene "bubble" greenhouses; nylon screen cages erect within fibre-glass greenhouses and inside large polythene tunnels.

For most circumstances, colonies containing 3 to 4 combs of brood and bees are adequate for pollination in cages. It can be disadvantageous to use colonies that are too large for the areas needing pollination. More bees inside pollination cage can badly damage the anthers, stigmas and carollas. Strength of bees should be according to the size of the cage. However, pollination results are not that satisfactory due to the difference in environment inside and outside the greenhouses. Therefore, large number of materials which can be used for making green houses need to be screened, so that the bees forage inside on if they were working in the open. The management of bees inside the greenhouses needs attention.

Difficulties in Pollination

The natural unfavourable conditions of climate like temperature, rainfall and wind are the main difficulties; under optimum conditions rate of pollination get increased. Other serious problem as discussed earlier is the insecticidal sprays. It can be solved by many ways:

(a) By using less harmful insecticides.

(b) Spraying must be done during evening when the bees are inside the hive.

(c) Queen gate of the colonies should be kept closed during spraying.

(d) Temporary transfer of bee colonies to other areas.

(G)

BEE DISEASES AND ENEMIES AND THEIR CONTROL

Like other living organisms honey bees are also attacked by their natural enemies and infections. Both the adults and broods of honeybees are susceptible to the attracts and bee diseases are classified into *brood diseases* and *adult diseases* depending on stage of development, as well as causative organism.

BROOD DISEASES

Honeybees develop in the cells of combs where temperature and humidity are controlled by the adult bees. Brood combs from healthy colonies typically have a solid and compact brood pattern. Almost every cell from the centre of the comb outwards contain an egg, larva or pupa.

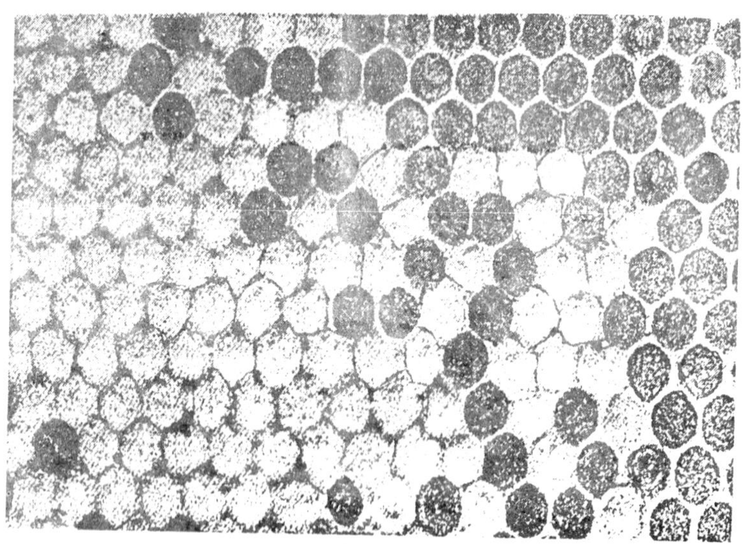

Fig. 3.62 : A part of the comb showing healthier brood.

Table 3.16 : Comparative symptoms of various brood diseases

Symptoms	American Foul brood	European Foul brood	Sacbrood/ Thai sac brood	Chalkbrood	Stonebrood
1	2	3	4	5	6
Causative agent	A spore forming bacterium, *Bacillus larvae*	A bacterium, *Melissococcus pluton*	A virus Morator *aetatulus*	A fungus, *Ascosphaera apis*	A fungus *Aspergillus flavus*
Appearance of brood comb	Sealed brood, discoloured, sunken or punctured cappings.	Unsealed brood, some sealed brood in advanced cases with discoloured, sunken or punctured cappings.	Scaled brood, scattered cells with punctured cappings often with two holes.	Sealed and unsealed brood. Affected larvae usually on outer fringes.	Some perforated and covered with a greenish layer.
Age of dead brood	Usually older sealed larvae or young pupae upright in cells	Usually young unsealed larvae, occasionally older sealed larlvae. Typically in coiled stage.	Usually older, sealed larvae; occasionally young unsealed larvae. Upright in cells.	Usually older larvae; upright in cells	Dead larvae in unsealed and sealed cells.
Colour of dead brood	Dull white becoming light brown. coffee brown to dark brown or almost black.	Dull white becoming yellowish white to brown, dark brown or almost black.	Greyish or straw coloured, becoming brown, greyish black or black. Head end darker.	Chalkwhite, sometimes mottled with black spots.	Green yellow, hard and shrunken.

Consistency of dead brood	Soft becoming sticky to ropy	Watery, rarely sticky or ropy; Granular	Watery and granular; tough skin forms a sac.	Watery to paste like.	Attacked cells can have greenish, mouldy appearance.
Odour	Slight to pronounced glue odour to glue pot odour. Putrid faint.	Slightly sour to penetratingly sour.	None to slightly sour.	Slight, non-objectionable	Mouldy in advanced stage.
Scale characteristics	Uniformly lies that on lower sides of the cell. Adheres tightly to cell wall. Fine thread like tongue of dead pupae may be present. Head lies flat, brittle, black.	Usually twisted in cells. Does not adhere tightly to cell wall, rubbery black.	Head prominently curled toward centre of wall, does not adhere tightly to cell wall, rough texture, brittle, black	Does not adhere to cellwall Brittle, chalky white, mottled or even black	None

The cappings are uniform in colour and are convex (higher in centre than at margins). The unfinished cappings of healthy brood may appear to have punctures, but since the cell are always capped from the outer edges to the centre, the holes are always centred and have smooth edges (Fig. 3.62).

AMERICAN FOUL BROOD

Beekeepers in temperate and sub-tropical regions all over the world generally consider American foul brood as perhaps the most destructive

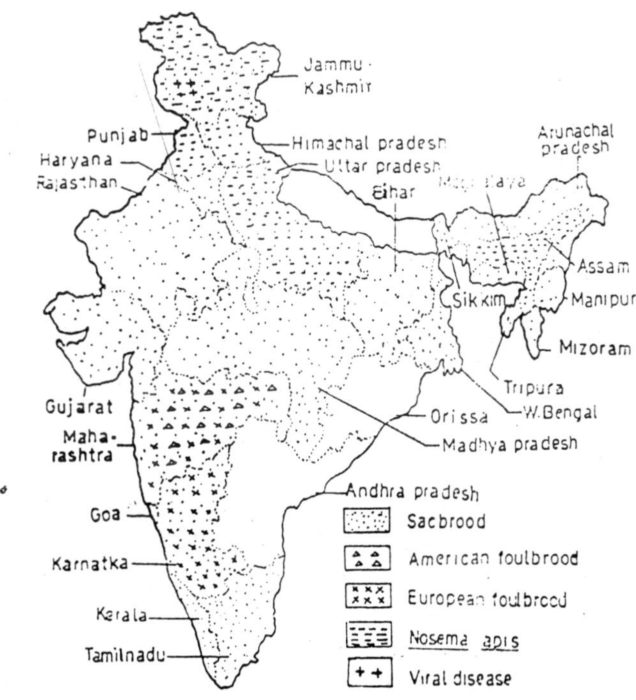

Fig. 3.63 : Map of India showing spread of honeybee diseases.

microbial disease affecting bee brood. The disease neither originated in nor it is confined to the America. It is widely distributed, wherever colonies of *Apis mellifera* are kept. In tropical Asia, where sunlight is abundant and temperatures are relatively high throughout the year, the disease seldom cause severe damage to bee colonies. Nevertheless, the spore of this bacterium are very resistant to heat, disinfectants and drying out and can remain viable for long period of about 35 years. American foul brood disease is not of much concern in India, although an isolated report of its occurrence was made in Jeolikote (Nainital), Uttar Pradesh in *Apis cerana indica* colonies. The reoccurrence however, is not reported from any part of India subsequently.

Cause : American foul brood disease is caused by a spore forming bacterium, *Bacillus larvae* (Fig. 3.64). *B. larvae* is slender rod with slightly rounded ends and a tendency to grow in chains. The rods vary greatly in length from 2.5 to 5.0 μ and are about 0.5 μ wide. The spore is oval and approximately twice as long as wide, about 0.6 × 1.3 μm. Approximately 2.5 billion spores are produced in each infected larva. It affects bee brood only; adult bees are safe from infection. At the initial stage of colony infection, only a few dead older larvae or pupae are observed and subsequently, if remedial action is not taken, the disease spreads within the colony. It can quickly spread to other colonies in the apiary as a result of robbing, drifting of workers, or contamination through the bee-keeper's hive manipulations.

Symptoms: At the initial stage of AFB infection, isolated capped cells from which brood has not emerged can be seen on the comb. The caps of these brood cells are usually discoloured, darker than the caps of healthy cells (Fig. 3.65), sunken and often punctured, whereas the caps of healthy brood cells are slightly protruding and fully closed. As the disease spreads within the colony a scattered, irregular pattern of sealed and unsealed brood cells can be observed which is different than compact pattern of brood cells in healthy colonies. The infected brood die at prepupal or late larvae stage. At first, the dead brood is dull white in colour, but it gradually changes to light brown, coffee brown, and finally dark brown or almost black. The consistency of the decaying brood is soft.

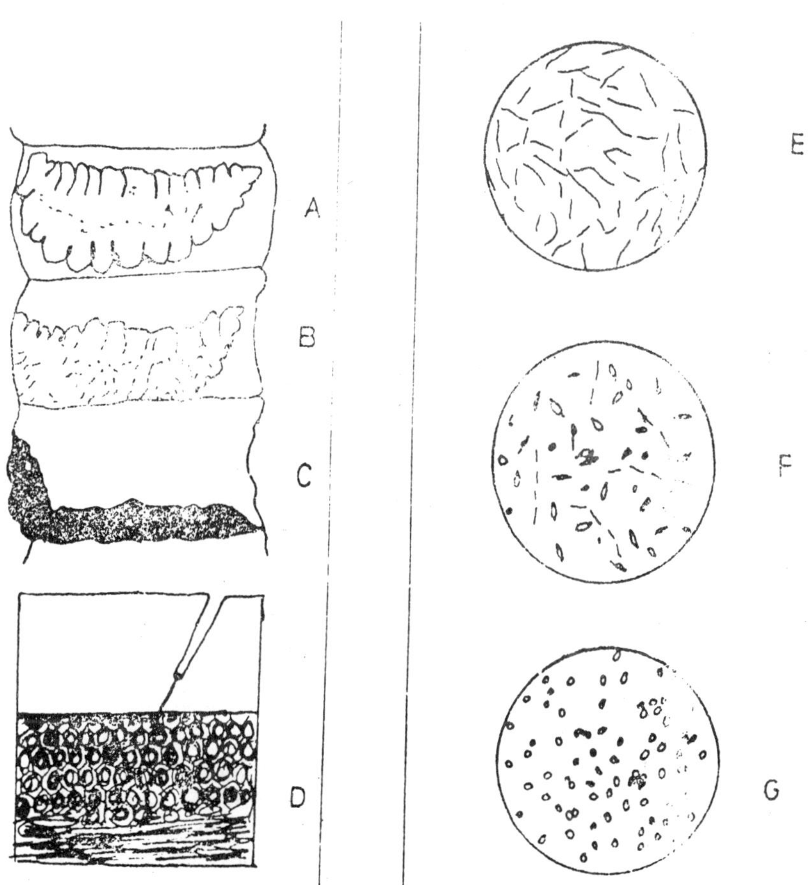

Fig. 3.64 : Showing (a) healthy larva (b) APB infected larvae in ropy stage (c) dried scale of dead larva (d) A dry stick inserted in the diseased larvae showing ropy stage, and bacterium that cause AFB (e) *Bacillus larvae* (vegetative cells); (f) *B. larvae* (spore formation), (g) *B. larvae* (spores).

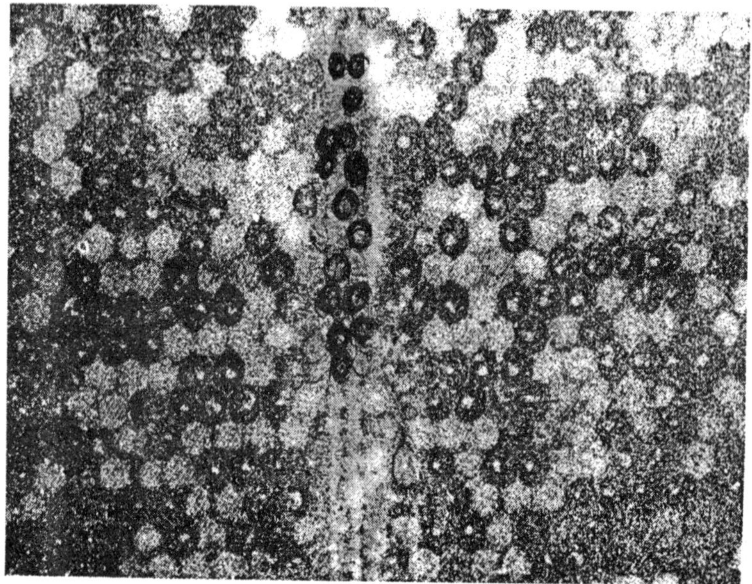

Fig. 3.65 : A part of the comb showing heavily infested brood with *Bacillus larvae* causing American Foul brood disease.

Diagnostic Procedure : A simple way of determining whether the death of the brood was caused by AFB is the "Stretch test". A match stick or tooth stick is inserted into the body of the decayed larva and then, gently and slowly, withdrawn. If the disease is present, the dead larval content adhere to the tip of the stick, stretching upto 2.5 cm before breaking and snapping back in somewhat elastic way. This "ropiness", typifies American foul brood disease, but it can be observed on decaying brood only (Fig. 3.64d). Once the dead brood have dried into scales, the test cannot be applied. The dry brood lie flat on the lower side of the cell wall, adhering closely to it. The scale is usually, black or dark brown in colour, is brittle, often a fine threadlike proboscis or tongue of the dead pupa can be seen protruding from the scale, angling toward the upper cell wall.

Control of American Foul brood disease

Following methods are employed for its management:

1. The management practices similar to those for TSBV could be adopted in AFB.

2. *Chemotherapic treatments* : Various chemicals have been recommended for the control of AFB which include:
   - Feeding of sodium sulphathiazole @ 0.1 g/litre in sugar syrup.
   - Feeding Oxytetracycline (Terramycin) 0.25 to 0.4 g in 5 litre sugar syrup.
   - Feeding streptomycine in sugar syrup @ 0.05-0.15 g/litre.
   - Dust Terramycin (TM50) in powdered sugar (1:20) @ 4 tea-spoon full on top bars of the brood frames.
   - Give antibiotics in the patties of pollen substitute/supplement.

The efficiency of drug treatments varies greatly. Chemotherapy has no effect on spores that contaminate the equipment. Chemotherapy is not advisable when the incidence of disease is low and can be economically managed by destruction of few colonies. Furthermore, the development of resistance of *B. larvae* to drugs is always a possibility. Nevertheless, chemotherapy can efficiently decrease infection if combined with suitable management practices and manipulations in the colonies.

Sterilization of Combs

The fumigation of empty combs and equipment is probably the safest method.

1. Formalin @ 150 ml/lt, of water at normal temperature can fully sterilize the combs. However, the main problem is the sterilization of capped cells and toxicity to bees and honey contamination.
2. Fumigation with chlorine gas is effective but it makes the combs brittle and corrodes the metal parts.
3. Knox *et.al.* (1976) recommended the use of ethylene oxide @ 1 g/lt for 48 hrs at 43°C in fumigation chambers.
4. Radiation from cobalt 60 (CO^{60}) or high energy electrons has been shown to inactivate spores of *B. larvae* in combs and honey from 0.1 to 1.0 Mrad have been suggested for effective control of AFB.
5. Hornitzky and Wills (1983) found that 99.9994 per cent spores of AFB were inactivated by a dose of one Megarad (Mrad) of Gamma radiation. *Liu et.al.* (1991) in a similar study reported that combs heavily contaminated by AFB when sterilized by gamma radiation @ 1.2 Mrad did not develop any infection.
6. Uchida and Nagata (1986) found that combs immersed in iodehole solution (100 times dilution) for 24 hours at 23°C

reduced number of spores from 5.7×10^8 to 1.4×10^5. Furthermore, no abnormalities in bees or brood rearing were observed.

7. Feeding of cinnamon oil (1.2 g) mixed with 1:1.5 honey and sugar semi-solid mixture (3 kg) to bees in six doses @ 0.5 kg per 10 day interval proved effective to control *B. larvae* infections in *A. mellifera* in Sardinia.

European Foul Brood

As is the case with American foul brood disease, the name of this bacterial bee-brood disease is inappropriate. The distribution of European foul brood disease is not confined to Europe alone but the disease is found in all continents, where *Apis mellifera* colonies are kept. In India, EFB has not been recorded so far in *Apis mellifera* colonies, however the disease was reported to occur in certain pockets of Maharashtra (*e.g.*, Mahabaleshwar) and in Karnataka around Castlerock near the Goa border. The rest of the country is free from EFB (Fig. 3.63).

Cause : The pathogenic bacterium of FFB is *Melissococcus pluton*. It is anceolate in shape and occurring singly in chains of varying lengths, or in clusters (Fig. 3.66). The bacterium is Gram-positive and does not form spores. While many strains of *M. pluton* are known, all are closely,, related, and they cannot be distinguished by serological methods. The bacterial cells measure 0.5-0.7 by 1.0μm. The disease is generally a stress related and less serious than AFB.

Symptoms : Honeybee larvae killed by EFB are in younger stages than those killed by AFB. Generally speaking, the diseased larvae die when they are 4 to 5 days old, or in the coiled stage. The colour of the larva changes from shiny white to pale yellow and then brown, as it decays. When dry, scales of larvae killed by EFB do not adhere to the cell walls but can be removed with ease. The texture of the scales is rubbery rather than brittle, as is AFB. A sour odour can be detected from the decayed larvae. Another symptom typifying EFB is that most of the affected larvae die before their cells are capped. The infected larvae appear somewhat displaced in the cells. Because of the lack of adequate number of nurse bees to feed them, they lose their lump appearance and appear underfed. When a scattered pattern of sealed and unsealed brood is observed in a diseased colony, this is normally an indication that the colony has reached a serious stage of infection and may be significantly weakened.

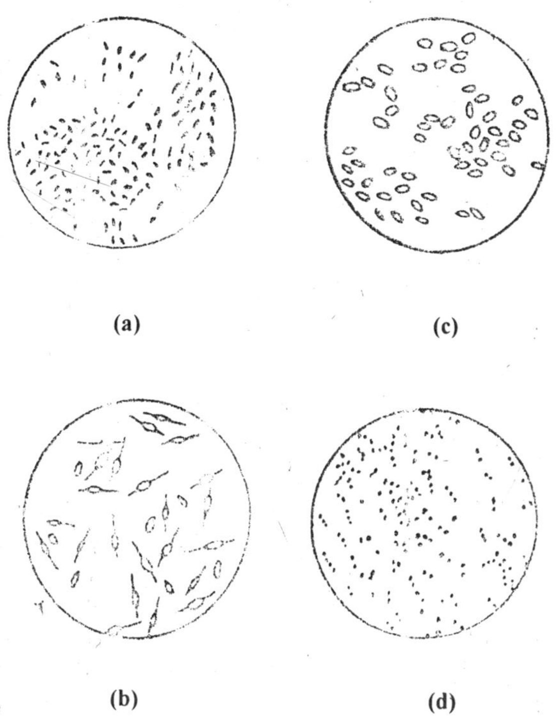

Fig. 3.66 : Bacterium that cause, European foul brood disease in honeybee (a) Melissococcus pluton (b) *Bacillus alvei* (c) *B. laterosporous* and (d) *Enterococcus faecalis.*

European Foulbrood Disease : Control Methods

The treatment of European Foulbrood is usually not required if the infection is light because the colonies can overcome the infection without assistance. A good steady nectar flow sometimes help elimination of the disease. In case of severe infection, treatment of EFB is relatively difficult as it has number of causal agents such as *Streptococcus faecalis* and *Achromobacter eurydice* besides number of bacteria that produce symptoms which mimic the disease. However, the following methods may be employed for the control of EFB disease:

1. Management practices are similar to Thai sacbrood virus disease
2. Sterilization of combs and honey could be done with formaldehyde and acetic acid. These fumigants have been reported to kill the resting stages of *Streptococcus pluton.*

Ethylene oxide and gamma radiation upto 0.8 Mrad proved ineffective.

3. *Chemotherapic treatments :* 0.5 or 1 g oxytetracycline (Terramycin) dissolved in 500 ml of concentrated Sucrose syrup sprinkled on bee clusters gave good result.
4. Breeding of strains resistance to disease can be a semi-long lasting measure.

Further research is needed to establish the pathology of EFB as the taxonomy of *S. pluton* and *A. eurydice* is still questionable and more research is needed to firmly establish the legitimate taxa.

FUNGAL BROOD DISEASES

Fungus diseases of bees appear less destructive than the bacterial diseases. These are chalk brood and stonebrood diseases.

Chalk Brood

The causative agent of this disease is *Ascophaera apis*. *A. apis* is a heterothalic fungus which develops the spore cyst when opposite thallic strains (+ and -) fuse. Spore cysts measure 47-140μm in diameter (Fig. 3.67). Spore balls enclosed within the cyst are 9-19μm in diameter, and individual spores are 3.0-4.0μm by 1.4-2.0μm. Diseases of fungal origin are generally more prevalent in damp and cool conditions. For this reason, chalk brood is more commonly found in spring and early summer. Colonies rarely die from the disease, but in some cases honey yields may be reduced. In India, chalkbrood disease has not been recorded so far. In temperate America and Europe, however, cases have occurred in which chalk- brood has caused serious damage to colonies and Indian bee-keepers should, therefore, be aware of the problem. Bailey (1967) opined that *A. apis* is an opportunistic pathogen and kills the individual larvae only when they are exposed to other stresses. Stress factors in various geographical areas should be examined in relation to the expression of the disease. Evidently, in this disease, dynamics of the pathogen and of the bee colony need to be explored in unison. The spread of the disease occurs through ingestation of spores which germinate within the gut, hyphae penetrate the gut walls and under favourable conditions all the softer tissues are attacked by mycelia.

Symptoms : As its name implies, it affects honeybee larvae or brood. Infection by spores of the fungus is usually observed in 3-4 days old larvae. Initially, the dead larvae are swollen to the size of the cell and covered with the whitish mycelia of die fungus. Subsequently; the dead larvae become mummified, hard, shruken and chalk like in appearance.

The colour of the dead larvae varies with the stage of growth of the mycelia; white at first, then grey, and finally black when the fruiting bodies are formed. When infestation is heavy, many sealed brood die and dry out within their cells and when such combs are shaken the mummified larvae give off a rattling sound. If only one strain (+ or -) of mycelium is present, the larva dries into a hard, shrunken, white chalk like mummy, thus the name chalk brood.

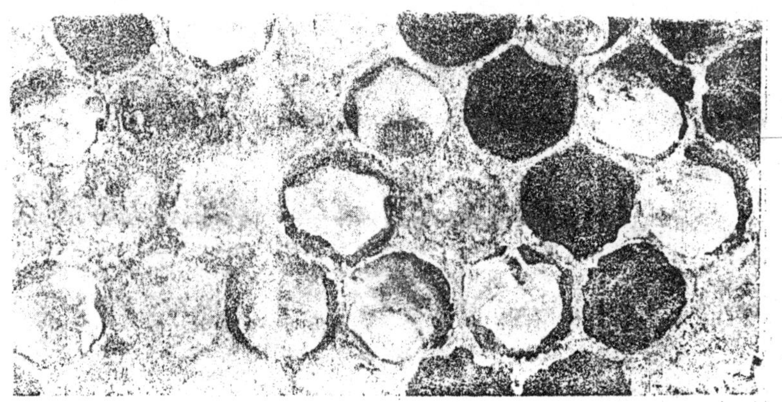

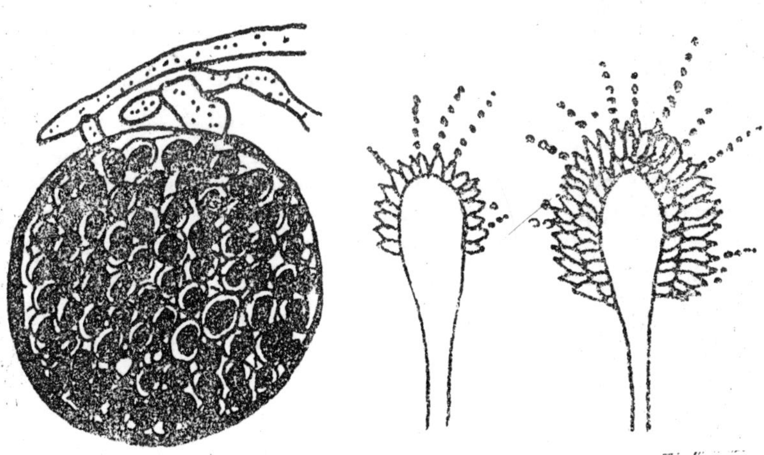

Fig. 3.67 : Spore cyst of *Ascosphaera apis* which causes chalkbrood disease in honeybee larvae, (b.) conidial head of *Aspergillus flavus* Link, causing stone brood disease in honeybees.

Stone Brood

Stone brood disease is usually caused by *Aspergillus flavus*, but other fungi belonging to the genus *Aspergillus* can also cause this disease (Fig. 3.68). These fungi are soil inhabitants that are also pathogenic to adult bees, other insects, mammals and birds.

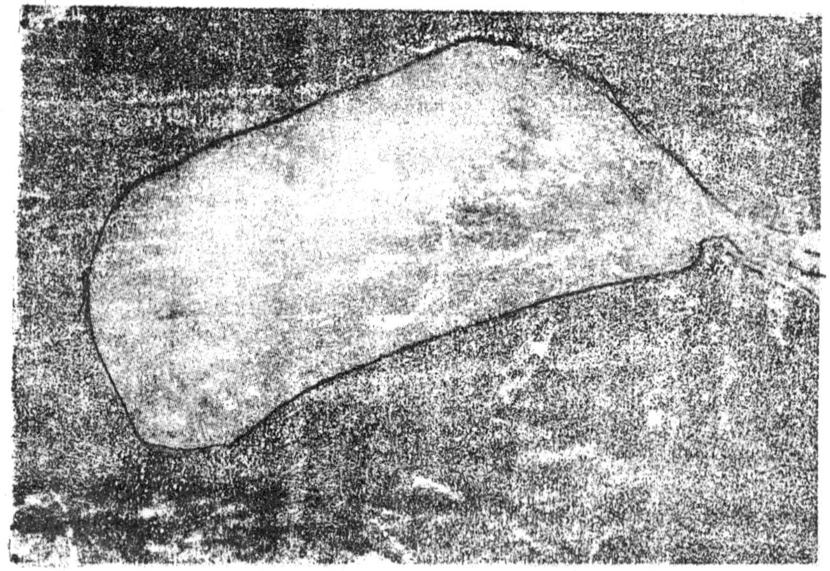

Fig. 3.68 : Typical prepupal sac showing Thai sac brood virus' disease.

Symptoms : As in chalkbrood, stone brood disease causes a mummification of brood. Larvae and pupae that are infected with *A. flavus* turn green in contrast to white or black in chalk brood. The green growth is powdery and can be readily seen with unaided eye. Fungus spores are found most abundantly near the head of the affected larvae and pupae. Stone brood diseased larvae are solid mummies and not sponge like as in chalkbrood.

Bailey (1981) suggested that the fungus grows rapidly in honeybees if they are weakened by other factors. The disease in encouraged under damp conditions and poor ventilation of hives This disease has not been reported from India so far.

Control Methods

Chalkbrood is a stress related disease as the starved and confined bees are more susceptible. Cultural studies have revealed that *Ascosphaera apis* contaminates more diverse substrates and survives better in bees and hives that maintain for hygiene. Following methods may be used for its control:

1. *Management practices* : Management practices similar to other brood diseases may be adopted in chalkbrood disease also.
2. *Manipulative treatments* : Moisture accumulation and poor ventilation should be prevented since cold and damp weather conditions encourage the development of chalkbrood disease.
3. Breeding of disease resistant strains is the best method to keep the colonies disease free.
4. *Chemotherapic treatments* : 0.7 per cent thymol has been reported to prevent the growth of Asocsphaera apis in culture. Several investigators have recommended the use of antimycotics and antiseptics for the control of chalkbrood disease. Giauffret and *Taliercio* (1967) reported that amphotericin B was most effective against chalkbrood disease. Actidione on the other hand exhibited high toxicity to bees and griseofulvin did not inhibit the growth of *A. apis*. The antiseptics such as cetylthmethyl ammonium were quite stable but more toxic to bees. Sorbic acid and sodium propionate fed to bees in pollen-sugar patties controlled infection within 7 days without affecting bees.

Gilliam (1985) found *Ascosphaea apis* is an opportunistic pathogen that kills the individual larvae only when they are subjected to other stresses. Stress factors in various geographical areas should be examined in relation to the expression of the disease. This is a disease in which the dynamics of the pathogen and of the bee colony need to be explored in unison.

Viral Diseases

So far, the world over, 18 viruses have been found to infect honeybees. Out of these viruses, Thai sac brood virus and *Apis* iridescent virus have been found to cause heavy losses to bee industry in India. The contiguity of land with Europe has been suspected to be the major factor for the spread of these diseases

Thai Sac Brood

Thai sac brood was first noticed in 1976 in Thailand from *Apis cerana F. colonies. Bailey et.al.* (1982) established it as a new strain of sacbrood virus and named it as 'Thai sac brood virus' (TSBV). This virus spread to India via Malaysia and Burma. After its first report in Thailand in 1976, the Thai sac brood disease' has traversed a long distance over the northern belt areas of India, viz., Meghalaya, Assam, Sikkim, north Bihar, north Utter Pradesh, Himachal Pradesh and to Jammu and Kashmir by 1985-86. The disease is caused by a virus (TSBV) which primarily infects the larvae.

Symptoms : In case of infections in *Apis mellifera* by SB mostly sealed larvae die after the cells are capped, whereas in *Apis cerana* indica late larval or mostly prepupal infections seem common, resulting in the death of larvae in the uncapped cells. The examination of the diseased colonies show irregularly capped brood with sunken and faded caps. The dead larvae in such cells are observed with their heads turned up, partly across the cell openings. Such larvae when picked up with forcep show nothing but entire larva/prepupa transformed into a 'sac' filled with fluid. Infected larvae turn pale yellow and finally brownish when dead.

Mode of spread : In an attempt of clearing the cells of dead brood, the young adult bees ingest infective virus and in turn contaminate healthy larvae, when these are fed by house cleaning bees. Even deserted swarms of infected colonies are capable of infecting newly developed young larvae reared in newly constructed combs. Drifting and robbing helps in spread of disease. Exchange of brood combs from infected to healthy colonies may spread infection.

Seasonal Occurrence : The infection generally takes place in the larval alimentary canal through ingested food. In Europe and Australia, the peak of the outbreak of SBV generally occurs in spring although its occurrence in other seasons is not very uncommon. But in India, as seen in Bihar, winter months seem to be favourable for the outbreak. Its continued occurrence in other seasons, especially in the beginning of summer in north-eastern states is also recorded.

DISEASES OF ADULT BEES

Apis iridescent virus

Another serious disease called as clustering disease is caused by Apis iridescent virus called as Iridovirus. *Apis* iridescent virus disease outbroke in mid fifties and again in the seventh decennial in Indian honeybee, *A. cerana indica* colonies and reports were published from different parts of the country.

Symptoms : The disease appears during hot months but *Mishra et.al.* (1980) correlated the appearance of the disease with dearth. The reports of disease incidence revealed that bees suffering from iridescent viral disease had reduced egg laying and brood rearing activity, the bees become sluggish and form clusters at the hive entrance and on the inner side of the hives many of which descend and crawl on the ground. The foraging discontinues resulting in depletion of honey stores, starvation of colonies and their dwindling/death.

Diagnosis : Infected adult bees can easily be diagnosed by illuminating their tissues, especially those in the abdomen and examining them with a hand lans, or in sunlight even with the naked eye. The exposed muscles under incident light under microscope look bluish, hence the name iridescent.

Shah and Shah (1988) reported that (AIV) infects various internal organs and is probably transmitted through glandular secretions in food.

Miscellaneous Viruses, Bacteria and Fungi

Honeybee species are also attacked by various viruses, bacteria and fungi. The world distribution of various viruses viz; black queen cell virus (BQCV); chronic bee paralysis virus (CBPO); Kashmir bee virus (KBV); acute paralysis virus (APV); Chinese sac brood virus (CSBV); Thai sac brood virus (TSBV); cloudy wing virus (CWV); deformed wind virus (DWV); Filamentous virus (FV); Apis iridescent virus (AIV); Arkanas bee virus (ABV); Japan bee virus (JBV); Slow bee paralysis virus (SBPV); bacteria viz. *Bacillis coaqulans*; *B. anthracis*, *Serratia marcescens* and bifid bacteria such as *Bifidobacterium asteroides*, *B. indicum*, *B. coryneforme*. *B. pulvifaciens*, *Pseudomonas apiseptica*; *Spiroplasma* spp.; *Bacillus para-alvei*, *B. laterosporous*; *Bacterium eurydice; Streptococcus* apis, *S. faecalis*, fungi viz. *Aspergillus flavus* and yeast like micro organisms causing melanosis. The data collected clearly reveals that these miscellaneous viruses, bacteria and fungi have rather limited distribution and occur in few countries of the world. However they cause destruction of honey bee colonies in the apiaries, wherever they occur.

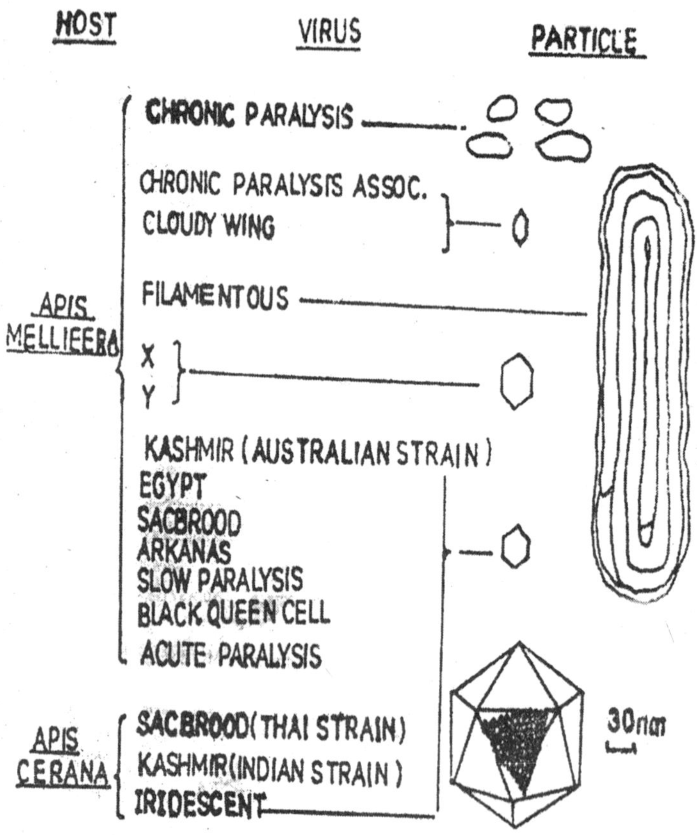

Fig. 3.69 : Characteristic feature of different viruses attacking honeybees.

Nosema Disease

Nosema disease has been reported from several parts of India. Kshirsagar *et.al.*, (1974) detected Nosema disease in *Apis cerana indica* colonies in Jeolikote, Uttar Pr.idesh. Rahman and Rahman (1994) have recorded Nosema disease in *Apis mellifera* colonies in Assam, The disease causes serious losses wherever it occurs.

Causative organism : The honey bees are infected by a single celled microscopic organism, a protozoan, *Nosema apis* Zanders. This organism enters the food canal of the adult bees in the form of spores which germinate in cells of the wall of the midgut. It multiplies rapidly at the cost of food absorbed from the cells of food canal of the bees. This constitutes an amoeboid stage. In a few days, further multiplication ceases resulting in the formation of spores. These spores are shed into the cavity of gut and then passed on to the rectum, the end part of the gut. In the rectum they are accumulated in masses and then passed on with excreta. These are picked up by other bees in the act of cleaning of cells etc. and these bees in turn get infected. The spores remain viable for several months in dried extereta on brood combs or hive parts. Workers, queens or drones of all ages are susceptible to this disease.

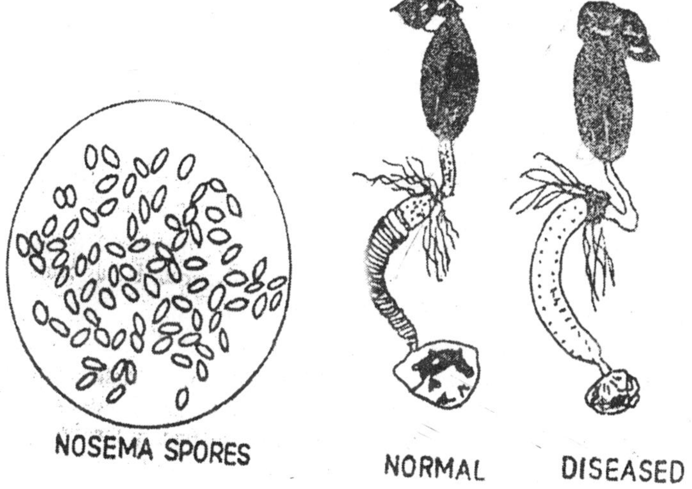

Fig. 3.70 : Showing (a) Nosema spores as they appear in wet mounts (x 400); (b) showing normal and (c) Nosema infected alimentary tract.

Symptoms and Diagnosis : Infected bees do not show any specific clinical symptoms of the disease. The disease may be present even if no signs of the trouble are seen inside or near the hive. The disease can be diagnosed with certainty only with the aid of microscope. The mid gut of infected worker bees become distended and on supturing large number of spores can be seen under the microscope. The Nosema infection, however is manifested in the manner, which may at times be the manifestation of certain other disorders also such as, dysentery, paralysis, acarine disease etc. The Nosema infected honeybees are unable to fly. They drop loose excreta on the combs or on alighting board or other hive parts. Number of crawlers can be seen in front of the hive on the ground after a play flight by the bees. Colony becomes weak as a result of death of foraging bees. Honey reserves get depleted. Infected nurse bees do not work normally due to deterioration of brood-food glands. Thus they cannot produce enough royal jelly. Nosema infected queens stop egg laying and occasionally queens also die.

Spread : The disease is spread through cleaning activity by nurse bees in infected colonies. Robbing, uniting, exchanging of hive parts etc. are other cause of spread of the disease in the apiary.

Control Methods

Management practices : Management practices similar to brood diseases can be employed for the control of adult bee diseases. The combs that contain no honey or pollen can be disinfected with formaldehyde but when the combs contain food the use of formaline is extremely poisonous for bees. *Lunder* (1972) found that acetic acid fumigation is the most safest method and does not harm the bees. *Cantwell* and *Shimanuki* (1970) on the other hand found that keeping the contaminated combs at 49°C and 50 per cent relative humidity for 24 hours greatly reduces the infectivity without damaging them.

Chemotherapy : The only successful drug so far found effective against *Nosema* is fumagillin. It has been reported to quickly supress the infection when fed to bees in concentrations ranging between 0.5 to 3 mg/100 ml sugar syrup without adversely affecting the bees. *Webester* (1994) found that two formulations of bicyclohexylammonium fumagillin B and Fumidil B were equally effective against *Nosema apis* infections.

HONEY BEE MITES AND OTHER ENEMIES

Honeybee Mites

Mites that affect honey bees or are found in bee hives may be divided into three group: parasites, phoretic mites and scavengers. The number of species of parasite mites are few, but some are very serious enemies and these are endo-and ecto-parasitic. Serious enemies are *Acarapis woodi* (endoparasite) and two ectoparasites *i.e., Varrea jacobsoni* and *Tropilaelaps clareae.* Phoretic mites are flower or leaf-feeding mites that use honey bees for transport from one plant to another and arrive

Table 3.17 : Brood and adult diseases of honeybees, their causative agents and occurrence in India.

Disease	Causative agent	Bee species affected	Symptoms/colour the brood	Locations
Brood diseases				
American foul brood (AFB)	Bacteria, *Bacillus larvae*	*Apis cerana indica*	Dull white dead brood becoming brown to white	Nainital (Uttar Pradcsh)
European Foulbrood (EFB)	Bacteria, *Melissococcus plulon*	*Apis cerana indica*	Dull white dead brood turning yellow to dark brown	Mahabaleshwar-Maharashtra Kamataka (around Castel rock near Goa border)
Sac brood disease (SBV)	Virus, *Morator aetatulus*	A. mellifera	Grayish or straw coloured becoming brown, grayish black or black or black head end darker	Not recorded from India so far.
Thai sac brood virus disease (TSBV)	Virus *Moralor aetatulus* (Thai strain)	*A. cerana indica*	Grayish or straw coloured becoming brown, grayish black or black or black head end dcirker	Whole india except Andhra Pradesh and Orissa
Chalkbrood disease (CB)	Fungus, *Ascosphaera apis*	*Apis cerana indica*	White chalklike mass sometimes referred to as mummy	Not recorcded from India so far.

Store brood (SB)	Fungus; *Aspergillus flavus*	*A. cerana indica* *A. mellifera*	The fungus forms a characteristic whitish yellow collor like ring near the head end of the infected larvae after death. Infected larvae become hardened and difficult to crush hence called 'Stone brood'	Not recorded from India so far.
Adult diseases				
Nosema disease	Protozoan, *Nosema apis*	*Apis cerana indica, Apis* mellifera	Shining swoolen abdomen	Uttar Pradesh, Himachal Pradesh, Jammu & Kashmir Punjab, Assam Orissa, Nagaland.
Amoeba disease	Protozoan, Malpigha-moeba mellificae	*A. mellifera*	Cysts in Malphighian tubules	Not recorded from India so far.
Bee paralysis	Filterable virus	*A. mellifera*	Black hairless shiny bees	Not recorded from India so far.
Septiceamia	Bacterium, *Pseudo-monas* apiseptica	*A. mellifera*	Destruction of connective tissue of legs, wings and antennae	Not recorded from India so far.
Clustering disease	Iridescent bee virus	*Apis cerana indica*	Bees leave combs and form clusters on the wall of hive or outside the hive become sluggish, queen stops egg laying and crawlers appear around the colony.	Jammu & Kashmir, Himachal Pradesh, Punjab Maharashtra.

accidentally in beehive. Scavengers are the species that feed on old provisions and a few species also feed on other mites. Mites rarely feed on stored pollen in the active hives, although large numbers of pollen-feeding mites are often found in stored combs. Overview of various types of mites, so far recorded from *Apis* species in India and neighbouring countries is given in Table 3.18.

Acarapis Woodi

Acarapis woodi is the only endoparasitic mite that infests *Apis mellifera* and *Apis cerana indica* honeybees.

This mite infests the tracheal system of adult bees; queens, workers and drones being equally susceptible. Sometimes some mites may also be found in air sacs in the head and addomen of bees.

Distribution

Presently this mite is reported from all over the world. In India it was first confirmed parasitizing *A. cerana indica* colonies in Katrain (HP) by Singh (1957) when colonies were noticed to be dwinding rapidly.

Fig. 3.71 : Adult *Acarapis woodi* mite.

Table 3.18 : Mites associated with honeybees (*Apis spp.*)

Mite	Hosts (Apis)	Mode of living	Habital
Acarapis woodi	*A. mellifera, A. cerana*	Endoparasite	Trachea of adult bee
Acarapis dorsalis	*A. mellifera*	External	Thorax region of adult bee
Acarapis externus	*A. mellifera*	External	Neck region of adult bee
Varroa jacobsoni	*A. mellifera, A. cerana*	Ectoparasite	Brood cell, adult bee
Varroa underwoodi	*A. cerana*	Ectoparasite	Brood cell, adult bee
Euvarroa sinhai	*A.florea*	Ectoparasite	Brood cell, adult bee
Tropilaelaps clareae	*A. mellifera, A. dorsata A. laboriosa, A. cerana, A. florea.*	Ectoparasite	Brood cell, adult bee
Tropilaelaps koenigerum	*A. dorsata, A. laboriosa*	Ectoparasite	Brood cell, adult bee
Pyemotes herfsi	*A. cerana*	Ectoparasite	Adult bee
Neocypholaelaps indica	A. cerana	Phoretic	Adult bee, pollen storage cells
Allodinychus flagellifer	*A. dorsata*	Not known	Not known

Biology

Adult female mites (Fig. 3.71) enter into the tracheae of worker bees within 24 hours after any emerge from bee brood cells. Single female mite lays 5 to 7 eggs after 3 or 4 days which hatch after 3 or 4 days. The males emerge on the 11th or 12th day and the females on 14th or 15th day. Infestation from one adult honey bee to another spreads when a mite leaves the bee via the first thoracic spiracle, it climbs a hair, usually on the thorax, and clings near its tip with one or both hind legs. It grasps a hair of another bee with its forelegs brushing past and descends to the surface of the new bee's body. Mites seem able to migrate only in this fashion and they are apparently unable to find a new host via an inert intermediate substratum such as honey comb or flowers and they will not pass through to wire-gauge screen separating infested bees from young uninfested individuals, even though the bees are able to feed one another through the gauze. It is believed that mites are attracted to the region of first thoracic spiracular openings by the puffs of air coming out of them which are caused by the respiratory movements or the abdomen. Since acarine disease is transmitted from one adult bee to another, robbing plays an important role in disease transmission. The disuse can be carried to healthy colony by robbers, drifting bees, or be a beekeeper who inadvertently transfers diseased bees to a healthy colony.

The thoracic tracheae which lead from the first thoracic spiracles are the main ducts for air in the thorax. Numerous mites in these ducts would partially suffocate the bee (Fig. 3.72) and at least impair its ability to fly, because oxygen is supplied to the flight muscles in the thorax by these ducts.

Irregular dark stains develop in infested tracheae, the whole of which eventually blackens. Such tracheae become brittle as compared to normal ones. Evidence of slight damage to flight muscle fibres and large nerve serving the wing and passing in close contact with the tracheae occurred. Mites sometimes live externally at the wing roots and may occasionally breed there in sufficient numbers to cause damage, perhaps even causing wings to fall off.

The life of overwintered infested bees in colonies is shortened and those old overwintered bees that were infested in both the tracheae died sooner than uninfested bees. *Acarapis woodi* infested honeybees have more bacterial infection, particularly in their haemolymph, than uninfested bees and it is probable that this causes more damages than the mites alone.

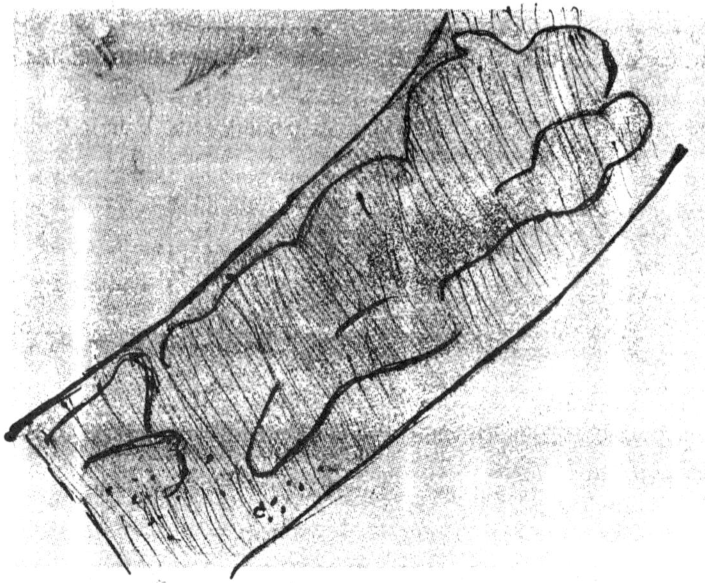

Fig. 3.72 : Eggs and adults of *Acarapis woods* in trachea.

Infestation with *A. woodi* occurred within two days of emergence of worker bees and 52-60 per cent got infestation within a week. Dying bee fell away from the clusters or left the hive, therefore, the spread of infestation from dead bee was of no significance. Sharma *et.al.* (1990) reported that most susceptible age group of *Apis mellifera* workers is between 4 and 7 days after which chances of infestation decreased. They spelled out that after 13 days of age, bees are practically immune to this mite.

Symptoms and Diagnosis

There are no specific outward signs of *Acarapis woodi* infestation. When the infestation is low, external symptoms are not noticeable but in the heavily infested colonies of honeybees below mentioned symptoms can be seen:

1. Large number of crawling bees (Crawlers) can be seen infront of the infested colony, climbing on the grass stems. These crawlers are young bees and fall on the ground when they make 'Orientation flight' in the afternoon of the day.
2. 'K' winged condition develops in workers bees due to the separation of fore and hind wings.

3. Yellow droppings may appear in some cases on hive parts and infront of the hives.
4. Bees have distended and shining abdomen.
5. Bees become sluggish and paralytic and do not cover the brood in the normal manner and formed scattered clusters.

Control Measures

Various chemicals like sulphur, Frow's mixture, methyl salicylate and acaricides like 'Dimite' and chlorobenzilate prepared for use under the trade names 'P.K.' and 'Folbex' strips are tested against *Acarapis woodi*. Out of these, Folbex fumigation is the tested control method for this mite all-over the world.

These Folbex strips are sheets of paper about 11 × 4 cm in size, impregnated with potassium nitrate and with about 500 mg of the acaricide, chlorobenzilate. Before using these strips, keep an empty super above the brood chamber after removing the inner cover. In the evening, when all the forager bees have returned to the hive, close the entrance cracks and crevices in the hive. Ignite the Folbex strip and allow it to smoulder. Hang the strip with the help of long pins of the side walls of the empty super chamber. Close the hive replacing inner cover and top cover. Keep the hive closed at least for 30 minutes and open the entrance thereafter. Always treat the honeybee colonies in spring season or in the beginning of breeding season in September when winters are not severe. Never use Folbex treatment in the lean period as queen mortality can be very high.

Mixture of nitrobenzene, methyle salicylate and petrol was effective but induced severe uncontrolable robbing and damaged the brood. Therefore usual treatment of the colonies with Folbex during active seasons was recommended. The fumigant action of insecticide/acaricide claimed to have given mite control. Sharma *et.al.,* (1983) claimed that 85 per cent formic acid @ 5 ml/day continuously for 21 days gave very effective control of the mite. All the treated colonies got cleared of mite infestation on 16th day. Chemical has no adverse effect on brood or the longivity of adult bees.

Southwick (1988) obtained an effective control of the mite with menthol and further observed that it has no honey-bee toxicity and hive contamination. He placed 50 gram packet of crystals on the bottom board of infested colony. Six weeks of continuous exposure to the menthol vapours was provided by introducing new packet every two weeks.

External Mites

Besides *Acarapis woodi* there are two other species of the genus *i.e., Acarapis externus* and *Acarapis dorsalis*. Former is also known as 'neck mite' and reported to cause wing loss or malfunction. Mites are found only in the area where the head and thorax join. *Acarapis dorsalis* causes very little damage and lives on the thorax in a groove between the mesoscutum and mesoscutelium. It may also be found at the base of the wings and on the propodeum and the fore part of the abdomen.

Varroa Jacobsoni

Distribution : V. Jacobsoni Oudemans was first detected by Dutch acarologist Jacobson on the Eastern honey bee, *Apis cerana* in 1904. It was first described by Oudemans in the *Apis cerana indica* in 1904 as parasite of the *A. cerana indica* in Java. While *A. cerana* has been recognised as the mite's native host. Delfinado (1963) collected specimens of *V. jacobsoni* from *A. mellifera* brood in Hong Kong in 1962. This was the first report of the utilisation of *A. mellifera* as an alternative host by *V. Jacobsoni.*

V. Jacobsoni has been anthropogenically spread to every continent of the world. Its wide dispersal has been the result of human carelessness in matters involving transcontinental shipment of bee stock. It has caused serious concern for beekeeping all-over the world, particularly in the queen rearing yards of *A. mellifera.* This ectoparasitic mite has been reported on the Indian and exotic honey bees from India but is of no major concern to both the *Apis* species.

Biology

Varroa jacobsoni is a large, 1.1 to 1.2 mm long and 1.5 to 1.6 mm, wide, dorsoventrally flattened, pilose, reddish brown mite and can be seen with the naked eye (Fig. 3.73). The dorsal shield covers the entire length of the idiosoma and almost completely hides the gnathosoma. It is an ectoparasite of *A. cerana* and *A. mellifera* adults and brood in the late larval stage. The shape of the mites body makes it easy for it to hold on to the bee, and it is also adapted in other ways *i.e.,* the base of each tarsus is modified into a lobed sucker, and the stiff hairs on the ventral side (which tangle with the those of the bee) make it almost immpossible to knock the mite off. The mites most commonly hold on to the bee between the first two abdominal segments, and are difficult to detect between the abdominal sclerities. They are also found between the head and thorax and between thorax and adbomen. These are places where the mite can easily penetrate the intersegmental membrane and gain access to the blood.

Varroa ingests small quantities of blood frequently. The bee may be harmed not only by the loss of blood but the wounds may allow microorganisms to enter the bee's circulatery system.

Life-cycle

The gravid female mite enters a brood cell shortly before this is capped and in the capped cell she starts to lay eggs as soon as the larva has finished cocoon spinning. Laboratory tests have shown that the female must feed on larval blood before she can lay eggs since she can live on the blood of adult bees, but cannot lay eggs. The female mite lays 2 to 5 eggs, at varying intervals and the eggs hatch after 24 hours and six legged larva develop into eight legged protonymphs after another 48 hours. The protonymphs are able to suck blood and they moult within 48 hours becoming deutonymphs and adults (Fig. 3.74) after another 3 days. The complete development of the female takes 8 to 10 days and that of the male 6 to 7 days. The male is considerably smaller and yellow to greyish-white. The mites mate in the capped cells. The chelicerae

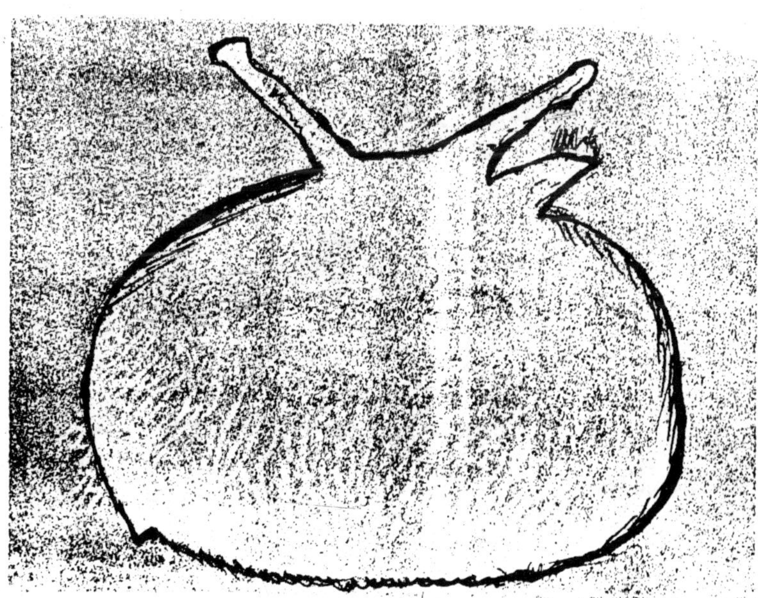

Fig. 3.73 : Dorsal view of female *Varroa jacobsoni*.

of the male are transformed for sperm transfer and the male therefore, cannot take food, and soon dies. The mother female and young females (now mated) remain in the brood cell until the adult bee emerges when they leave the cell sticking to the body of the bee.

The adult bee is usually an intermediate host and means of transport for the mite. The most serious parasitization, nearly always occurs on the older larva. Drone larvae are preferred to worker larvae. Queen larvae are infested only in cases of heavy infestation. The number of mites in an *A. mellifera* colony may exceed 10,000 individuals. An infested worker bee may act host to as many as five mites and a drone to up to about 12. Mite populations in an individual brood cell are reported to be as high as 12 in worker brood and 20 in drone cells.

The young female mites seek brood cell again after 4 to 13 days and the mejority lay eggs in only one cell. They live for about 2 months in summer and 5 to 8 months in winter. The mites cannot reproduce during winter period when there is no brood in the colony in temperate climates, however under Indian conditions mite infestation can be observed since winter is brood rearing period.

Symptoms

1. Adult mites can be seen on bee's body

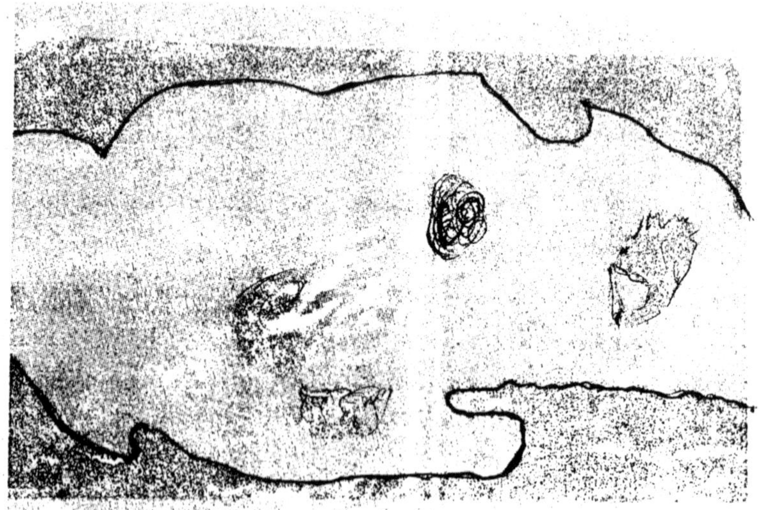

Fig. 3.74 : Female Varroa mites on bee pupa.

2. Spotty brood pattern.
3. Dead larvae, pupae, malformed workers and drones appear at the hive entrace.
4. The parasitized larvae which complete their metamorphosis emerge as crippled individuals with shortened abdomen, deformed legs and wings and are seen crawling at the hive entrance.
5. White droppings are seen on the walls of the empty cells.
6. Some larvae die in the prepupal stage with characteristic raised heads which might be mistaken for sac brood disease.

Control Measures

Chemical control : Both chemical and non-chemical methods have been advocated for the control of *Varroa disease*. Chemical control agents used are in the form of sprays, powders, evaporation agents, fumigants and systemic agents. List of various chemicals used for the control is given in Table 3.19.

Out of the above mentioned chemicals used against *Varroa*, Folbex Forte and recently introduced Apistan/Fluvalinate strips are most effective. Folbex Forte is available in the form of paper strips, impregnated with bromopropylate. Method of application is same as in a. carine disease with Folbex strips.

Table 3.19 : Chemicals used to control Varroa disease.

S. No.	Name of chemical	Mode of action
1.	Kelthane	spray
2.	Synecar	Powder
3.	Malathion	Powder
4.	Thymol	Powder
5.	Formic acid	Evoporation agent
6.	Danikoroper	Fumigant
7.	Varrostan	Fumigant
8.	Phenothiazine	Fumigant
9.	Bromopropylate (Folbex forte strip)	Fumigant
10.	K-79 (Chlorodimeform hydrochloride)	Systemic agent
11.	Pluvalinate (Apistan or Mavrik strip)	Contact
12.	Amitraz	spray and fumigant.

Apistan is a veterinary drug based on fluvalinate. Apistan is presented in the from of plastic strips into which the active substance is incorporated during manufacture. The delivery system ensures sustained and constant release of the .active substance. Apistan acts exclusively by contact of bees with the Strips. Through their social behaviour, the bees spread the active substance to the whole colony.

Non-Chemical Treatment

The Varroa mite depends on bee brood to complete its development cycle. Since the mite prefers drone brood to worker brood, frames of drone cells are given to the colonies which will rear drone brood in them. When the cells are sealed, the frames containing the mites applied inside the cells, can be removed and destroyed.

The other method involves caging the queen for 21 days to make the colony broodless. Thereafter the activities of queen are restricted to a couple of combs using queen excluder. The queen lays in these combs. The adult *Varroa* mite will enter the brood cells to deposit their eggs. When most of the brood is sealed these frames are removed and destroyed, in some countries, this "mite trapping approach is combined with hive fumigation, since it reduces the number of acaricide applications. These methods are, however, labour intensive and time consuming and likely to affect the productivity of the colony to a great extent.

Varroa under-woodi

Varroa underwoodi is a new species of *Varroa* and recorded and described for the first time from *Apis cerana* colonies in Nepal by Delfinado-Baker and Aggrawal (1987). This species is very similar to *V. jacobsoni*; it is much smaller in size with general morphology and naetotaxy as in *V. jacobsoni. Varroa underwoodi* is a ellipsoidal, light chestnut brown mite; female 780 mm long and 1168 mm wide. The male morphology is similar to that of male *Varroa jacobsoni.* The body of the male is rounded, very weakly sclerotized with light tanning on legs and setae. The length is 678 mm and width is 598 mm. Only deutonymphs are known. Characters that readily distinguish the *Varroa underwoodi V. jacobsoni* are the smaller size of body and the long female lateral marginal setae radiating outward. The mite presumably feed on bee brood and are capable of becoming a pest of honeybees. No further information is available in literature regarding this new mite.

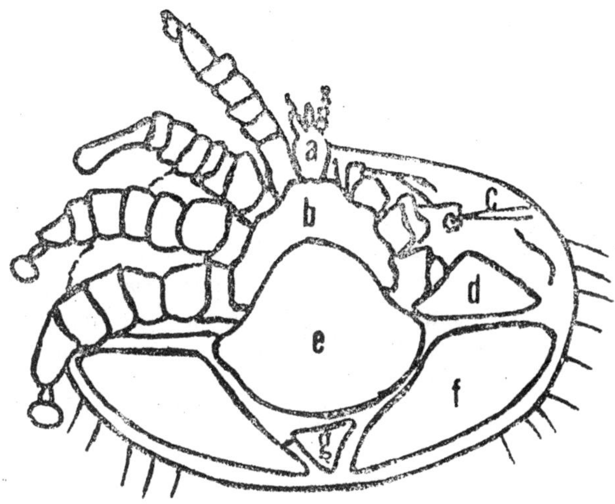

Fig. 3.75 : *Varroa jicobsoni* (ventral view).

Tropilaelaps Clareae

Modern beekeeping with *Apis mellifera* in tropical and sub-tropical Asia frequently encounters the problem caused by infestation with *Tropilaelaps clareae*. The mite is a native parasite of the giant honey bee, A. dorsata, widely distributed throughout tropical Asia and mite infestation of the colonies cannot be avoided. Situation becomes more complicated and serious when *T. clareae* and *V. jacobsoni* infests the colonies simultaneously. In case of dual infestation, the number of both mites often were observed to exceed 25:1 in favour of *T. clareae*.

Biology

T. clareae is an ectoparasite of *A. mellifera* having separate sexes. Mites of both the sexes are oval or elliptical in shape (Fig. 3.76). The female mite is reddish brown in colour having palpus directed anteriorly. It is very active and can be seen running on the brood combs with naked eyes. The male is of light brown colour and somewhat smaller than the female. The female is 0.990 ± 0.25 mm long and 0.510 ± 0.013 mm wide. The male is 0.917 ± 0.035 mm long and 0.483 ± 0.18 mm wide. In both the sexes, gnathosoma is completely hidden from above by the dorsal shield. Chelicerae of the male are toothed. The movable chela is modified into a long sinuous spermato-dactyl having a spiral coil. It is used to transfer sperms to the female genital aperture. A number of plates are present on the ventral side of the idiosoma. The sternal plate is rectangular and the opisthogenital plate is long, blunt posteriorly. Both are reticulate and fused together in the male. The opisthogenital plate overlaps anal plate in the female but is separate in the male.

The life cycle of *T. clareae* as a parasite of *A. mellifera* appears to be similar to that of *V. jacobsoni.* According to researchers, newly hatched adult female mites mate inside the brood cells as no formal mating was observed on the combs. Rath *et al.,* (1991) had elaborated the mating process very nicely. According to them, the mating took place in a cell with a young bee ready to emerge. No premating activities were observed. The male usually reacted with a quick grasping movement, clasping and then mounting the female, or the male clinging to the ventral side of the female. A male riding on the dorsum of a running female would quickly move to the ventral side of the female when it came to a stop, then the insemination sequence followed. The mating lasts approximately 15 minutes. The insemination is done by podospermy, *i.e.,* the sperm material is transferred to the female spermatheca through the external openings by means of male chelicerae. The external openings are located on either side of the female body between coxae III and IV. The movable digits of the male chelicerae of *T. clareae* are modified into spermatodactyls for sperm transfer. One female insemination may serve to fertilize the entire eggs.

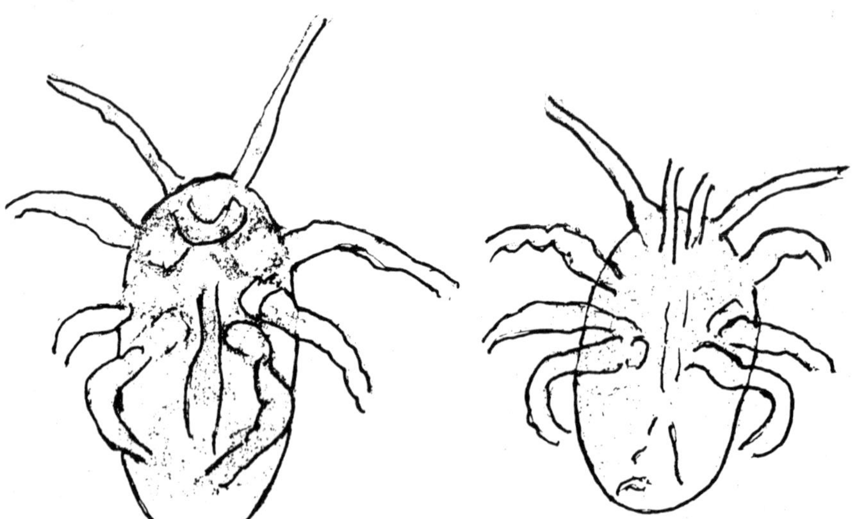

Fig. 3.76 (a) Adult female of *T. clareae* (b) Adult male *T. clareae*.

The gravid female mites come out alongwith the emerging worker bees. These mites run swifly on the brood combs and enter the cells before they are capped. Then these mites move to the bottom of the cell and some got attached to the developing larva. They feed on the body fluids of the late instar larva and pupa. Oviposition frequently occurs on the side walls of the brood cells, body of prepupa and also on the underside of the capping. Only gravid female mites were seen in the sealed cells, after 3rd day of sealing. Eggs were recorded only on the 4th day after sealing. The developmental stages of the mites are: eggs, six-legged larva, protonymph, deutonymph and adult.

Eggs are almost round in shape and measure 0.277 mm, long and 0.238 mm wide and appear milky white to the naked eye. The duration of egg stage is 1-1.25 days. The egg hatches into a six legged larva. The larva is milkly white in colour with somewhat swollen idiosoma. Its body is egg shaped with three pairs of legs. Larva remains inside the egg. The length and width of the larva is 0.464 mm and 0.354 mm, respectively. The duration of larval stage is 0.67-0.83 days. The larva hatches into an eight legged protonymph, The protonymph is egg shaped with four pairs of legs. It is an active and feeding stage. It is milky white in colour with swollen idiosoma. Chelicerae and pedipalps are formed. It attaches itself to the body of the worker pupa and feeds on the vital fluids. Its body length is 0.765 mm and width is 0.474 mm. This stage lasts for 2.5-2.8 days. The deutonymph is oval in shape. Its body resembles that of adult mite except that there is no chitinization. It is an active and feeding stage. Freshly formed deutonymphs are white in colour, but later on turn light yellow. Rudiments of opisthogential and anal plate appear.

Symptoms

Brood pattern in an infested colony is scattered and gives irregular and spotty appearance (Fig. 3.77) as compared to healthy brood which is uniform and compact. Infested brood cells have depressed/sunken cappings. Cell with infested pupae are uncapped and in some pupae partly eaten by the bees can be seen. Reddish brown adult *T. clareae* mites are quite often seen running on the combs freely. Hive debris contains number of dead mites. Partly eaten pupae and worker bees with deformed/mutilated wings and stunted abdomens can be seen infront of the hive which are discarded by nurse bees. In heavily infested colonies, crawlers can be seen crawling on the ground. Such crawlers have normal wings and abdomen, but their longevity has been reduced due to mite infestation. In such colonies, queen often stops egg laying and colony become broodless. Food reserves are more quickly consumed in *T. clareae* infested colonies as compared to healthy ones.

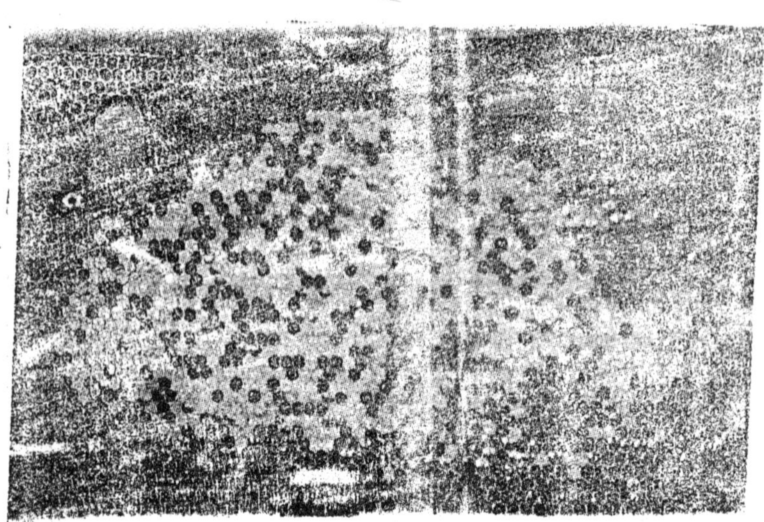

Fig. 3.77 : *Apis mellifera* brood infested with *T. clareae*.

Control Measures

Chemical control : T. clareae infestation at present is restricted to south east Asia and moreover due to overriding interest in *Varroa jacobsoni*, not much work has been done on the control measures. Both chemical and non-chemical methods have been tried but work has not been extensive. *Savilla* (1963) and *Mathews* (1973) recommended sulphur dusting to control *T. clareae* infestation. *Atwal* and *Goyal* (1971) found feeding terramycin and fumigation with chlorobenzilate as ineffective. *Laigo* and *Morse* (1960) too reported chlorobenzilate ineffective in eradicating *T. clareae*. *Garg et.al.* (1984) established a high effectiveness of fumigation with 5 ml of 85 per cent formic acid/day continuously for 21 days. They further reported that formic acid fumigation has no deleterious effect on the brood and adult worker bees. *Garg* and *Sharma* (1988 b) studied the efficacy of chlorobenzilate and menthol strips against *T. clareae* and found them ineffective due to non-residual effect of these chemicals. *Woyke* (1987 a) tested, the efficacy of phenothiazine and amitraz and reported amitraz to be more effective than phenothiazine. *Nyein* and *Zniarlicki* (1982) recommended depriving *A. mellifera* colonies of brood by either removal of all brood or by caging of the queen for at least a full 21 days and subsequent fumigation of colonies with phenothiazine. Lubinevski *et al.*, (1988) controlled *T. clareae* infestation under subtropical and tropical climates with Mavrik TM and reported that it is highly effective and had no bad effect on the developing or adult bees nor has any effects on the egg laying rate of the queen. *Garg* and *Sharma* (1988 c) reported that 10 gm of thymol crystals per colony are also effective against *T. clareae* mite. *Burgett* and *Kitprasert* (1990) evaluated Apistan for the control of *T. clareae* under tropical conditions and reported it to be very effective against this mite. Recently, Kumar *et al.*, (1994) reported that amitraz and fluvalinate fumigation had no deleterious effect on the brood and longevity of adult *Apis mellifera.* Sulphur dusting, formic acid fumigation and use of Apistan strips is normally recommended for the effective control of *T. clareae* infestation in *A. mellifera* colonies in India.

Non-chemical control measures : Non chemical control methods involve the manipulation of the brood rearing cycle of the infested colonies in such a way that the mites are deprived of sealed and unsealed brood for at least three days. These methods are based on the studies of *Woyke* 185), according to which *T. clareae* cannot survive on adult *A. mellifera* bees for more than one to two days. He reported that after removal of brood from *A. mellifera* colonies, all *T. clareae* died within 48 hours.

There are several means of creating broodless situation in infested colony. One can simply remove the brood frames; both sealed and unsealed, from this infested colony and put them in a new hive. Before the new larvae-hatch, the hive so manipulated will be short of brood for two to three days which is enough time to starve off mites. The new hive with the removed brood frames is given mated queens, but these are caged for 14 days, a period that allows most of the brood to emerge, while no new brood is reared because the queen has been confined. When the drone population of colonies is relatively high, and one wishes to increase the number of colonies, he can give these newly reared, capped queen cells instead of mated queens to the new colonies. By the time the virgin queens emerge, mature, mate and are ready to lay, most of the brood will have emerged; the rest can be destroyed before egg laying begins. There will thus be sufficient time to starve most of the mite population in the colonies. The best time of the year to apply these techniques is during a heavy pollen-flow season, enabling the .colonies to rear brood after the period of brood deprivation.

Another method is to combine chemical treatment with brood dcprivation technique. In this approach, all sealed brood is removed from the mite-infested colonies, which are then fumigated. The adult female mites, having no capped brood cells to hide are killed by the fumigant and reduces the number of fumigations required.

Phoretic Mites

Phoretic mites are flower or leaf feeding mites that use honey bees for transport from one plant to another and arrive accidentally in beehives. *Neocypholaelaps indica* Evans is only phoretic mite reported from *Apis cerana indica* honey bee in India. The mites are very minute, having their idiosoma 0.47 mm broad and 0.63 mm long. This mite has been observed on the workers of *A. c. indica* in the month of December. This mite was found to be phoretic and foragers carried 33-60 mites.

There is only one report on the chemical control of this mite. As suggested by *Percy et al.* (1968), fumigation with smaller dose of Folbex strip (Chlorobenzilate) in the evening, when all foragers have returned to hive is effective. It killed the mites, but next day again the returning foragers carried more mites.

Scavengers

Scavenger mites are those which feed on the old provisions and few species also feed on other mites. Presently, there is no record of any scavenger mites from any Apis species. Systematic studies are required on these mites in India.

ENEMIES OF HONEY BEES

Apart from various parasitic mites, there are enemies which pose serious threat to honey bees in India. Sometimes these enemies take a heavy toll of bee life. A large number of desertions occurs due to their nefarious activities. A list of major and minor enemies of honey bees is given in Table 3.20.

Wax-moths

There are two types of wax-moths, which are of major concern to the beekeeping industry. The greater wax moth (*Galleria mellonella L.*) is by far the most serious problem in the combs, whereas lesser wax moth (*Achroia grisella Fabr.*) is a minor pest.

Table 3.20 : Enemies of honey bees in India.

Common Name	Scientific Name	Class	Order	Status
Wax moths	*Galleria mellonella, Archroia qrisella*	Insecta	Lepidoptera	Major
Wasps	*Vespa mandarinia, V. tropica, V. Velutina, V. basalis, V, orientalis, V. aurana, V.* cincta, *Polistes hebraeus*	Insects	Hymenoptera	Major
Assassin bugs	*Acanlhaspis siva*	Insects	Hemiptera	Minor
praying mantis	*Odontomantis micans*	Insecta	Dictyoptera	Minor
Beetles	*Protaetia sp., Anomala sp.*	Insecta	Coleoptera	Minor
Black ants	*Componotum sp., Dorylus sp. Monomorium sp,*	Insecta	Hymenoptera	Major
Spiders	*Nuphilia sp.*	Arachnida	Araneida	Minor
Birds	*Merops apiaster*	Aves		Minor
	M. orientalis	Aves		Major
	Cypselus spp.	Aves		Minor
	Apus spp.	Aves		Minor
	Dicrurus sp.	Aves		Minor
	Lanius sp.	Aves		Minor
	Picus sp.	Aves		Minor
	Honey guides	Aves		Minor
Frogs and toads	*Rana tigrina, Bufo* sp.	Amphibia	Anura	Minor
Lizards and snakes	*Calotes sp., Hemidactylus* sp.	Reptilia	Squamata	Minor
Bears and pine martins		Mammalia	Carnivora	Minor

Biology

The adults are brownish grey in colour with a length of 10 to 18 mm and wing expanse of 25 to 40 mm, Thc female is larger than the male. The colour and size of adults varies a great deal in accordance with the food eaten during the larval period. The outer margin of the front wings of males has a semi-lunar notch, whereas that of the female is smooth.

The greater wax-moth is active from March to October and passes the winter mostly as hibernating larva and sometimes as pupa. The moths emerge during March-April and mate outside the beehive. The, females re-enter the hives usually at night, when bees are not active. They lay spherical, smooth and creamy white eggs in cluster in cracks and crevices of the hive or are found right on the combs. A single female may lay about a thousand eggs during its life span of two weeks. The eggs hatch in 7.18 days and the young caterpillars feed on gnawed pieces of comb or other debris. As a protection against the attack of bees, they make silken tunnels in the comb or on the bottom board, wherein they feed unhindered. They pass though 5 to 7 stage (Fig. 3.78) till full grown in about 4 weeks in summer. The full-grown larvae spin thick silken cocoons in which they pupate. The pupal stage lasts for about a week and the life-cycle is completed in 6-7 weeks during the active period and it passes through several overlapping generations in a year.

The damage is caused by the caterpillars which eat combs and interfere with brood-rearing by making silken galleries (Fig. 3.79) through the cells. In case of severe infestation, the whole comb becomes a mass of webbings in which excreta of the caterpillars is enmeshed. Severe damage is caused in bee colonies which are week and some combs are not covered by bees. The infested-colonies often abscond. In the off season when the combs are stored, the caterpillars damage them.

Preventive Measures

Preventive measures include ensuring that the colonies, whether of *A. cerana indica* or *A. mellifera* are strong and have adequate food stores, reducing the hive entrance and sealing cracks and crevices in the hive wall, protecting the colonies against pesticide poisoning and controlling pests and diseases that might otherwise weaken them, and removing wax debris accumulated on the bottom boards of the hives.

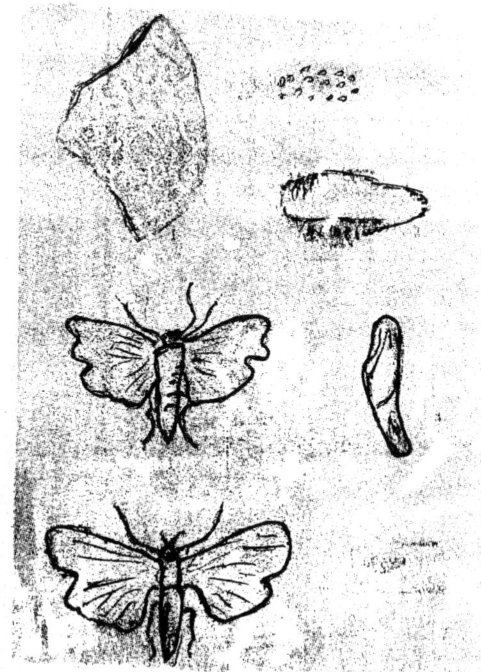

Fig. 3.78 : Development stages of *Galleria mellorella.*

Fig. 3.79 : Comb preavily infested with wax moth.

Chemical Control Measures

Measures can be taken to prevent or control the wax-moth infestation of stored combs, Fumigation is often necessary. Among the most commonly used fumigants are PDB (paradichlorbenzene) crystals, which evaporate slowly. Aluminium phosphide, naphthalene, ethylene dibromide and methyl bromide are the other fumigants used for wax-moth control. Ramachandran and Mahadevan (1951) recommended sulphur fumes for the control of wax-moth in stored combs. Kapil and Sihag (1983) reported one tablespoon of EDB very effective to kill all stages of wax-moth in stack of eight supers.

Biological Control

Spore forming bacterium, Bacillus thuringienis Berliner has been used as a biological agent to control wax-moth. A large number of treated colonies remained unaffected; and wax-moth pathogen and its toxin had no effect on honey bees (*Kapil* and *Sihag*, 1983). *Battu* and *Singh* (1977) too obtained good control of *Galleria mellonella* larvae by spraying 16 ml thuricide in 4 litters of water. The bacteria can also be sandwiched in the comb foundation sheets.

Lasser Wax-moth

The adults, larvae and pupae of this moth are smaller than those of the greater wax-moth. Adult *Archroia grisella* is silver-grey in colour with a distinct yellow head. Normal body length of adult female and male are about 13 and 10 mm respectively. The egg stage lasts for two to four days, larvae from 34 to 48 days, pupal from 5 to 12 days and the adults live for about a week The female moth lays from 300-500 eggs during its eight day long life. There are three to four generations during the active season.

Infestation by the lesser wax moth usually occurs in weak honeybee colonies. The larvae prefer to feed on dark comb with pollen or brood cells; they are often found on the bottom board in the wax debris.

Black ants

Various species of ants like *Componotus compressus; Dorylus labiatus* and *Monomorium* spp are the common predators of honeybees in India. Being highly social insects, they attack the hives *enmasse*, taking virtually everything; dead or alive adult bees, the brood and the honey. Apiaries of *Apis mellifera*, under the attack become aggressive and difficult to manage and weak colonies will sometimes abscond. Absconding is also the defence of *A. cerana* against frequent ant invasions.

These ants live in underground colonies. Their nests should be destroyed with insecticides like BHC, Chlordane and aldrin. Bee colonies can be kept free from ants by placing the hives on stands with their legs in earthern cups containing water or by smearing the legs of its stand with greese or tar. Tape soaked in corrosive sublimate and band round the legs of stands is a good repellent but needs replacement once or twice a month.

Wasps

Wasps of genus Vespa are the troublesome enemies of honeybees, especially in hilly areas. Colonies of both *A. cerana* and *A. mellifera* are frequently attacked. Wasp invasion of *A. cerana* colonies generally causes the bees to abscond, and similar behaviour is reported from weak *A. mellifera.*

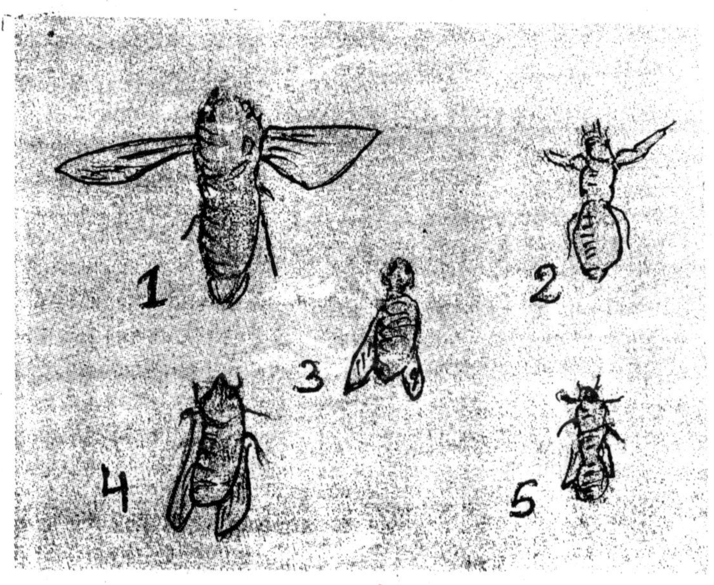

Fig. 3.80 : Different species of honeybee predatory wasps. 1. *V. mandarinia* 2. *V. tropica* 3. *V. basalis* 4. *V. velutina* 5. V. *orientalis.*

Biology : The wasps remain active during the warm summer months and hibernate as fertilized females during the winter, in cracks and crevices, in the ground or in other places of protection. *Vespa mandarinia* makes its nest in the hollows of tree trunks and is the largest wasp in north India. It is dark brown and hairy and has a robust look. Its head is bright orange with very strong mandibles. Its 5-7 individuals can finish a bee colony within one hour. *Vespa basalis* or black wasp makes its papery nest atop of the trees. It has a brown coloured head and abdomen is purely black with hairs. Its movements, are like that of a drunken man and 7-8 wasps attack a colony collectively. It also feeds on the dead bees lying in front of the hive. *Vespa tropica* or the yellow banded wasp is a fast flier and mostly catches the forager bees. It makes its nest in the ground. *Vespa velutina* or brown wasp is a very notorius pest of honey bees. It is rust-red and hairy. It catches the loaded bees coming to the hive. It also sits on the alighting board to catch a bee and flies back to its nest. It makes papery nests atop of the trees. These four species of Vespa are very troublesome for beekeeping in hills.

Polistes hebfeaus and *Vespa orientalis* are the species found in the plains. The former is uniformly yellow and makes simple round nests in the ceilings of Varandahs. The latter is larger and deep brownish with yellow bands across the abdomen which give a shiny appearance. It is very common on the sweetmeat shops. It makes irregular simple nest, generally in hidden places in the walls or in the hollow of trees. *Polistes* spp. are rarely seen attempting to prey on bees. *V. orientalis is also* not serious since they are in smaller number and make long effort to catch a bee. Bees are also able to defend against these species.

In all the species, the first brood in the spring is raised in old or new nests by the queen alone which lays eggs in the cells. The foragers which emerge take over the duties of foraging, nest building and defence etc. from the queen which then confines only to the nest and continue laying eggs. From then onwards, the brood is reared more or less continuously. The young larvae are fed on masticated animal tissues obtained by predation on insects. There are several broods in a year and the last brood reared in the beginning of the autumn consists of both male and fertile females which mate, leaving behind the fertilized queens for overwintering. During the cold season, all the workers and males die, only the fecundated Females survive till the next spring. These queens hibemate in the cracks and crevices during winter.

The damage caused by these wasps is of various types. All of them feed on ripe pears and other sweet fruits, sometimes causing heavy damage. The mountainous species are very serious predators of honeybees and become a limiting factor in the success of beekeeping in certain areas.

Control Measures

Due to relatively large body size, *Vespa spp.* can reach considerable distances from their nests, which may have populations of many thousands of individuals. For this reason the destruction of entire wasp nests, although useful when it can be done, is sometimes difficult to achieve. When a nest can be found and is within the reach of the beekeeper, destroying the entire nest is recommended.

An interesting technique for finding nests, known as 'Wasp lining' has been developed in some areas of south-east Asia where hornet brood is considered a delicacy. A foraging hornet is captured alive and a string, about 15 cm long is tied around its thorax; it is then released. When the hornet flies off toward its nest, the hunter follows, using the string as a cue. Killing of wasps attacking bee colonies in an apiary is generally practised with fly flappers all-over India.

Various other methods are also recommended by the researchers for the control of wasps. *Subbish* and *Mahadevan* (1957) recommended destruction of wasp nests and elimination of alighting board. Kshirsagar and Mahindre (1975) tested different baits for wasps and found that candy was more attractive than meat or fruit. *Singh* (1972) observed that *A.c. indica* have defensive behaviour against the attack of hornets. He further reported that bee mortality was highest when few hornets were attacking; as the number increased the bees showed organised defence with 60-200 guards, and fewer bees were killed. Destroying *V. auraria* nests with kerosene torches, fumigation with calcium cyanide or spraying, with carbaryl were recommended by *Sharma et al.,* (1979). Flapping for 30 minute reduced the number of wasps for the following 3 hours. According to them, compared with untreated meat (100%), meat poisoned with 0.25 per cent trichlorfon had 59 per cent and meat +0.5 per cent trichlorfon had 20 per cent attractancy. With lindane and chlordane the attractancy was low. Three *Vespa* species preferred decaying fish to decaying meat and *V. basalis* preferred rotten apples. They also could induce defensive behaviour in the bee colony be putting injured wasps at the entrance and consequently wasp visits decreased. A new method of mass poisoning was extensively tested in the laboratory and field by *Mishra et al.*, (1989). In this method poisoned jaggery (a product of sugarcane juice) in gelatin capsule cups by glucing them to the thoraces of trapped foragers was sent to the nests of *V. cincta*. The poisoned jaggery is shared by the inmates in the nest and only half to one dozon loads are sufficient to finish the colony.

Birds

Many insect species are preyed upon by birds, and honeybees are no exception. The heavy traffic of bees flying out of the hives in apiaries provides an exceptional opportunity for insectivorous birds. Fry (1983) has given nice account of honeybee predation by birds.

Birds which have been listed as taking honey bees are *Merops orientalis*, *M. apiaster, Dicrurus adsimilis*, *Cypselus* spp., Appus spp, *Lanius* spp., *Picus* spp. and honey guides. Bee shrikes honey buzzards, swifts, swallows, king bird are professional predators. Honey buzzards have morphological adaptations to protect them from bee sting but jacamars bee eaters and some shrikes have behavioural adaptations; they cause bees to discharge venom and eject their sting by rubbing off the insect to and fro.

Merops orientalis (Fig. 3.81) is of size of the house-sparrow and is found throughout Indian subcontinent. It is grass green, tinged with reddish brown on head and neck. It has a slender, long, slightly curved bill and central pair of tail feathers have blunt extremities.

Fig. 3.81 : Common Bee-eater, *Merops orientalis*.

The common green bee-eater is a resident or a local migrant. It inhabits open country, and is found near gardens and woodlands. It is often seen sitting on electric or telephone wires snapping at honeybees and-other wild bees, flies, wasps and large number of other insect species. It makes a horizontal or oblique tunnel-like nest in the side of earth cuttings, sides of hillocks or in green ground. The bird is extremely destructive to domesticated honey-bees and is the cause of failure of beekeeping in certain areas. In the plains the green bee eater poses serious problem from May through August when it does not find other insects to prey upon during summers. They also attack in large numbers during overeast weather during winters.

Fig. 3.82 : Black Drongo, *Dicrurus* sps.

Dicrurus adsimilis (Fig. 3.82) is a slim and agile glossy jet black bud, about as big as a bulbul, with a long deeply forked tail. It is commonly seen in the open countryside and around cultivation, perched on fence-posts, bush-tops or telegraph wires etc. From these lookouts the bird swoops down to the ground to pounce on honeybees. The nest is a flimsy-bottomed cup of fine twigs and grasses cemented together with cobwebs. It is built in the fork of an end twig in an outhanging branch of a large tree, usually standing by itself amidst cultivation and providing an unobstructed view of the surroundings. These birds visit apiaries occasionally on cloudy days, and prey upon bees. They have otherwise bright record as friends of the farmer.

The level of damage caused by apivorous birds varies. An attack by a single bird or by a few together rarely constitutes a serious problem, but when a relatively large flock descends upon a few colonies or an apiary, a substantial decline in the worker population in some or all the hives may be observed. The degree of damage to commercial apiaries caused by predatory birds depends largely on the number of the predators and the intensity of the attack, the mere presence of a few predators in apiaries engaged in queen-rearing can inflict serious losses.

Control

Locating the burrow of common green bee eater and destroying the eggs may give some relief. The birds can also be trapped but reduction of their numbers is not desirable since they act as important biological control agents of large number of crop pests. Shooting and killing of these birds can also not be advocated. Various methods such as scaring, recording and playing the distress voice at high volume, restricting bee movement/flight, using reflective tapes etc. have been evaluated (*Mishra* and *Kaushik*, 1993) but none of the measure works. However, the birds are unable to take flight and catch the prey if the colonies are kept under a thick canopy of trees such as poplar and mulbery and their predation is minimized.

Enemies of Minor Importance

Besides the above mentioned pests there are also other enemies of honey bees in India. Mahadevan (1951) reported a bug, *Acanthaspis siva* predating on *A. indica*. This bug pierces the body of the worker bee to suck body fluid. Rao et al. (1972) observed preying mantis (*Odontomantis micans*) as a predator of honey bees. These wait on the flower heads and inflorescences of coconut and caught the worker bees as they foraged. The spider, *Nuphilia kuhlii* makes webs on shrubs during monsoon and autumn, and trapped 10 forager bees in the webs daily. The spider could be controlled by spraying water with spray jets in August. Some beetles were reported feeding on stored pollen in the colonies. Rangarajan (1964) found *Protaetia aurichalcea* F. feeding on pollen in brood combs of *A. indica. P. impavida* and *Anomala dimidiata* were also found feeding on stored pollen in *A. mellifera* and *A. indica* hives in Himachal. Slug (*Limax sp.*) was also reported eating cells of combs, and bees got stuck to the slug.

Damages to honeybee colonies was also reported by bears and pine martins (Thakur et at. 1981), particularly in autumn and winter. They suggested cheap walled enclosures or hives in the walls of dwelling houses if few colonies were to be kept. *Rao* (1968) advocated suspending of hives from horizontal branches of trees against bear's damage. Murthy and *Venkataramanan* (1985) reported pseudoscorpion: (*Ellingsenius indicus*) associated with *A. c. indica* bees. They do not harm the bees, but used them phoretically for dispersal. Occasionally cockroaches come to hive in weak colonies and nibble away small bits of combs. The death's head moth Achrontia stags enters the hives at night and drinks honey. Robber flies and dragon flies catch bees and queens on their mating flights. Blood suckers, lizards, frogs, toads etc. pick up bees at hive entrances. Rats and mice-are the enemies of stored combs or empty combs in a colony particularly in winter. They are tempted to enter a colony because of its warmth. They also eat dead bees, honey etc.

(H)

PESTICIDE BEE POISONING

The use of pesticides results in large economic and health benefits to the society. Their use enhances and stabilizes crop yields, protects the nutritional integrity of food, facilitates storage and ensure year round supplies and provides for attractive and appealing food products, but the potential negative aspect of pesticidal use include environmental impacts on the quality of water and wildlife habitat, pesticide resistance in targeted species, applicator/harvestor safety problem and consumer exposure to pesticides residues in food.

Pesticide poisoning of honey bees has been a problem since the advent of synthetic, brood spectrum insecticides. The destruction of honeybee colonies by agricultural insecticides is well documented. First published account of honeybee losses due to insecticides was given over 116 years ago during on application of paris green to a pear tree in blossom. Since then toxicity of pesticides to bees has been studied extensively and it has been observed that almost all synthetic pesticides are hazardous to honeybees. Therefore, it became essential to use the pesticides judiciously to avoid losses to honeybee. Protection of bees, the efficient pollinators, against commonly used insecticides has been accepted on one of the important technological methods for the production of agricultural crops and maintenance of their high yields. It has been estimated that 80% or more of pollination of fruits and food crops is accomplished by honeybees in the united states as well as other countries.

Symptoms of Poisoning

The first sign of poisoning is a build-up of pile of dead and dying worker bees at the hive entrance. When poisoning is severe, thousands of bees will accumulate in front of a hive each day. This can be noticed easily when fenvalerate is dusted or sprayed on pigeonpea in a nearby field. If it is an organophosphorus compound, many of the workers will die with their tongues extended, and the mass of dead and dying bees becoming wet and sticky with nectar regurgitated from their honey sac. If the Insecticide is fast acting, only few of the affected bees will return to the colony, the rest will die in the field and remain unnoticed. Large colonies are often more severely damaged than small ones, because they have large foraging force working in the contaminated fields.

Poisoning cause the bees to become agitated and aggressive. Other signs include stupefaction, paralysis and abnormal jerky or spinning movements. Bees often behave as if they are chilled. Crawling around in front of the hive is an almost sure sign of carbaryl poisoning. Similar effect is produced by diedrin. In such cases, bees quickly lose the ability to fly and die only after 2-3 days. In severe poisoning, broods die due to lack of attention from workers or by direct poisoning through storage and transfer of insecticides. Contaminated bees sometimes become hyperactive around their colony. Foragers contaminated with pesticides are often rejected on return to the hive. Lack of coordination in body organs such as legs, wings and distension of abdomen, laying of eggs by queen in irregular fashion or stopping it altogether, broods emitting foul smell, supersedure etc. are some other symptom of poisoning due to insecticides.

INDIRECT EFFECTS OF PESTICIDE USAGE

Pesticide application under field conditions may produce several indirect effects. It may result into reduced foraging activity due to repellency or because of reduced foraging force owing to poisoning. Repellency could also be due to irritant effect on honeybees. It is also hypothesised that the so called repellent effect of insecticides can be due to sub-lethal toxicity resulting into transitory inhibition of activity. Sub-lethal doses can also influence other bee behaviour patterns *i.e.,* dance rhythm, flight velocity, walking speed, wing beat frequency, etc.

Pesticides can also cause physiological injury to bees when their applications are repeated. In such cases it is not the bee mortality which is significant, but it may reduce longevity. Pesticides gradually accumulate in combs as a result of absorption from stored pollen, honey/nectar. Drift of pesticide droplets can also cause this phenomenon. Small doses may cause chronic poisoning under stress conditions and sub-lethal pesticide dose application could be a significant contributing factor. *Fiedler* (1986). *Lensung* (1986; 1987), *Nation et.al.*, (1986) observed reduced egg laying and brood rearing due to small doses of pesticide application. *Atkins* and *Kellum* (1987), *Czoppelt* and *Rembold* (1988), *Ruijter* and *Steen* (1987) observed a morphogenic effects in delayed and abnormal development.

Pesticide contaminated food/nectar may cause bees to cease feeding or there may be reduced consumption and collection of nectar. Pesticide application can also change the physiology of nectar or pollen producing plants. Mishra and Sharma (1988a) have reported significant effects of growth regulators on pollen production, nectar volume and insect foraging in *Brassica* crops. This phenomenon results into changes in attraction of bees to flowers and consequently affected pollination and yield. Pesticides may also affect pollen viability and there are reports of similar effects of organophosphorus insecticides on tomato (Gentile *et. al.*, 1971) and of fenvalerate and phenthoate on apple pollen (Mishra *et. al.*, 1987). Moreover, germination of pollen on contaminated stigma can also be affected. All these indirect effects of pesticide usage are serious to pollination potential and honey production, yet pesticides may be overlooked as the cause.

Factors Influencing Bee Poisoning

1. *Pesticide Formulation :* Dusts are highly hazardous to bees because of their tendency to drift to considerable distance and particles remain adhered to plant surface for long. So is the case with wettable powders which also remain unabsorbed on the plant surface for longer periods than emulsifiable concentrates. The emulsifiable concentrates are relatively safer. The granular insecticides are the safest. However, granular insecticides with systemic action may contaminate nectar and may result in losses to bees, foraging upon them. The insecticide with fumigant effect may also be hazardous to bees. Micro-encapsulated granules applied on flowers are sometimes collected by bees and stored in the hives, where they may be eaten by adult bees or fed to brood causing high mortality besides causing long term contamination of hive part and hive products.
2. *Selectivity of Pesticides :* There are some insecticides which have little effect on honeybees when applied as sprays *e.g.*, endosulfan, phosalone, pirimicarb, fluvalinate or trichlorfon. High toxicity of carbamates to honeybees has been considered due to their very low level of phenolase enzymes. It has also been claimed that the greater acetylcholinesterase concentration in young honeybees enables them to tolerate malathion. Similarly, high tolerance of trichlorfon by honeybees has been correlated with relatively high pH of the body.
3. *Period of Application :* Insecticides when applied to crops in flower may be hazardous to bees or when pesticides are applied to a non-flowering crop but having large number of attractive

flowering weeds or hedges in the fields or in the adjoining fields. The bees are also affected if they pass through a field treated or sprayed with pesticides. In mango orchards bees attracted to honeydew secreted by mango hoppers are killed in large numbers when insecticides have been applied to the trees.

4. *Time of Application* : Little foraging takes place early in the morning or late in the evening. Application of pesticides during late evening or early morning provides relative safety. This avoids direct deposition of pesticides on the bee body and even residues on the treated surfaces are rendered less harmful, especially in case of short residual pesticides. Species variation has also been noticed. Whereas *Apis cerana indica* became active at 7-10°C ambient temperature, *Apis mellifera* never became active unless the ambient temperature reached 13-16°C or above. During honey-flow season of litchi in Bihar, non-foraging period could be from 1900 hr to 0600 hr next morning and spraying pesticides beyond these hours could be extremely hazardous to honeybees. In case of sunflower foraging is greatly reduced in the afternoon and blooming crop can be treated in the evenings. This tip can also be followed in crops like lucern where the flowers close in the afternoon.
5. *Attractiveness of Crop* : Some crops, notably rape and mustard are extremely attractive to bees which will forage from colonies upto 3 km away or more. These crops remain attractive until the very end of flowering. It remains attractive even in cool and dull weather when they may not visit other crops.
6. *Weather* : Warm and sunny weather is conducive to foraging by bees. When there is a prolonged dull weather the foraging activity is reduced considerably and insecticides can be applied to field crops.
7. *Method of application* : Aerial application of pesticides has been regarded more hazardous to bees than ground application. Bees get less time to escape the drift of insecticides. Systemic insecticides applied on the blooming crop may cause hazards to the bees. Fine sprays are safer than coarse sprays.
8. *Colony strength* : Populous colonies always suffer greater losses than small colonies, because more foragers are exposed to the pesticides.
9. *Age and body size of honeybees* : Newly emerged bees are more susceptible to insecticides than older bees. Smaller bees likewise

are more susceptible to insecticides than the larger bees because their body surface area is larger in relation to their body weight.

10. *Distance of colonies* : Honeybee mortality is inversely proportional to the distance of colonies from treated fields. Farther the crop is from the colony the less likely it is to attract large number of foragers.

Prevention of Bee Poisoning and Care of Poisoned Colonies

Insecticides safest to the bees should be applied only when their use is justified. Hazardous insecticides could be applied only when the crop is not in bloom and does not contain flowering weeds, and preferably when there are few or no flowers in surrounding hedges. Insecticides having selective action and selective formulation (causing low toxicity to the bees) should be used. Safer formulations of the pesticides should be used. Sprays of undiluted technical material may be more toxic than diluted sprays. When using pesticides hazardous to bees, the beekeepers should be notified so that they may provide protection to their colonies. The time of application of pesticides and location of colonies are important considerations. Treatments made when bees are foraging in the field are usually the most hazardous. Treatments during hot weather when bees are clustering on the outside of-the hive may cause severe losses. Treatments with insecticides which break down within a few hours, made during late evening, night or early morning before bees are foraging, are the safest. Colonies located near the treated field may sustain more losses than colonies kept beyond treated area. Farther the colonies from the treated area the less critical is the treatment time. Treating large areas and repeating applications may cause greater damage and bee losses.

Insecticides should not be applied when unusually low temperatures are expected afterwards because residues would remain toxic to bees for a much longer time. Integrated pest management programmes which rely upon biological and cultural methods and which tend to minimize the use of chemicals should be advocated.

Honeybee Repellents

Protection of honeybees from pesticideal hazards has been a challenging task. Application of honeybee repellents either in combination with pesticide formulations or separately on the crops to be sprayed with pesticides is one of the alternatives to protect honeybees from the pesticidal hazards. In the search for repellent compounds to reduce bee poisoning numerous substances have been tested. The idea of using a honeybee repellent to reduce the harmful effects of insecticides by preventing exposure of honeybees to toxic chemicals is not new. Probably the first suggestion to use a chemical addition in insecticide sprays was made by *Callbreath* (1900) who recommonded the use of carbolic acid. Later on, the repellents like cresol compounds, carbondisulphide, nicotine sulphate, naphthaline, milkol etc. were used as gustatory honeybee repellents. Tests on effectiveness of chemical repellents to honeybees were confined to loboratory and almost all had failed in the field tests as repellents. In field conditions efficacy of the compounds is limited due to their short persistence on the crop. To increase longevity certain other chemicals have been found suitable which can be exploited commercially to repel honeybees from pesticide sprayed field for desired duration. The area of chemical repellency is very rich for further research. It is actually beginning of understanding the insect's chemical repellency. There are many opportunities for investigation and possibilities for exploitation, some of which probably have not yet been investigated. Mechanism of action of perception of repellent compound by honeybees has not been discovered so far.

Except the protection of honeybees the repellents may be used for insect control as on alternative to organic pesticide. These chemicals may be used for communication disruption, mate finding, host finding (feeding, oviposition) and self defence. The greatest value of these investigations may lie in knowledge gained about the structure activity relationship, mechanism of perception and the process that integrate the information and control the behavioural responses. This knowledge should lead to new, safer, environmentally benign and more effective methods of insect control.

(I)

ECONOMICS OF BEE-KEEPING

It is a matter of common knowledge that the major industry of India is agriculture and most of the agriculturists do remain idle for at least six months in a year—a major loophole in the Indian rural economy. Thus if he knows about the utility or otherwise of the life around him and also the industries based on it, he may utilise the off-season months to plug this loophole. Looking to the appreciable benefits in return to the least physical labours and casual attention, farmers are increasingly attracted towards bee-keeping activity. The role of bees in agriculture is significant as we have discussed in previous topics and farmers should adopt planned bee pollination programmes for boosting of crop yields and also for better quality of the fruits and seeds.

Bee-keeping is a gainful activity and bee-keepers in developed countries like U.S.A., U.S., England, Australia, Hungary, Israel, Germany treat it as a major occupation. All this is possible because farmers and bee-keepers have acquired scientific knowledge and have adopted new techniques in bee-farming. Some bee-keepers keep bees to produce honey and some rent hives to farmers who grow plants that need pollination.

In India bee-keeping still it remains as a subsidiary one. Indian bee-keepers need to develop scientific outlook and better management techniques to make it most successful and gainful occupation.

The average honey yield per bee colony in India is 5 kg. In India there are bee-keepers maintaining 50 or more bee-colonies and are earning Rs. 7,000 to 12,000 annually. Individual family can easily maintain about 25 to 30 bee colonies where abundant bee pasturage is available. It requires minimum supervision work and any member of a family can inspect the colony at the leisure time.

It is estimated that to start with bee-keeping, the initial investment will be around Rs. 32,000 to purchase and construct equipments, labour charges etc. The annual income in the form of honey, wax and other products is about Rs. 45,000, plus bonus benefit of bees as a pollinating agent. In India, bee-keeping industry produces 9000 tons of honey per year, the rate of honey in market is 120 to 180 Rs. per kg; you can imagine the transaction of money from the honey, additionally the role of bees in pollination also increase the production about 15 to 20 %. If you sum up both, the transaction will be in thousand millions. This net profit is with minimum working hours and is as positive part of this activity.

Training Courses

Before starting a apiary one should have a general idea and interest otherwise it will become a difficult job for beginners. To gain the basic knowledge, join your local bee-keepers association, visit local bee-keeping supplier. They are usually enthusiasts who will discuss what you need to buy from them and when. Try the bee-journals that are available and look through the manufacturer's catalogues. Many local authorities and some agricultural colleges run bee-keeping evening class courses which enable you to discuss plans and problems with an expert, and to meet other beginner bee-keepers.

At International Level IBRA (International Bee Research Association, U.K.) which is an educational and scientific charitable trust, funded annually by subscriptions from its members and by income from the sale of publications and other services. IBRA provides the world's most comprehensive information and advisory service on all aspects of bees and bee-keeping—thereby helping to promote bee-keeping and bee research worldwide.

At national level all India co-ordinated project on honey bee research and training was sanctioned by Indian Council of Agricultural Research (ICAR), New Delhi. ICAR has settled various centres and objectives of the centres are :

1. To increase in number of bees-colonies and improvement in their performance.
2. Assess the utility of Indian honey bees in planned pollination.
3. Quality control and processing of bee-products.
4. Conducting graded training courses in bee-keeping.

Some of the important centres are as follows :

1. Central Bee Research Institute, Pune.
2. Indian Agricultural Research Institute, New Delhi.
3. Indian Institute of Horticultural Research, Chethalli.
4. J.N. Krishi Vishwa Vidyalaya, Indore.

5. Andhra Pradesh Agricultural University, Hyderabad.
6. Assam Agricultural University Jorhat.
7. Punjab Agricultual University, Ludhiana.
8. Orissa University of Agriculture and Technology, Bhubhaneshwar

Additionally, as discussed earlier, a statutory body, KVIC (Khadi and Village Industries Commission) initiated the; planning organisation and implementation of programmes for rural development. The KVIC has taken up the programmes of bee-keeping on a wide scale since 1953-54. The appreciable progress has been achieved with the active co-operation of the various *State Khadi and Village Industries Boards*, social service bodies and co-operatives. In Maharashtra, the State Khadi and Village Industries Board and, Khadi and Village Industries Commission are involved in research, extension and training. These institutes organise different refresher courses which include theory cum practicals for field; staff and laboratory technicians. It helps in updating their knowledge by learning the improved techniques and advanced researches in this subject.

In Maharashtra there are number of training centres, important ones are in Western Ghats *i.e.,* Mahabaleshwar, Panchagani and Pune. They are conducting some short term training courses of 8 days to 3 months. Additionally centres of Khadi and Village Commission has settled training centres in tribal areas as well as in Urban areas at Mumbai, Rajpur (Ambegaon), Bordi (Dahanu). In last some years in Western Maharashtra, there are some farmers, who knows the role of bees in pollination started bee-keeping at local level and credit goes to Dr. T.B. Nikam who has played a key role to diffuse, spread the information, importance of bee-keeping in Nasik District. Recently agricultural schools related to Yashwantrao Chavan Open University, Nasik has accepted to introduce bee-keeping as a full term course in their curriculum. But in general it is observed that there is no whole-hearted participation of Agricultural Universities in bee-keeping. We hope and trust that the picture will be changed in coming some years and bee keeping will become a productive agro-industry.

POLICIES OF GOVERNMENT AND OTHER ORGANISATIONS

Beekeeping infact should have been included in the schedule of agriculture and only then it would have acquired the needed impetus. At present it is neither considered as an industry nor an agriculture activity. As a result there is lack of sufficient financial help from government and other lending institutions for the development of beekeeping. It also requires long term loans at easy rates of interest. This was the procedure adopted by China to take up beekeeping on a commercial scale. Beekeeping is a long term developmental activity and needs to be given some tax incentives for people to take it up in a big way. Furthermore, it is a high risk activity, depending upon favourable weather conditions for heavy production. Therefore, beekeepers need financial support during seasons of bad honey harvest to sustain the colonies for next season.

The Government of India in 1988 removed the excise duty on equipment manufactured for use in beekeeping industry (Verma, 1980) which is a good sign for the development of Indian apiculture. In USA beekeepers are supported through a non-recourse loan programme with a national loan rate of 53.77 cents/pound of honey. Support loans help the beekeepers with interim finance and help them to market honey in a more orderly manner (Singh, 1993).

In India there is a general lack of consumers awareness of uses of honey and bee products. It is mainly used as medicine and its value as food by the general consumers has not been realised. The per capita consumption of honey in India is merely 3 gm. General interest and awareness about beekeeping and bee products can be created amongst the consumers by beekeeping societies and national bodies by organising honey festivals, honey beauty queen competition and award to beekeepers for high honey production. To make beekeeping a success, we have to make bee products available to common man at affordable price which can be achieved by increasing the production rather than the price.

At present we do not have any legislation for harmonizing and regulating the use of pesticides harmful to bees, as a result of which many times there are heavy colony losses due to pesticide poisoning. Indiscriminate use of pesticides also destroys a number of valuable naturally occurring pollinators. Thus there is an urgent need for legislation for judicious use of pesticides to avoid mortality of honey bees and other wild pollinators which are helping us with needed pollination of cross pollulated crops.

From the research point of view, honeybee research in India is being undertaken by entomologists and there is no separate apiculture department exclusively engaged in research on honeybees. Beekeeping needs the same status as dairy or poultry for its proper development. Indian Council of Agricultural Research in 1980 felt the need for multilocational research in apiculture and started an All-India Coordinated Project on Honeybee

Research and Training (AICRP) which at present is having nine different centres in the potential areas of the country but most of these centres are lacking in having specialist scientists in the field of apiculture. It is suggested that apiculture research in the country is planned and directed in such a manner that different research groups are established in different fields of apiculture (Mishra and Sihag, 1987). The important research groups can be in honeybee botany, bee management, bee genetics and breeding, bee pathology, bee products and engineering.

Need for a Central Body (Bee Board)

For the development of any industry an integrated approach is needed and therefore in order to boost commercial beekeeping it is imperative to bring research, extension, training and marketing under one umbrella. To attain this goal well coordinated efforts and cooperation is needed from organisations such as KVIC, Agriculture and other Universities having beekeeping centres, state departments of agriculture/ horticulture, cooperative societies and beekeepers associations. There have been suggestion for setting up of a bee board on the lines of dairy development board, oil seed development board or silk board and it is hoped that this board will be able to provide the much needed forum for the coordination among different agencies in research, marketing to achieve desired results. At present there is beekeeping development board in the ministry of agriculture but it is not a statutory body. It has no autonomy and technical cell/infrastructure of its own.

CORRECTIVE MEASURES AND STRATEGIES FOR BALANCED GROWTH OF APICULTURE

For the balanced growth of apiculture following measures are suggested:

1. Apiculture should be considered as a special discipline at the national level for generating strong know-how. The existing National Honeybee Board should be upgraded and created on the lines of other statutory boards to look after all aspects of the industry starting from hive to market.
2. Apiculture activities should be diversified by producing other bee produce like royal jelly, pollen, beeswax, bee venom and propolis etc. Appropriate market for these products should also be worked out.
3. To create interest in beekeeping, more beekeeping societies and national bodies be formed and different competitions and awards, should be encouraged.

4. Beekeeping with appropriate bee species should be taken up according to local floral and agro-climatic conditions and available technical know-how.
5. Selection and breeding of queens of high yielding races for increased honey production should be given top priority. Nurseries of superior genetic stock in different regions of India should be maintained so that beekeeper may get best genetic stock. Bee colony multiplication units should be established in each beekeeping area/state from where the colonies and equipment should be available to the beekeepers at reasonable rates.
6. Beekeeping should be advertised as a recreation and employment generating activity.
7. Maintenance and conservation of all the species of honeybees .should be taken up and their value as pollinators should be assessed.
8. Some national law needs to be enacted for legalising the activities and functioning of bee-keeping industry which include strict quarantine measures within the country as well as legislation on safe use of pesticides harmful to honey bees and wild pollinators
9. Mass propagation of bee flora should be undertaken with the help of different agencies on wastelands, road side and along railway tracks to make beekeeping a profitable venture.
10. There should be a clear distinction between centrifugally extracted and squeezed honey; while the former may be used for table purposes, the latter can be used in industry.
11. There should be appropriate quality control of honey since for export market fresh, light coloured unifloral and liquid honey fetch good price.
12. Cooperative societies be formed for processing and sale of honey because it is not possible for an individual beekeepers to look after all these aspects as these are too expensive and time consuming.
13. Loan should be made available to the beekeepers at easy rates of interest and tax benefits be extended to beekeeping industry.
14. People engaged in traditional beekeeping be motivated to adopt modem beekeeping by providing technical guidance and financial support.

15. There is a necessity to work out carrying capacity of different migration centres during honey flow season so that beekeepers do not overcrowd their colonies in these areas.
16. Grants and subsidies provided by the Government under different schemes for beekeeping should be made publicly known to avoid misappropriation or favouritism in disbursement of these funds. The present policy of subsidies needs to be reviewed; the subsidy should be linked with the performance since the existing pattern has failed to raise new beekeepers.
17. Due to escalating cost of bee hives, beekeepers are finding it difficult to purchase bee hives, it is, therefore, suggested that some element of subsidy be added on the cost of bee hives and other bee equipments. The hive manufacturers can be given wood logs at subsidised rates and price of finished equipment should be controlled.
18. Different research groups should be established in the country in different fields of specialization so that technical expertise is generated, particularly on bee diseases and bee enemies.
19. Adequate man power be develop so that we may have trained field workers in beekeeping at village, block, tehsil, and district levels. Graded training should be imparted in different organizations. Universities should set up separate departments of apiculture so as to get a regular flow of specialists in apiculture to man the research and extension in the country.

4

SERICULTURE

1. INTRODUCTION

Bombyx mori commonly called as *Chinese* or mulberry *silkworm moth* which is well known for the pure silk. The moths are reared for silk. The industry of obtaining the silk from the silkworm by artificial rearing is called "Sericulture".

HISTORICAL BACKGROUND

Historical evidence shows that silk was discovered in china and that the industry spread from there to other parts of the world.

The earliest authentic reference to silk is found in the chronicles of chocking (2200 B. C.) where silk figured prominently in public ceremonies as a symbol of homage to the emperors. According to some sources the first country after China to learn the secret was Korea, where Chinese immigrants started sericulture in about 1200 B.C. The industry later spread to Japan. During the latter part of the 19th century Japan gave serious attention to the development of the industry, introducing the use of modern machinery and improved techniques and carrying out intensive research in Sericulture. The industry is said to have spread to Tibet when a Chinese princess, carrying silkworm eggs and mulberry tree seeds in her headdress, married the king of khotan in Tibet. From Tibet the industry spread slowly to India and Persia. According to some, mulberry tree cultivation had spread to India through Tibet by about 140 B.C. and the cultivation of mulberry trees and the rearing of silkworms began in the areas flanking the Brahmaputra and Ganges rivers.

According to some Indian scholars silkworms were first domesticated in the foot hills of the Himalayas. There is also evidence in ancient Sanskrit literature that certain kinds of wild silks were cultivated in India from time immemorial.

When the British came to India they found a flourishing silk trade. The British East India Company exploited the industry and developed silk centres in many parts of the country. The company exported large quantities of the raw silk produced in West Bengal to England. Other major silk-producing states namely Mysore, Jammu and Kashmir took steps to develop the industry at that time.

The main silk-producing countries of the world today are Japan, the China, the U.S.S.R., Korea and India. Next in importance come Italy, Bulgaria; Brazil, Iran, turkey and Thailand, where sizeable quantities of raw silk are produced. Other minor countries where sericulture is practised include Rumania, Spain, Greece, Hungary, Yugoslavia, Taiwan etc.

2. TAXONOMIC CLASSIFICATION

PHYLUM—ARTHROPODA

They are bilaterally symmetrical, metamerically segmented and body enclosed in a tough, chitinous exoskeleton. The segmented body bears paired and jointed appendages. Blood vascular system is open type, respiration by gills or trachea or book lungs, excretion by green glands or by malpighian tubules. Fertilization internal and development includes metamorphosis.

CLASS—INSECTA (HEXAPODA)

They are air breathing, mostly terrestrial and aerial and rarely aquatic. Body is divisible into head, thorax and abdomen. Head consists of 6 fused segments, and bears a pair of compound eyes, a pair of antennae and mouth parts adapted for different food habits. Thorax has 3 segments, each bearing a pair of legs and two pairs of wings. Abdomen is divided, into 7-11 segments without appendages. Respiration by trachea. Excretion by malpighian tubules.

SUB-CLASS—PTERYGOTA (METABOLA)

Wings are usually present. Abdomen has no appendages except genitalia and cerci. Metamorphosis simple or complex.

DIVISION—ENDOPTERYGOTA

Wings develop internally. Metamorphosis complete including pupal stage.

ORDER—LEPIDOPTERA (SCALE WING)

Small or large sized insects with 2 pairs of wings. Wings membranous, covered with minute or overlapping scales. In adults, mouth parts are of sucking type, maxillae modified into spirally coiled proboscis. Metamorphosis complete. Larvae typical caterpillars with biting mouth parts. Three pairs of thoracic legs, 4 to 5 pairs of prolegs on the abdomen which often with silk glands. This order includes moths and butterflies including the silkworm moth.

SUPER FAMILY—BOMBYCOIDEA

Maxillary palpi and tympanal organs absent. Fraenylum almost always atrophied or vestigial, proboscic rarely developed. Chaetosoma absent. Antennae pectinated, especially in male.

FAMILY—BOMBYCIDAE

This family includes economically important moths which produce natural silk.

Genus—*Bombyx*

Species—*mori*.

TYPES OF SILKWORMS

There are four kinds of natural silk which are commercially known and produced. Along them mulberry silk is the most important and contributes as much as 95% of world production, therefore, the term "silk" in general refers to the silk of the mulberry silkworm. Three other commercially important kinds are : *Eri* silk, *Tasar* silk and *Muga* silk.

Mulberry Silkworms

The insect producing mulberry silk is a domesticated variety of silkworm. The caterpillar feeds on mulberry leaves: The Mulberry silkworm may be further classified and identified as of Japanese, Chinese, European or Indian origin based on geographical distribution, or as *Univoltine, Bivoltine* and *Multivoltine* depending upon the number of generations produced in a year under natural conditions; or as *Tri-moulters, Tetramoulters* and *Pentamoulters* according to the number of moults during larval growth; or finally even as Pure strains, and as hybrids which may be either *Monohybrid* when two strains are involved or *Polyhybrids* when more than two strains are involved in the hybrid.

Bombyx mori produces super quality silk which is due to its shining and creamy white colour. It also produces cocoons with continuous silk filament and therefore can be industrially reeled to produce raw silk.

Eri Silkworms

These belong to family Saturniidae and consists of two species namely *Phiosamia ricini* and *P. cynthia*. *P. cynthia* is a wild species-while *P. ricini* (also called as a castor silkworm) is a domesticated one reared on castor oil plant leaves so as to produce a white or brick-red silk popularly known as eri silk. Since the filament of the cocoons span is neither continuous nor uniform in thickness, the cocoons cannot be properly reeled.

Tasar Silkworms

The tasar silkworm belongs to family *saturniidae* and genus *Antheraea* and they are all wild silkworms. There are many varieties such as the Chinese tasar silworm *A. pernyi* which produces the largest quantity of non-bulberry silk in the world, the Indian lasar *A. mylitta*, next in importance, and the Japanese tasar silkworm *A. yamamai* which produces green silk. The Indian *tasar* silkworm feeds on leaves of Arjan, Oak, Sal and various other plants.

Muga Silkworms

It also belongs to family *Saturniidae* and genus *Anthraea assama*, produce unusual lustrous golden-yellow silk thread which is very attractive and strong. These are only found in state of *Assam* and feed on some (*Machilus bombycina* and saalu (*Litsaea polyantha*), champa, Cinnamon etc., leaves. The quantity of muga silk produced is quite small and is mostly used in the state of Assam itself.

The mulberry, silkmoth never occurs in the wild state, it is completely domisticated. *Bombyx* is extensively cultivated all over the world and usually produce a single brood in a year *i.e., Univoltine*. Some strain, however, produce 2-7 broods a year (*Polyvoltine*) and are cultivated in warm climate.

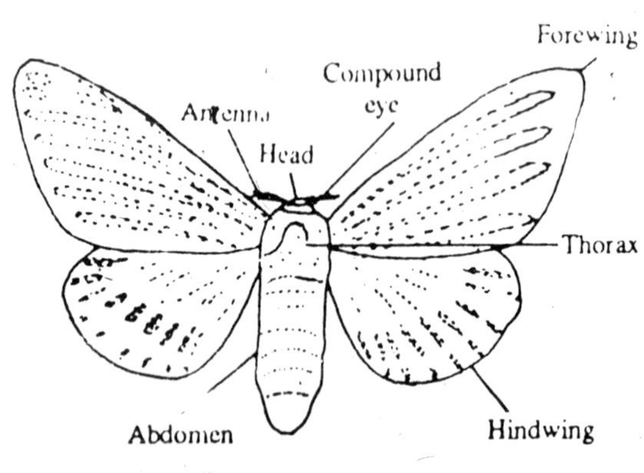

(a) Moth (*B. mori*)

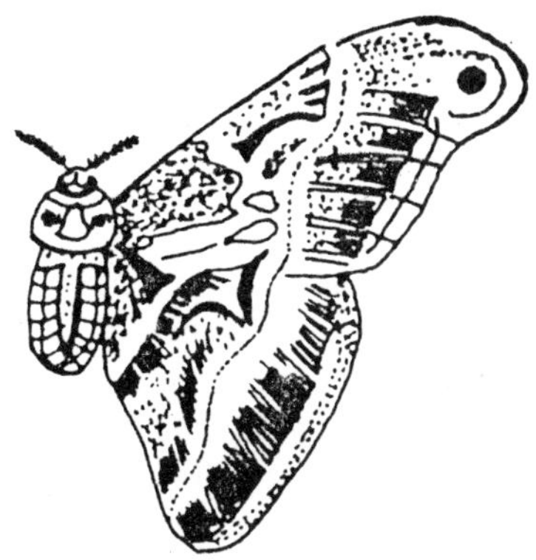

(b) Eri Silkmoths

(c) Tasar silkmoth *Antherea paphia* (*after Lefroy*).

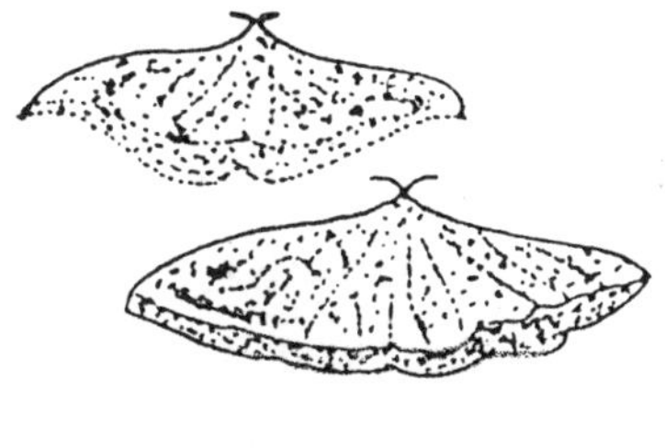

(d) Muga silkmoth

Fig. 4.1

3. EXTERNAL MORPHOLOGY AND LIFE CYCLE

Male and female moths copulate in tail to tail position. Life cycle pass through a complete metamorphosis (Holometabola) *i.e.,* comprising four stages

1. Egg,
2. Larva,
3. Pupa,
4. Adult.

1. Eggs

Soon after fertilization, each female lays about 300-500 eggs in clusters upon the leaves of mulberry tree. They are fixed to leaves with a gelatinous secretion. The eggs are tiny, smooth and spherical measuring 1 to 1.3 mm in length and 0.9 to 1.2 mm in width and when laid they are yellowish white but become darker later on. The female dies within 3 or 4 days after egg laying. In the univoltine race the eggs may hatch in months. In polyvoltine race the eggs hatch in a few days.

Fig. 4.2 : Eggs of Silkworm.

2. Larva

The Larva is called *Caterpillar* or *silkworm*. The caterpillar larva which hatches from the eggs measures 5-7 mm in length. These caterpillars move on the leaves in a characteristic looping manner. Their body is rough, wrinkled and greyish in colour. They are made up of 12 segments which is distinct into three parts : head, thorax and abdomen.

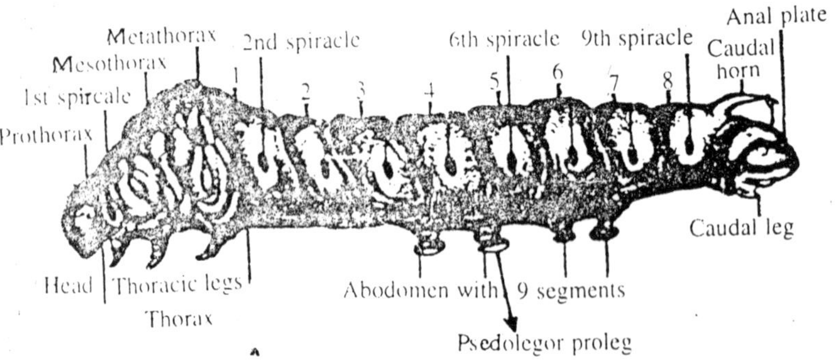

Fig. 4.3 : Morphology of Silkworm Larva.

Head consists of six body segments fused together with a cranium. There are six pairs of ocelli or larval eyes which are located behind and a little above the base of the antennae. Head also bears mandibulate type of mouth parts which are adapted for mastication.

The maxillae on the ventral side of the mouth, consist of cardo, stipes, maxillary lobe and maxillary palpi. Maxillary lobe and maxillary palpi discriminate the taste of food. The labium is located ventrally carrying a big-sized, lightly chitinized mentum. The prementum is chitinized and black. Distally the prementum carries a median process or spinneret through which silk is expelled from the silk gland to form the silk bave or thread. The sensory labial palpi are found on both sides of the spinneret.

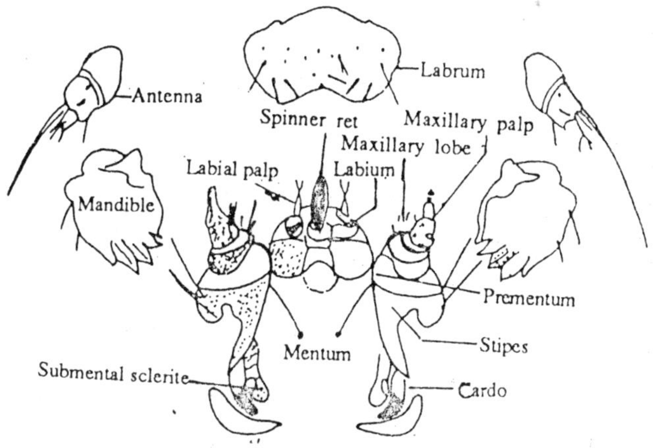

Fig. 4.4 : Mouth parts of Silkworm Larva.

The thorax is 3 segmented and all the segments bear a pair of true 5 jointed legs ventrally with claws which help in holding mulberry leaves while feeding. The abdomen is 10 segmented and bears 5 pairs of unjointed, stumpy prolegs or psedolegs. These prolegs are present in each segment 3rd, 4th, 5th, 6th and 10th. A short dorsal *anal horn* is present on 8th segment. The *prolegs* of on the last segment turn backward to form the *claspers*. There is a row of small respiratory pores, the spiracle, on either side of the abdomen.

The larvae feed voraciously upon the mulberry leaves and grow very quickly. They stop feeding, become inactive after 4-5 days, and then 1st moulting (ecdysis—shedding of old skin) take place. The 2nd stage larvae resemble the 1st stage larvae except that they are slightly bigger in size. They also eat voraciously for next 7 days, then moulting takes place and 3rd stage larvae are formed. The larvae repeats this process for 4 times. The maturity is achieved in about 45 days, since the time of hatching. The matured caterpillar now measures 10-17 cm in length. By this time the formation of a pair of salivary glands is completed. Since these glands secrete silk, they are called *silk glands*.

3. Pupa

The pupal stage is generally called the resting, inactive stage of the silkworm when it is incapable of feeding and appears quiescent. But this does not suit here. The pupal stage is a transitional phase during which definite changes takes place. During this period of biological activity the larval body and its internal organs undergo a complete change and assume the new form of the adult moth. The mature silkworm larva passes through a short transitory stage of pre-pupa before becoming a pupa. During the pre-pupal stage the dissolution of the larval organs takes place and this is followed by the formation of adult organs during the pupal stage. Soon after pupation the pupa is white in colour and soft but gradually turns brown to dark brown and the pupal skin becomes harder.

The prominent morphological parts visible are a pair of large compound eyes, a pair of large antennae, fore-and hind-wings and the legs. Ten of the abdominal segments can be seen on the ventral side, but only nine of them are visible on the dorsal side. Seven pairs of spiracles can also be discerned in the abdomen; the last pair is non-functional. Sex markings are prominent and it is much easier to determine sex in the pupal stage than in the larval stage. The female has a fine longitudinal line on the eighth abdominal segment, whereas such a marking is absent in the case of the male. (Fig. 4.5).

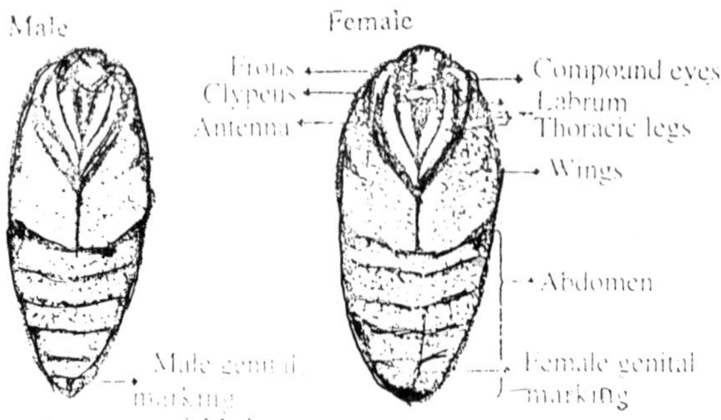

Fig. 4.5 : Drawing of Male and Female Pupa *Bombyx mori*.

During pupation, larva secrete the sticky secretion from the silk glands through a very fine pore called spinneret situated on the hypopharynx. The secretion is continuous and after coming in contact with the air the sticky secretion is converted into a fine, long and solid thread of silk. The thread is made up of five filaments stuck together by sericin secreted by two other glands. The thread becomes wrapped around the body of larva forming a cocoon. This process continues for 3-4 days. The cocoon serves as a comfortable house for the protection of the caterpillar for further development.

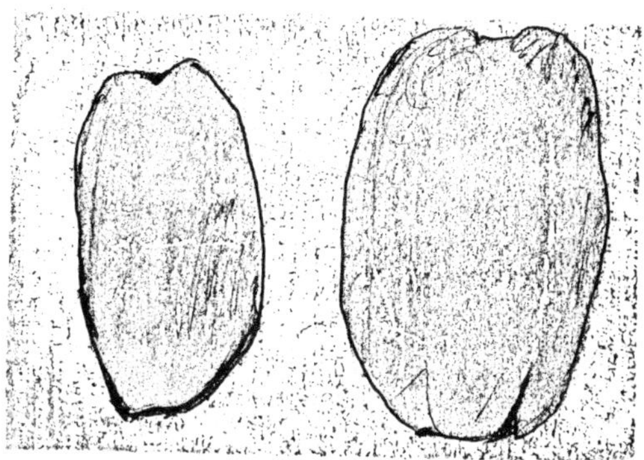

Fig. 4.6 : Male and Female Pupa *Bombyx mori.*

Cocoon is a thick, white or yellow and oval capsule. The outer or initial filaments of the cocoon are irregular, but the inner one that form the actual bed of pupa is one long continuous thread about 300-400 metres long round in ring. The silk thread is secreted at the rate of 150 mm per minute. Within 15 days the caterpillar is transformed into a brownish pupa or chrysalis (as described above). Later on, larval organs such as prolegs, claspers, anal horn and mouth parts are lost. The adult organs develop within 12-15 days and the adult is formed.

4. Adult (Imago)

The adult secretes an alkaline fluid that moistens and softens on end of cocoon. The cocoon breaks at this end and the adult comes out. The adult moth emerging from the pupa is incapable of flight, it does not feed during its short life span. The silk moths mate, lays eggs and die.

The body of the moth is composed of three distinct segments : head, thorax and abdomen. Body surface completely covered by scales like butterflies. The compound eyes are situated on either side of the head. The ocelli are absent. The antennae are conspicuous large and bi-pectinate.

The Thorax consists of three segments namely pro-, meso-and meta-thorax as in the larva. The meso thorax is the largest and is pentagonal. There are three pairs of thoracic legs one pair on each of the three thoracic segments. Each of the thoracic legs is composed of five segments. The meso- and meta-thorax bear two pairs of wings, the front pair overlapping the hind pair when the moth is in the resting position.

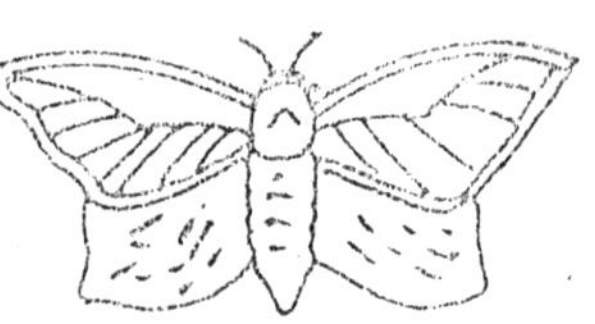

Fig. 4.7 : Adult Female.

In the male eight abdominal segments are visible; in the female seven. There are six pairs of spiracles present laterally on either side of the body.

Morphologically the female and male can be easily distinguished in the adult stage. The female has comparatively smaller antennae, its body and the abdomen are fatter and larger and it is generally less active than the male moth. At the caudal end, the male moth has a pair of hooks known as harpes (Fig. 4.8) whereas the female moth has a knob-like projection with sensory hairs (Fig. 4.9). These differences help to a large extent in separating the sexes for preparation of hybrid eggs.

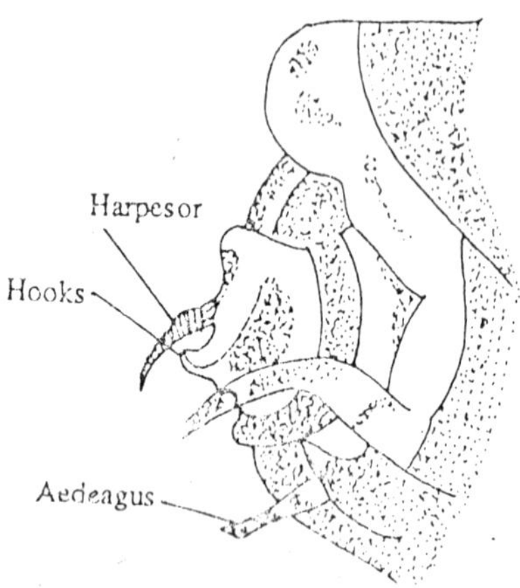

Fig. 4.8 : Abdominal end of male moth.

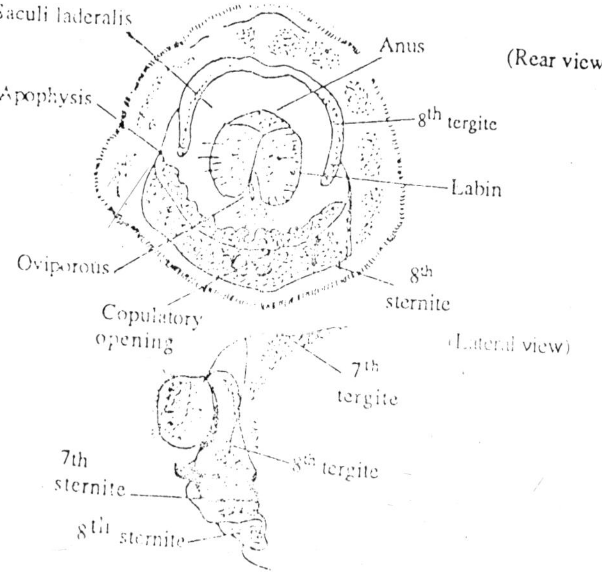

Fig. 4.9 : Abdominal end of female moth.

(A)

CULTIVATION AND HARVESTING OF MULBERRY

Mulberry forms the basic food material for silkworms and the bulk of the silk goods produced in the world are from mulberry silkworms. Scientific sericulture is the meeting place for agriculture and art and industry, ancient culture and civilization, the rich and the poor and it reflects the interdependence of these.

Production of mulberry leaves on scientific lines is essential for organizing sericulture on sound economic lines. It is estimated that one metric ton of mulberry leaves is necessary for the rearing of silkworms emerging from out of one ounce of eggs, which will yield about 25 to 30 kg. of cocoons of international standard. The cost of leaves works out to be 60% of, the total cost of production of silk. One hectare of fertile land can produce about 15 to 40 tons of mulberry leaves over a period of 1 year. This, however, depends on the favourable climatic conditions prevailing in the region; in the temperate and subtropical region of Europe and Japan, about 15 to 20 tons of leaves per hectare is produced in a year whereas in the tropical parts of India, under intensive cultivation practice, about 30 tons per hectare could be harvested in a year.

Morus is the latin word for mulberry. Mulberry plant is exploited in different ways for commercial production of silk, as mulberry is the chief food for Bombyx mori. Mulberry leaf protein is the source for the silkworm to bio-synthesize the silk which is made up two proteins, fibroin and sericin. Nearly 70% of the silk proteins produced by a silkworm is directly derived from the proteins of the mulberry leaves.

Mulberry is grown as a bush in tropical countries and as middlings and trees in temperate countries like Japan where silkworms can be reared only in three seasons a year. With an object of increasing the production of mulberry and reducing the cost of production, attention has been paid to intensive cultural operations including application of economics dosage of fertilizer and adoption of suitable irrigation schedules.

With efficient use of mulberry leaf, coupled with pruning programmes, it has been possible in sericulturally advanced countries like Japan to rear silkworms thrice a year utilizing the second leaf harvest. In India by following bottom pruning of the crop it has been possible to harvest more and better quantity leaf. This has also made possible artificial hatching of biovoltine silkworms eggs and rearing of polyvoltine silkworms in a commercial manner in India.

The next scientific advancement in reducing the labour in harvesting leaf is introduction of mechanization of farming practices. Use of machines for cultural practices and for harvesting of leaves have helped to reduce the cost of production of mulberry leaf.

ORIGIN

Mulberry is believed to be native either of India or China and it is believed to have originated on the lower slopes of the Himalayas. Towards the year 2800 B.C. Chin-Nong, one of the successors of Empheras Fo-Hi taught cultivation of mulberry in China. Silk-industry took its origin in the province of Chang-tong. According to western historians mulberry culture spread to India from China through kotan (Tibet) by about 140 B.C.

DISTRIBUTION

A global survey of sericulture industry reveals that there are at least 29 countries where mulberry is cultivated (as shown in Fig. 4.10) and major countries are (1) Japan, (2) China, (3) Korea, (4) U.S.S.R., (5) India, (6) Brazil, (7) Italy, (8) France, (9) Spain, (10) Greece, (11) Sri-Lanka, (12) Bangladesh (13) Cyprus etc. There is no doubt

Fig. 4.10 : Distribution of Mulberry in the world.

that mulberry can grow and flourish in many other parts of the world and this aspect is being explored by the FAO of the united Nations.

CLIMATIC CONDITIONS

Mulberry can be grown under various types of climatic conditions, ranging from temperate and tropics.

TEMPERATURE

The optimum temperature for best mulberry growth 23.9 to 26.6°C. The climatic conditions in India are favourable for luxurient growth and rearing of silkworms throughout the year. The temperature of Karnataka state, major silk producing state, ranges from 21.1°C to 30.0°C. Climatic conditions in Kashmir are favorable to rear silkworms during may to October.

SUNSHINE

In temperate countries—5.0 to 10.2 hrs/day-in tropics it grows well with 9.0 to 13.0 hrs/day.

ELEVATION

A range upto about 700 m above sea level is considered desirable for luxuriant growth of Mulberry.

SOIL

Should be deep, well drained, clayey loam to loam in texture, friable, porus and with good moisture holding capacity and fertile.

RAINFALL

Mulberry can be grown in a rainfall range of 635 mm to 1500 mm. Under low rainfall conditions, the growth of mulberry is limited due to shortage of moisture, resulting in low yields as prevailing in Karnataka State under low rainfall conditions, supplemental irrigation is necessary.

HUMIDITY

Range of 65 to 80 % is ideal for its growth.

Mulberry belong to the genus Morus (Fig. 4.11) which includes more than twenty species. It is deciduous tree or shrubs and perennial one. The plant is of branching type and the branching character varies considerably due to the influence of different factors, such as the type of cultivation, mode of training, fertility of the soil, rainfall etc. The important species grown in India are *Morus indica* (Mysore local, Kanva-2 and Berhampore variety), *M. nigra* and *M. serrata*; other species distributed throughout the world are *M. arabica*, *M. mizuno*, *M. bombycis*, *M. kagayamae*, *M. rubra*, *M. alba* etc.

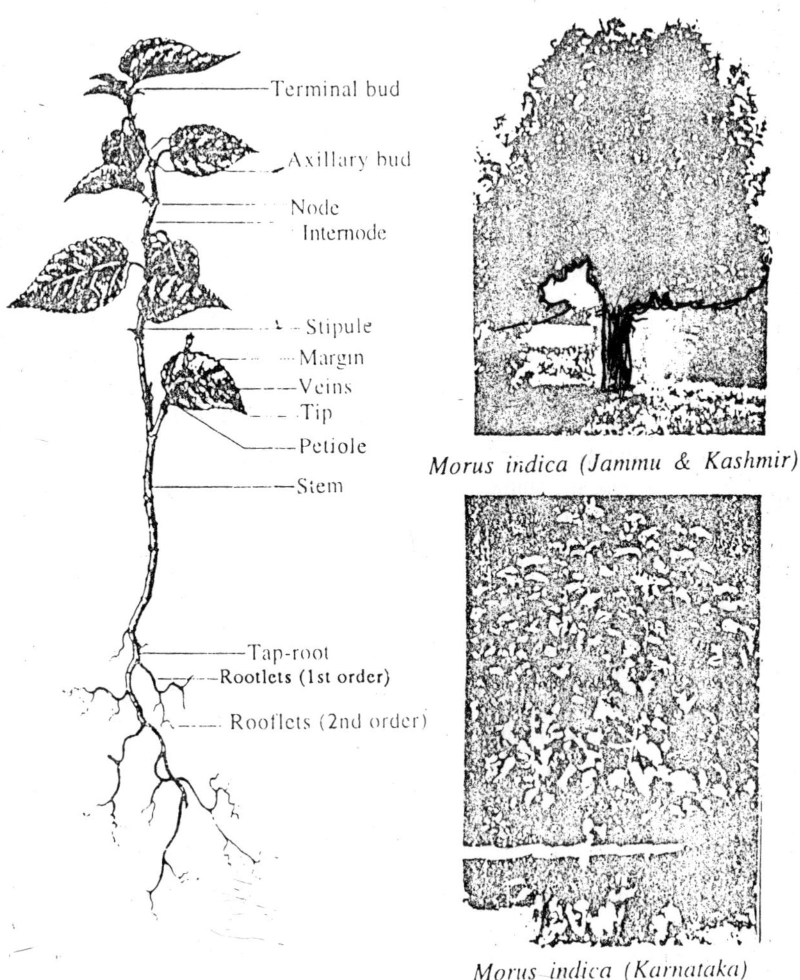

Fig. 4.11 : Morus indica

PLANTING OF MULBERRY

Mulberry plantation should desirably be established close to the rearing house for quick transport and immediate use of the leaves for feeding otherwise they will loose moisture and wither during summer. After selecting a flat and fertile soil for cultivation, it is necessary to prepare the land in the proper manner. The field is to be levelled and the fertility level improved. Mulberry land is perennial and arboreous and its roots grow widely and deep in the soil. Hence a thorough preparation of the soil with a deep ploughing or digging is a prerequisite for mulberry cultivation, whether it is raised as a rainfed dry crop or as an irrigated crop.

For proper establishment of mulberry plantations, early spring and late autumn seasons are best suited. Planting in winter and summer seasons are to be avoided. In India, planting season in Karnataka (July-August), in West Bengal (November-late Autumn).

DIRECTION OF PLANTING

Depending upon the light intensity (sunshine hours) and wind direction mulberry secdings are to be planted in rows either north-south or east-west directions, making the rows parallel to the direction of wind. In tropics, where the sunlight is not a limiting factor, mulberry rows can be planted in any direction. In sloppy lands the rows should be parallel to the contour lines.

PLANTING DISTANCE

The planting distance depends upon the agro-climatic conditions (sunshine, precipitation, temperature etc.) soil fertility level, intensity of cultivation practices adopted including the training and harvesting methods and also the variety of mulberry planted. In most parts of India, Mulberry is grown with very close spacing as a pure crop. In Karnataka, in rainfed areas, it is raised under pit systems with wider spacing. In some parts of Karnataka, such as kolar division, where intensive cultivation is practiced with heavy manuring and irrigation, closer spacing is adopted. In West Bengal, where the rainfall is heavy the 'strip system' of close planting is practiced.

Table 4.1 : Spacing for Mulberry Cultivation in India.

Name of system	Spacing between rows	Spacing between plants
1. Pit system (rainfed Mulberry)	0.9 to 0.75 m	0.9 to 0.45
2. Row system (irrigated)	0.45 to 0.60 m	0.45 to 0.60 m
3. Strip system	0.3 to 0.45 m	0.15 m

MULBERRY PROPAGATION

There are various methods of propagating mulberry plants. Generally, plant is raised either from the seed or through vegetative propagation from stem cutting or layering and grafts. The vegetative is the most common method of propagation because of various advantages like maintenance of particular character of the plant, relative speed in raising samplings in large numbers for plantation, the adaptability to a particular habitat, to develop resistance to pests and diseases and to modify the growth of plant. Propagation through seeds has got some limitations. Triploid plants can not be propagated due to their inability to produce viable seeds. It is not possible to reproduce true to the type from a seed due to biparental origin. Different countries depending upon climatic and soil conditions follow different modes of vegetative propagation,

In India the most common method of propagating mulberry is through *cuttings* in multivoltine regions like Karnataka and west Bengal. Exotic varieties (imported) which do not come up by cutting are propagated through root grafts.

Bud grafting (budding) is followed only when the scion (young shoot) material is scarce. Whenever a larger mulberry plant is to he obtained in a shorter time then the method of *layering* is followed.

In univoltine areas like Kashmir, the mulberry is propagated through seedlings and the exotic varieties through root grafts.

1. SEEDLING PROPAGATION

Propagation through seeds is mainly to bring about a varied population for the purpose of selection and hybridization. Since mulberry flowers are open for cross-pollination, the seeds thus collected are mainly used to obtain stock material tor grafting.

For this process of seedling nursery is maintained to protect the seeds from excessive light and temperature as well as handling the processes carefully.

SOWING OF SEEDS

Seeds with viable embryos are soaked for a day in water to soften the hard testa for easy and successful germination. They are broadcast in seed beds or sown in holes drilled in a line or sown in a line with the help of rope. The seeds should not be sown deeper than 2½cm in the bed. The holes are then covered with soft earth. Water is applied through watering, small quantity of insecticide can also be applied if necessary to protect from attack of insects. Seed beds are protected from severe sun by covering them with mats made up of bamboo strips.

GERMINATION OF SEEDS

Temperature and light have a marked effect on the germination of seeds. Seeds germinate better in bright light than in total darkness. The germination is delayed with lowering of tempcrature for instance six days of 33°C or 36°C, eight days at 30°C and ten days at 27°C.

SEEDLINGS

When the seedlings attain 3.5 to 5 cm height they are thinned by picking from dense areas and planted in thinner areas. Direct sunlight which is essential for young plants is allowed to fall on them during the cool hours of the day and on cloudy days. Transplanting of seedling is done first after three months with a distance of 22.5 cm between them. They are allowed to grow for one to two years, when they are removed either to raise tree plantations or utilize the stock for grafting.

CUTTING

In India propagation through cutting is the most common method. It is restricted in varieties which are fully acclimatized to local condition. Plants which conform to the qualities chosen for multiplication such as nutritious leaf, higher yield, quick growth, resistance to diseases and insect pests and drought resistance are selected.

PREPARATION OF THE CUTTING

Shoots of proper maturity and thickness with active and well developed buds are cut from the selected varieties. Cuttings taken from parts with high carbohydrate content root more readily and profusely than from a cutting selected from parts rich in nitrogen. Accordingly the portions of the shoot which are too tender at the top and over mature at the base are rejected. Generally cuttings taken from young seedings root more rapidly and profusely than those taken from old and mature plants. Cutting of 7 to 10 cm usually of pencil thickness with three or four active buds are prepared out of the central portion of the clone with the slanting cut. These cutting are planted in the field directly or in nursery beds. When kept in nursery all precautions should be observed in not allowing the cuttings to dry up. After two or three months, sprouted cuttings are transplanted into the field depending upon the type of plantation to be raised.

FORMATION OF ROOTS

Adventitious root develop from the basal end of the cutting as also from the root angles of the bud. The roots develop endogenously. Those varieties which do not ordinarily produce roots from cutting are induced to root with the application of requisite quantity of root hormones like Indole Acetic acid, Indole Butyric acid, rootone etc. by direct application of the powder, or soaking in dilute solutions or concentrated solutions or application as a paste in lanolin.

GRAFTING

Grafting consists of inserting a small branch of a plant into rooted plant of the same or allied species in such a way as to bring about an organic union between the two and finally make them grow as one. The branch that is inserted is known as the scion and the plant to which the scion is inserted is called the stock. The scion grows with the help of nourishment supplied by the stock. The stock is generally of an indigenous variety which is well acclimatized to the local conditions. Grafting thus facilitates the propagation of variety which has the desirable qualities and which cannot be propagated by other means.

There are three types of grafting namely, shoot grafting, root grafting and bud grafting. All these methods are pretty popular throughout the temperate zones. In Japan the best season for grafting is the spring.

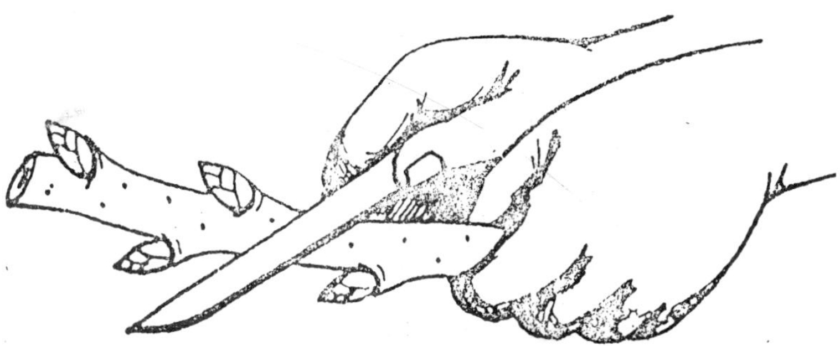

Fig 4.12 : Preparation of Stock & Scion.

Selection of stock and scion is very important as the success depends upon their quality. The stock should be preferably got from a one or two years old seedling. The size of the stock should be either slightly bigger or at least of the same size as that of the scion. Shoots of one year old and medium size are used as scions. The middle parts, cutting away the top and basal portions are selected. The leaves from these plants should not have been plucked during the previous year. In Japan these scion are stored before early spring, keeping them in a refrigerator at

5°C and burying them under wet rice bran or sand, covering with straw mat or digging a ditch in a cool place covering it with fine soil and straw mat enclosing the scions. In Japan the seedlings of *Morus latifolia* are generally used for stock. It is desirable to use such varictics which are strong having a great resistance to the damages caused by the diseases or insects and are vigorous in rooting. On both the stock and the scion a slanting cut should be made at both ends. The scion is well inserted in between the bark and wood of the stock. They are carefully placed under controlled climatic conditions and planted in the soil. The union between the stock and the scion will be established within two to three months.

ANATOMY OF GRAFT UNION

When the scion is inserted into the stock, organic union is established between them by the formation of secondary vascular tissues. Hence the cambium layers of the stock and scion are brought in close contact with each other. In case the cambium layers do not match, the union is delayed. When they are extremely mismatched union may be prevented. Cell division occurs rapidly for the formation of callus tissue. Proper temperature preferably 20°C to 40°C is conductive to rapid cell growth. It is also of importance to maintain high humidity to prevent the thin walled turgid cells from dessication and death.

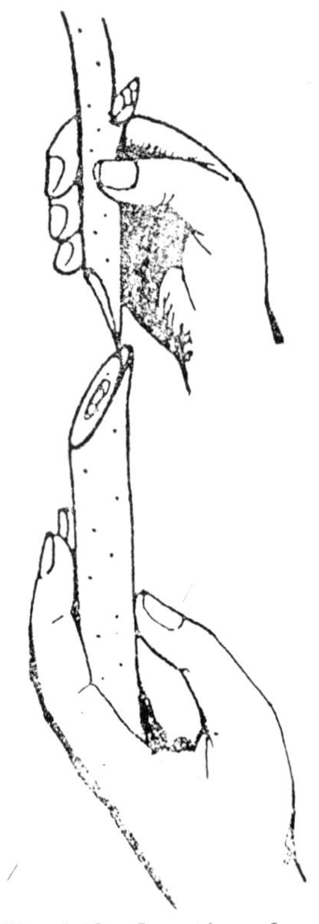

Fig, 4.13 : Insertion of Scion into stock.

As a second step interlocking parenchyma cells (callus tissue) are produced from the stock and the scion. The parenchyma of the phloem rays and immature parts of the xylem rays produce new parenchyma cells filling the gaps between the stock and the scion. Thus, the thin walled parenchyma cells serve the conduction of water and nutrients from both the components. This is the callus tissue whose surface layers get suberized and become bad conductors of heat.

Along the edges of the original strips of cambium, cells of the callus also convert into meristematic cells. These cambial cells grow inward across the callus till they meet. Thus a continuous ring of cambium is formed. The newly formed cambium cuts off secondary xylem towards inside and secondary phloem towards outside along with the original cambium of the stock and the scion. These new vascular strands soon establish their connection with the original xylem and phloem.

The union between the stock and the scion is from the cells which develop after the actual graft union has been made. There is no fusion of cells or cell contents. Cells produced by the stock and by the scion each maintain their own distinct identities.

There are different types of grafting followed in different countries:

(i) Wedge and crown grafting

(ii) Whip grafting

(iii) Root grafting either ordinary type of "In Situ";

(iv) Bud grafting

WEDGE GRAFTING

Wedge grafting is followed to renovate the old plant. The plant is pruned at a height of convenient level and incision is made at a cut surface in the form of letter "V". The basal end of the scion is also cut obliquely downward to fit in the incession made on the stock. The scion is inserted into the stock and the grafting wax or grafting clay is applied at the region of contact.

GRAFTING WAX

A mixture of tallow part, bees wax one part and resin four parts, melted together and baked with a small dough under water.

GRAFTING CLAY

Clay two parts, cow dung one part and some finely chopped hay mixed with water.

CROWN GRAFTING

When more than one scion is inserted, it is called crown grafting.

To a stock of about 1.2 to 2.5 cm thickness, sloping cut is given to a length of 3.5 to 5 cm. The scion of the same thickness is also cut in a similar way so as to fit in the stock exactly, it is inserted into the stock and tied firmly with soft fibre. The wound is then covered with grafting wax.

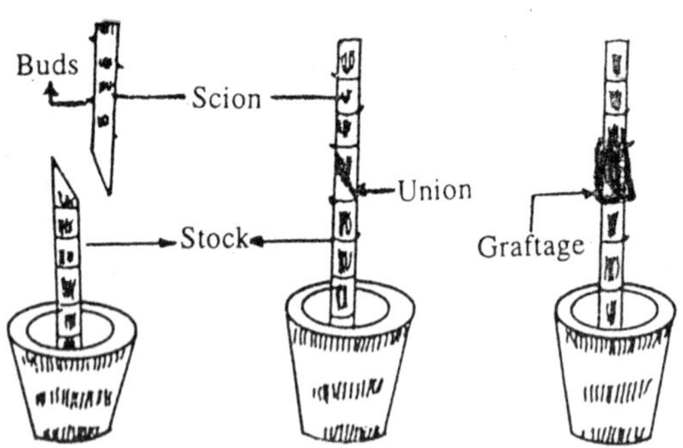

Fig. 4.14 : Whip Grafting.

ROOT GRAFTING

When the root is used as a stock instead of shoot the grafting is called root grafting. Mysore Local variety is used as a stock material. Root is selected to the thickness of 0.6 to 2.5 cm from a one year old seedling and cut into bits of 5 to 7.5 cm in length. The top end of the root bit is cut obliquely. The scion bit of four to five inches long with two to three active buds are prepared. The basal end of the scion bit is cut obliquely and inserted at the pointed end of the stock between the wood and the bark. The bark at the pointed end of the scion which goes inside the bark of the root is removed before insertion in such a way that the cambial layers of stock and scion are in close proximity with each other. Organic fusion is formed between the stock and the scion, and the plant having the characters of the scion is developed (Fig. 4.15). As bandaging is not done, in this case, this type of grafting is very economical for multiplication.

The advantages claimed in this method are:

(i) percentage of success is more in this type of grafting,

(ii) an individual can prepare 800 to 1,000 grafts in eight hours,

(iii) out of a two year old seedling, five to six stocks can be had and therefore five to six grafts can be prepared, thereby cost of labour is minimized.

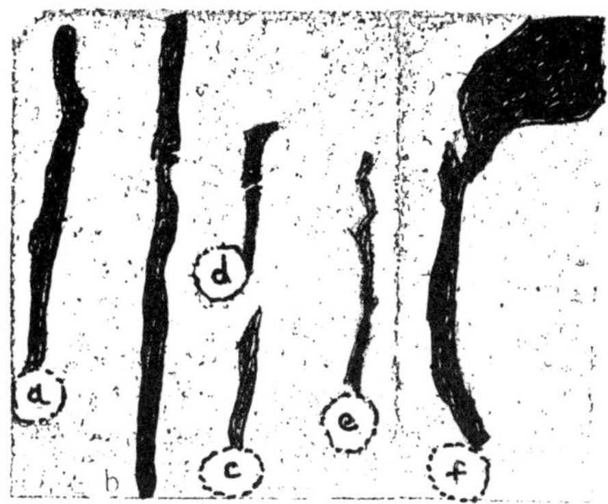

Fig. 4.15 : Root Grafting.

"IN SITU" GRAFTING

In this type of grafting, the seedling which is used as stock is cut below the ground level including the transitional level and the grafting is done as in the case of root grafts but without disturbing the root from its original place in the soil. Though a more vigorous plant is obtained in this case, only one graft is developed from a seedling. While making "In situ" grafting, it is necessary to take proper care in cutting the root below the transitional zone as otherwise there will be chances of sprouting from the region of the local stock.

BUD GRAFTING (BUDDING)

This method is followed only when the scion material is scarce. Bud grafting involves removal of only one bud and a small section of the bark without wood and grafting it to another clone. Since the physiological processes are the same as in grafting, it is called bud grafting or budding. The bud piece usually consists of the periderm, cortex and phloem. This bark piece is laid against the exposed xylem of the stock. Callus strands develop from terminal cells of broken phloem rays and adjacent young secondary phloem cells on the cut surface of the inner side of the bud piece. Correspondingly, the cambium and the newly formed derivatives on the exposed surface of the stock also divide and form callus strands. Similarly from the side also callus strands develop. With the joining of all these callus strands a continuous callus bridge is formed. Short areas of cambial cells appear along the callus tissue. Secondary xylem and phloem are formed due to the activity of the newly formed cambium and the union is established between the stock and the scion. The vascular strands get connected to the original xylem and phloem.

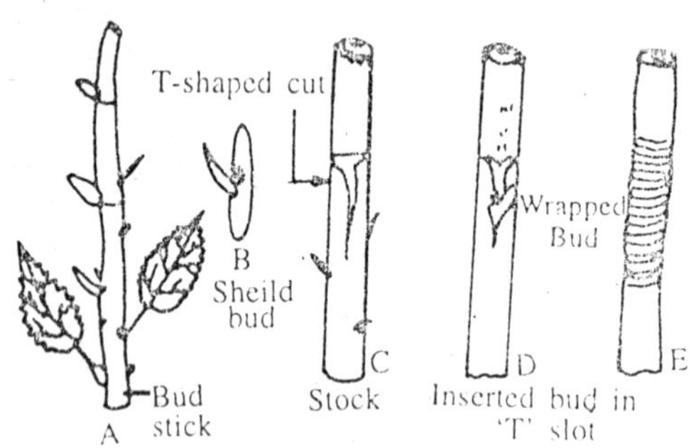

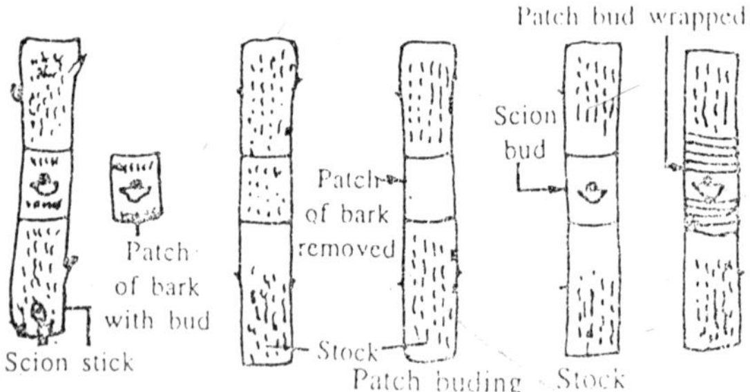

Fig. 4.16 : Bud Grafting.

There are three types of budding followed in mulberry :

(i) Patch budding,

(ii) T-budding and

(iii) Flute budding.

PATCH BUDDING

A portion of the bark with an active bud is removed from the bark as scion correspondingly a portion of the bark on the bud is removed from another stem to which bud grafting is to be made as stock. To this portion a bud of the scion is inserted and properly bandaged with a soft fibre and covered with grafting wax or grafting clay. Generally, dried fibre of the petiole of plantain is used for bandage because of its softness, easy availability and moisture retaining capacity.

T-BUDDING

A "T" shaped incision is made on the bark of the stock at the nodal region and a slit is formed by opening the cut ends. A bud cut out clean from the selected plant is inserted in this slit and properly bandaged.

FLUTE BUDDING

The bark of the plant is cut to a length of 2½ to 3¾ cm round and removed as a sheath. Similarly, the bud of the scion with the same length of the bark is removed and inserted in the stock from where the bark is removed and bandaged properly.

The following precautions are to be observed for preparing bud grafts :

1. While removing the bud from the scion, no woody portion should stick to the bark.
2. Bandage should be done in such a way that water and air should not enter the budded area.
3. Sprouting of other buds on the stock should be inhibited.

LAYERING

Layering is another type of propagating mulberry. It is the development of roots on a stem while it is still attached to the parent plant. The rooted stem is then detached to become a new plant growing on its own roots. The layered stem is known as a layer.

The principle involved in the formation of roots during layering is due to interruption of the downward translocation of organic material, carbohydrates, auxins and other growth factors from the leaves or growing shoot tips. These substances accumulates near the point of treatment and induce rooting in that area even though it is still attached to the plant. This method is considered to be safe and cheap. At the same time not many plants are formed by this method.

Layering does not depend upon the length of time that a severed shoot can be maintained before rooting occurs. Due to this reason, layering is more successful in many mulberry varieties than those propagated through cuttings. This method is usually followed to obtain a large sized plant in a shorter time than it would have taken from a cutting.

There are different types of layering such as simple layering, air layering or gooting, trench layering, etc. A lower branch of a bush mulberry is bent down, a ring of bark is removed 2.5 to 5 cm in length around the stem and this portion is pushed into the soft earth, made for the purpose, keeping the upper portion of the bark erect, the bent portion is covered with soil and a stone or brick is kept on it (Fig 4.17). When the roots have developed, usually within two to four months, the branch is cut off from the mother plant and is grown separately. This is simple layering.

Fig. 4.17 : Process of Layering.

AIR LAYERING OR GOOTING

In this method, 1.2 cm circular bark is removed from the stem, where peat mass or well decomposed organic manure with a little quantity of root promoting hormones is placed at the region and tied securely with a polythelene cover. Water is regularly poured at the region. After two or three months, roots develop from the cut end. When the connection between the mother plant is severed it can develop into a new plant (Fig. 4.18).

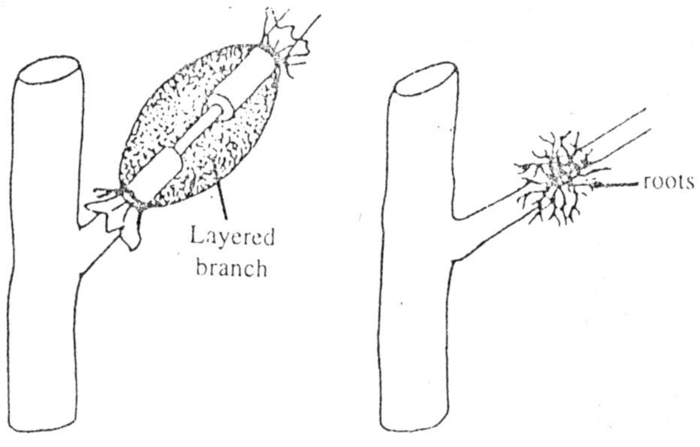

Fig. 4.18 : Air Layering.

TRENCH LAYERING

This is mostly followed in Japan. It consists of growing a plant or a branch of a plant in a horizontal position in the base of a trench and filling in the soil and manure around the new shoots as they develop. Roots develop from the base of these new shoots under the same conditions as is done in simple layering. New shoots which develop from the buds along with the stem grow upwards through the soil with roots at the base resulting in the formation of a number of layers from a single branch. (Fig. 4.19).

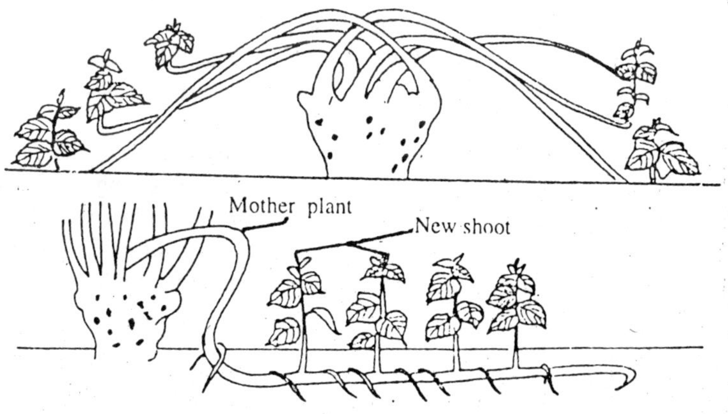

Fig. 4.19 : Ground Layering (Trench Layering).

PRUNING

Principles of Pruning

Pruning is the methodical removal of certain branches of a mulberry plant with the object of giving the trees a convenient shape and size, to increase the leaf yield and to improve its feeding value. Pruning is essential to improve the yield in certain agricultural and horticultural plants.

In mulberry, pruning is practiced solely to improve the yield of foliage and to maintain the shape of the plant for early harvest of the leaf for silkworm rearing. Pruning of mulberry plants is also useful in adjusting the production period to synchronize with the leaf requirement for silkworm rearing, and also to extend the leaf production period in all the three seasons, viz. spring, summer and autumn. Pruning is one of the cardinal principles of mulberry leaf production technology. Trimming or shaping of plants into some definite cut forms is essential in mulberry cultivation for easy harvest and inter-cultivation. Pruning helps to divert the energies of the plant for optimum production of foliage. Due to the profuse branching habit, more branches than required for uniform distribution and optimum production may persist, thus creating a competition and struggle for existence. Many of the branches in adverse position may not get required nutrition and sunlight thus leading to wastage of energy. Systematic pruning helps in diversification of energies to the selected branches for higher yield.

Usually more branches grow on the top of any mulberry plant than can persist, therefore there is overcrowding. By pruning, the residual branches receive greater proportion of energy, make quicker growth and yield more and better leaves. Pruning is eventually a thinning process and is one of the cardinal practices in the growing of mulberry plants, similar to tilling.

In Japan, silkworms are reared in two or three seasons of the year, *i.e.,* in spring and autumn, in summer and autumn, or in spring and early and late autumn seasons. Consequently, pruning of mulberry trees is done to suit the rearing seasons. Summer pruning is practised for harvest in autumn and spring, while the spring pruning for the harvests in summer or autumn.

CUT FORMS

In order to get a guaranteed leaf harvest, every year, it is desirable to make a certain form or shape of a plant and maintain this plant form. This is achieved by adopting suitable pruning schedules. If the plant is kept in a certain form, routine pruning and management works become easy. The method by which the shape and. form is maintained is called a "cut form". The plant form is kept in order by annual pruning. In Japan the plant forms are classified by the height of the main stem, into three forms, viz., (i) cut including semi-low cut, (ii) medium cut, (iii) high cut. The length of the main stem is maintained below 50cm in low cut form one metre medium cut and above one metre in high cut forms (Fig. 4.20 & 21).

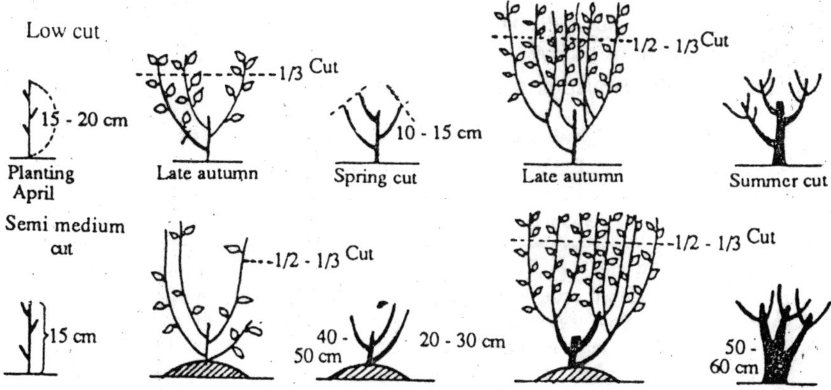

Fig. 4.20 : Low-cut and semi-medium cutforms.

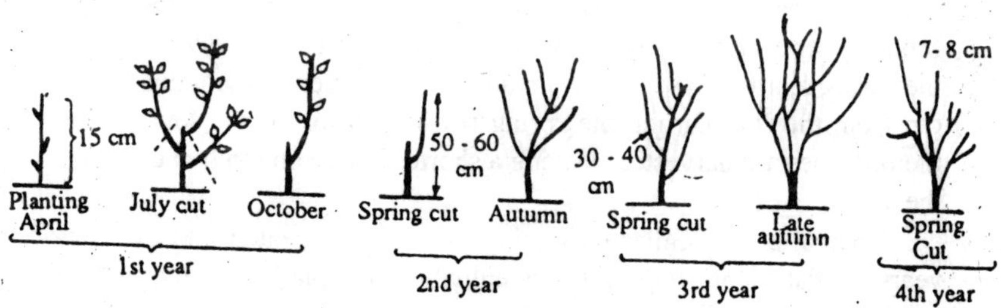

Fig. 4.21 : Medium cutform.

When the branches coming forth on the main stem are pruned every year, the part at which the cutting is repeatedly practised gets thick and shows a fist shape in a few years. This training by which the branches are cut back every year to form a fist shape is called "fist form". This fist form facilitates pruning and many branches are put forth on the fist.

Another form in which two or three branches are raised from the main stem from which secondary branches are allowed to develop is called "non-fist form". The plant trained by the non-fist form occupies a wider space and also assures an increased harvest.

TRAINING

Low Cut Form and Pruning

At the time of planting, the seedlings are cut to a height of a 15cm, above the ground level. From this plant three or four branches arise. Prior to sprouting in the following year, stout branches are chosen and these are pruned so as to have five to ten cm length above the ground level. These branches are termed primary branches.

Table 4.2 : Training schedule for low cut and semi-low cut forms practised in Japan.

Cut forms	I year	II year	III year
1. Low cut (30cm from base). 2. Semi-low cut (30-50cm from the base)	Planting in April. Cut 15-20cm from the base. In the late autumn season cut 1/3 top shoot.	Primary Pruning before budding (March) 10-15cm. above the first cut (25-35cm) from the base). In late autumn 1/2 to 1/3 of shoot is cut and used for late autumn rearing.	Spring leaf harvest by shoot cut method followed by pruning in late May or early June and is known as summer pruning and is followed every year.

The thinner branches at the base are removed. On this primary branch, three secondary branches are retained. Therefore, during the second year each plant has about nine branches. In the third year, during the spring rearing season, all the branches are pruned at a length of. three cm from the fist. During the spring rearing season of the fourth year, the branches are harvested, leaving a shorter length one to two cm from, the fist.

Thereafter, a Similar harvesting process is repeated and in a few years, a "fist shape" appearance eventually gives place to'the growth of ten to fifteen cm of long stout branches. This form of training is designed for harvesting mulberry leaves for the spring rearing.

By adopting summer and spring pruning schedules it is possible to harvest mulberry leaves for spring, summer and early or late autumn rearings, from a plantation which was formerly used only for a spring harvest. After the leaf harvest for spring rearing the branches are pruned in May-June, When the plants grow fast, and this pruning is called "summer pruning" (Fig. 4.22). From summer pruned plants, leaves are harvested for autumn and spring rearings. On the other hand, for harvesting of mulberry leaves for summer and autumn rearings, the branches are pruned in March, before spring sprouting of buds, and this is known as "spring pruning" (Fig. 4.23).

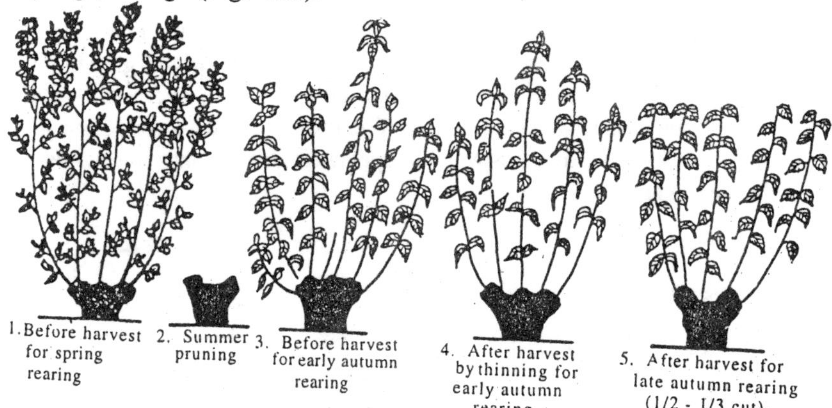

Fig. 4.22 : Summer Pruning.

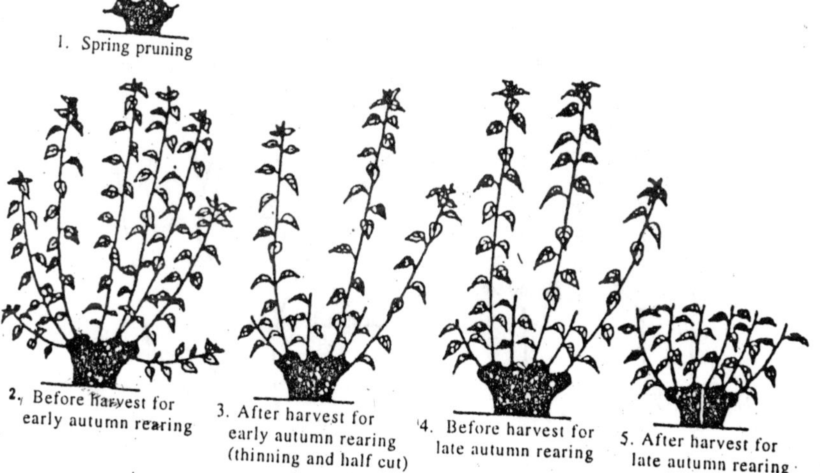

Fig. 4.23 : Spring Pruning.

TYPES OF PRUNING

The medium cut forms include the fist form, non-fist form and various other forms of pruning.

1. Fist Form

The branches are cut at the base, maintaining a constant height. Fist style pruning promotes growth of latent buds, besides normal bud, thus making the plant grow into a small tree. By this method pest and disease control becomes easy (Fig. 4.24).

2. Non-Fist Form

Branches are cut leaving basal parts to some extent and in this style the trees become taller every year. This is better for the growth of mulberry trees, because by this pruning method normal buds grow with many branches. However, retaining old growths leads to retention of injurious pests and diseases which perpetuate, multiply and cause severe damage (Fig. 4.24).

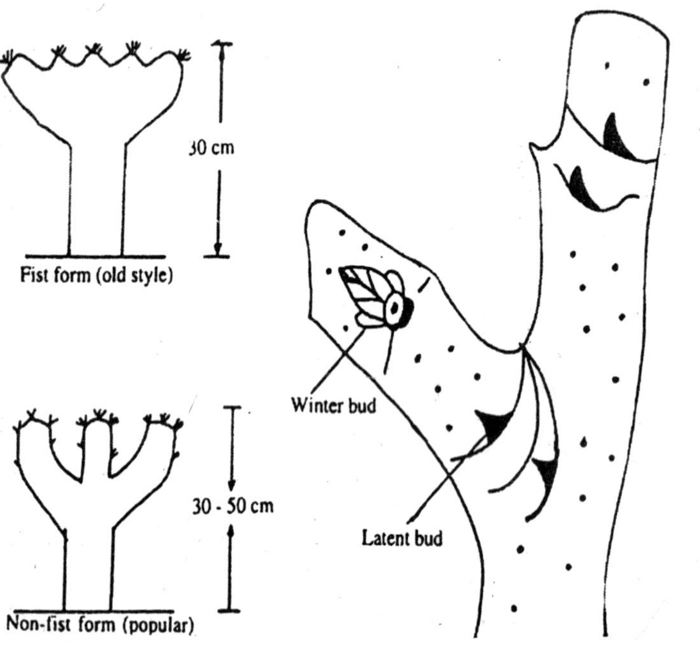

Fig. 4.24 : Pruning Technique (Fist and non-fist forms).

In pruning system, it is necessary to maintain the mulberry plant in proper form. Even in grown up trees a suitable pruning method should be adopted to maintain a desired shape.

In the tropical climates of India, where mulberry sprouts all the year round (except in Kashmir), pruning season depends upon the rainfall conditions, and the methods of leaf harvest. Under rainfed dry farming conditions, the plant is pruned once a year, during July-August, from 10 to 15 cm above the ground level and this is known as bottom pruning (Fig. 4.26).

In some cases "middle pruning" is also practised. Middle pruning is a method of cutting the branches of bush mulberry 45-60 cm above the ground level, during December-January. Middle pruning stimulates sprouting of the lower buds in the bush mulberry, during winter season (Fig. 4.25).

Fig. 4. 25 : Middle Pruning. **Fig. 4.26 : Bottom Pruning.**

Under the "Kolar system" and "Strip system" of cultivation the whole shoot is cut to the ground level at each harvest, (Fig. 4.27). Thus the pruning and harvesting are combined together. In all, five harvests are made in a year and thus the plant receives five prunings. This is suitable only for tropical climates where mulberry sprouts all the year round, with no dormant period for bud sprouting. This type of severe pruning, however, needs heavy fertilization and irrigation.

Fig. 4.27 : Shoot Harvest to the ground level.

In Kashmir, the pruning method is the same as that of Japan, since the climate of Kashmir is similar to that of Japan.

HARVESTING OF LEAVES

The method of harvest of mulberry leaves depends on the rearing practice in vogue in the locality or nation. The leaves are fed as a whole or as bits and in some cases the entire shoot or branch is used for feeding the worms. Furthermore, over the past several centuries the method of harvest must have become modified to suit the availability of labour and intensity of rearing practices. Thus, in Europe, the U.S.S.R., Japan and parts of India, the entire shoot is cut and fed to the older worms so as to save, labour cost whereas for feeding very young worms one has necessarily harvest individual tender leaves, prepare them into bits and feed.

There are three methods of harvesting mulberry leaves viz.:

(i) leaf picking,
(ii) branch cutting, and
(iii) whole shoot harvest.

LEAF PICKING

In this method the leaves are picked individually from the plant. After the leaves are harvested from the main stem, the terminal bud is removed and the auxiliary buds on the main stem allowed to develop (Fig. 4.28). This results in a rapid development of the lateral shoots which. would have otherwise remained dormant. Subsequently the leaves on the secondary branches are picked as a second crop.

When individual leaves are picked from the plant, the tender ones are fed to the young silkworms and as the worms advance in age more mature leaves are picked and fed to them. In India the bushes are stripped six to seven times a year whereas in Japan and the U.S.S.R. only three pickings are possible. In these two countries because of high labour cost individual leaves are picked only to feed the young larvae during spring, summer and autumn months.

In India, the leaf picking starts about 10 weeks after bottom pruning and the subsequent pickings at an interval of about seven to eight weeks, thus obtaining six to seven harvests in a year, after which the bushes are pruned or cut down almost to the ground level. While picking the individual leaves from the plant, they are cut with or without petiole, the former being an easy and less labour consuming practice and the latter having no specific advantage over the former.

BRANCH CUTTING

In this method the entire branch with leaves is cut and fed to the worms after the third moult. It is practised in Kashmir, West Bengal and parts of Karnataka in India and in Persia and Japan. In Kashmir it goes under the name of "Batchi" system and in Japan it is called "Jassoiku". The advantages of this method are:

(1) It is easy in low and medium cut bush plantations.
(2) It saves labour in collection of leaf, distribution of feeds, bed changing and spacing.
(3) When cut branches are fed the leaf is utilized by the worms to the maximum.
(4) Since the branch rearing is usually practised on a shelf or on the floor rearing equipment trays, shelves, etc., are dispensed with.

(5) It helps in maintaining hygienic conditions in rearing the silkworms.

(6) The feed quality of the leaves are better as they are attached to the branches maintaining prolonged succulency.

WHOLE SHOOT HARVEST

This system is practised in Kolar region of Karnataka state and in Malda district of West Bengal in India. The branches are cut close to the ground level and the top is fed to the worms settled for fourth moult. Thus topping helps uniform maturity of the leaf left over on the plant. The effect of top clipping is that the energy which would otherwise go to the formation of new leaves is redirected to the leaves left behind on the plant making them more uniformly mature. The shoots are generally

Fig. 4.28 : Individual Leaf Picking.

preservation, in addition to the loss of water due to transpiration. Thus, the leaves become poor in their nutritive value. The decomposition of molecules and consequent reduction in their nutritive value are in direct proportion to the changes in storage and atmospheric temperatures. Transpiration of water is inversely proportional to the relative humidity. Hence, care must be taken to keep the place where the leaves are preserved as cool and moist as possible. The ideal condition is below 20°C atmospheric temperature and over 90 per cent of relative humidity. Heaping up leads to fermentation and high storage temperature. To avoid this the leaves should be spread loosely in their layers and covered with wet gunny cloth or alkathene sheet.

(B)

SILKWORM REARING

Silkworms are reared for the sole purpose of obtaining cocoons which form the raw material for producing raw silk, which in scientific way called *sericulture.*

PREPARATION OF EGGS FOR REARING

It is well known that there are two types of silkworm eggs, hibernating and non-hibernating.

In hibernating eggs the embryo develops only half-way undergoes stage of dormancy called diapause, and hatches out the following spring. In non-hibernating eggs, on the other hand, the embryo develops without undergoing diapause and hatches out in the normal way. Under fixed environmental conditions, this characteristics depends upon the hereditary factor of the silkworms namely whether the egg produced is hibernating or not

Silkworms laying hibernating eggs in spring, which do not hatch out till the following spring, so producing only one generation in a year, are called *univoltines*. When the first generation of moths lays non-hibernating eggs and the second generation lays hibernating eggs which, hatch out only in the following spring so that there are two generations in a year, then such silkworms are known as *bivoltines*. Silkworms which lay non-hibernating eggs and so are able to produce many generations in a year are called multivoltines. However, the character of voltinism is more or less susceptible to environmental conditions, especially the temperature and illumination during the period of incubation of the eggs. Accordingly, it is possible to control the conditions during incubation in order to obtain hibernating leggs even from the first generation of bivoltines. Under high temperatures (over 25°C) and light conditions during incubation, hibernating eggs are obtained, whereas hon-hibernating eggs are obtained under low temperatures (such as 15°C) and dark conditions. When it is necessary to permit the larvae to hatchout from hibernating eggs, one way change the hibernating eggs to non-hibernating ones simply by employing an artificial hatching method using HCl treatment, within 24hrs of egg laying. In case of farmers, silkworm rearing is usually performed on a large scale 3 or 4 times a year, from spring to autumn, during which time mulberry leaves are available. The farmers buy eggs from silkworm—egg—producers, which take special care not only over the production of qualified eggs of guaranteed breeds but also over the production conditions and preservation of the eggs. In small scale, rearing such as in the laboratory, it is desirable to have eggs that are ready to hatch whenever they are needed. By the combination of HCl treatment and preservation of eggs at various temperatures, it is possible to obtain eggs which are hatchable at almost any time of the year.

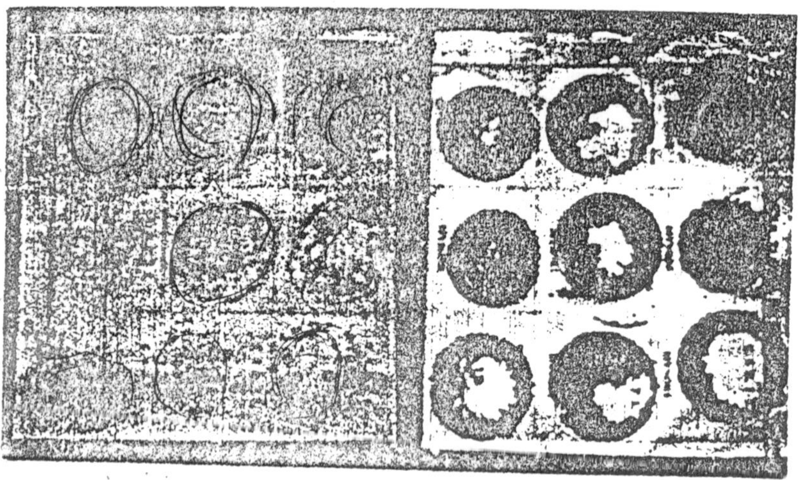

Fig. 4.30 : Hibernating and Non-hebernating Silkworms eggs.

When over-wintering eggs are kept at natural temperatures, the larvae do not hatch simultaneously but over several or more days, in April or May. In order to ensure'uniform hatching on a specific, hoped for day the eggs, are usually kept in an incubation room. Thus, incubation entails the hatching of silkworm, eggs under controlled environmental conditions, such as of temperature, illumination, and humidity.

Over-wintered eggs are taken from the cold store, kept at 15°C for 2 or 3 days, then transferred to an incubation room kept at 25°C at a humidity of at least 90%. The incubation room is illuminated for 16 hrs or more per day in order to prevent the appearance of non-hibernating eggs. The humidity during incubation affects the rate of hatchability and the unirormity hatching.

Before incubation, it is recommended to disinfect the egg surface by soaking the eggs in 2 or 3% formaldehyde solution for 30 min. The eggs are then washed in water repeatedly to eliminate formaldehyde, and finally dried. When non-hibernating eggs, prepared by means of acid treatment are to be incubated after cold storage, they must be kept at 15°C at least for half a day before transfer to the incubation room at 25°C.

In order to ensure hatching simultaneously, eggs are usually kept in a dark place or in a dark box one day before hatching, and transferred into the light as soon as hatching begins.

REARING PROGRAMME

In sericulture areas which enjoy temperate or subtropical climates, silkworm rearing is carried out twice a year in the spring and the autumn or three-times, in the spring, the summer and the autumn, coinciding with the growth and production of mulberry leaves. In tropical areas, however, where mulberry growth is continuous throughout the year silkworms are reared five to six times a year. Whether rearings are conducted *frequently* or for few times a year it is always beneficial to have a planned programme. It is advisable to rear a greater number of laying at a time than to rear many batches at very short intervals.

APPLIANCES FOR SILKWORM REARING

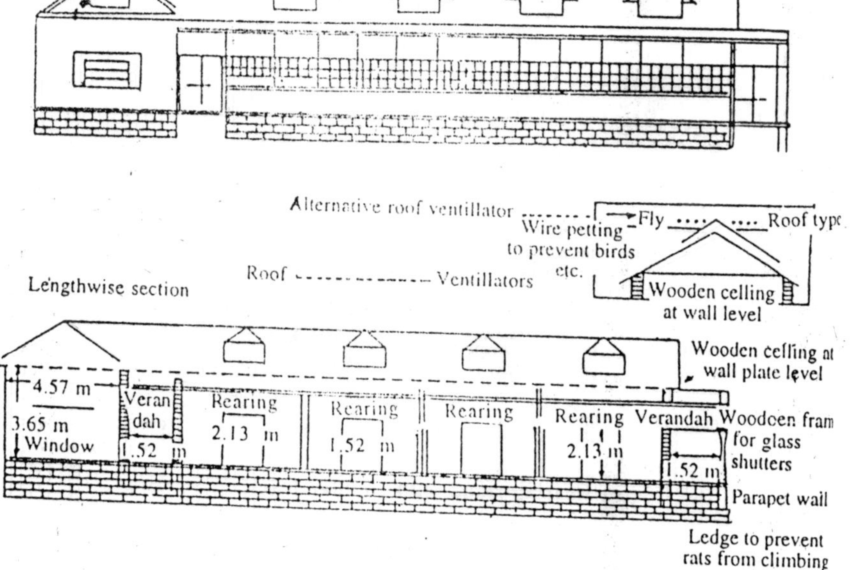

Fig. 4.31 : Rearing Houses.

The equipments required for a model rearing house for shelf rearing is shown in Fig. 4.31. The rearing house in each of the rooms 200-250 lavings or 800-1000 layings in the entire rearing house under shelf rearing.

REARING STAND

Rearings stands are made of wood or bamboo and are portable so that it is easy to move them from place to place (Fig. 4.32). A rearing stand should be 2.5 m high, 1.5m long and 1m wide and should have 10 shelves with a space of 20cm between each shelf. The trays are arranged on the shelves and each stand can accommodate ten rearing trays. Six stands are required for each rearing room.

(a) Rearing stand with rectangular wooden trays.

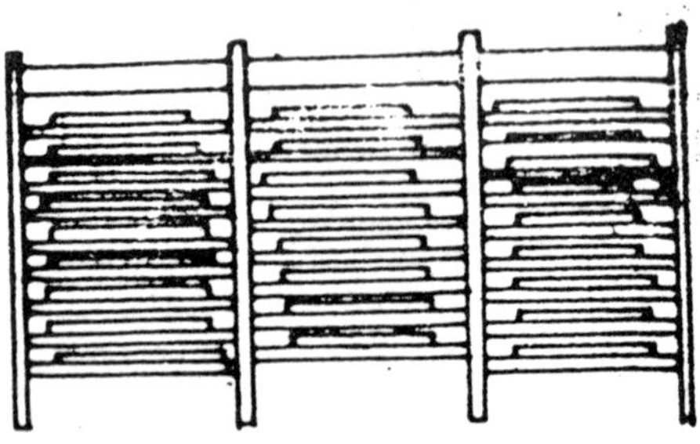

(b) Rearing stand with circular trays.

Fig. 4.32

ANT WELLS

Ant wells are provided to stop ants crawling onto the trays and attacking the silkworms. They are made of concrete or stone blocks 20cm square and 7.5 cm high with a deep groove of 2.5 cm running all round the top. The legs of the rearing stands rest on the centre of the block and water is poured into the groove to stop the ants reaching the rearing trays; each stand leg must rest in a well.

REARING TRAYS

These are used to rear silkworms and are usually made of bamboo so that they are light and easy to handle. They are either round with a diameter of 1.2 - 1.4 m and a depth of 7.5cm or rectangular measuring 0.7-0.9m × 0.9-1.2 m. On the basis of area of rearing seat required, an adequate number of trays should be provided.

RECTANGULAR WOODEN TRAYS OR BOXES

These are used for rearing early-stage larvae (Fig. 4.32). They are made of light wood and of convenient size for easy handling. A tray 0.9 m long, 0.7 m wide and 7.5-15 cm deep with the bottom covered may be found convenient. About 80 boxes are required for rearing 800 to 1000 layings up to the second instar.

PARAFFIN PAPER

This is thick craft paper coated with paraffin wax with a melting point of 55°C. It is used for rearing early-stage silkworms and prevents withering of the chopped leaves and also helps to maintain proper humidity in the rearing bed.

FOAM RUBBER STRIPS

Long foam rubber strips 2.5 cm wide and 2.5 cm thick dipped in water are kept all around the silkworm rearing bed during the first two instars to maintain optimum humidity. As a substitute, newspaper folded into convenient strips and dipped in water may be used.

CHOP STICKS

These are made of bamboo approximately 17.5cm-20cm long and thin and tapering to one end. Direct handling of the young age worms is to be avoided for hygienic reasons and to prevent damage to young worms. A pair of chop sticks is used to pick early-stage larvae. (Fig. 4.33).

FEATHERS

Birds feathers, preferably white ones, are used for crushing the delicate newly hatched worms on to the rearing bed (Fig. 4.33).

CHOPPING KNIVES

Chopping knives are used for cutting the mulberry leaves. They are usually 0.3-0.5 m long with a broad knife blade and a wooden handle. In technologically advanced countries such as Japan machines, either manual-operated or power-operated, are used for chopping leaves.

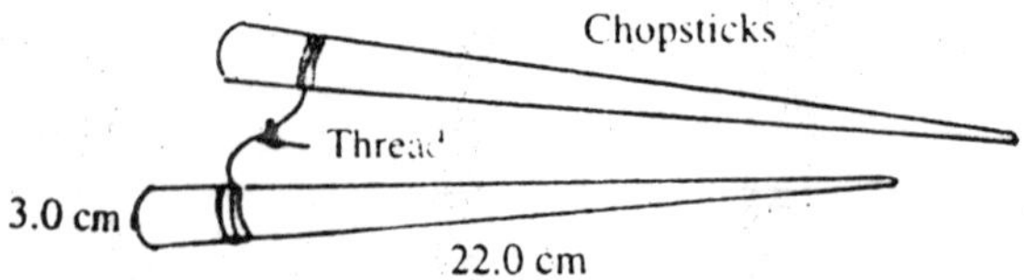

(a) Chopsticks

(b) Feather

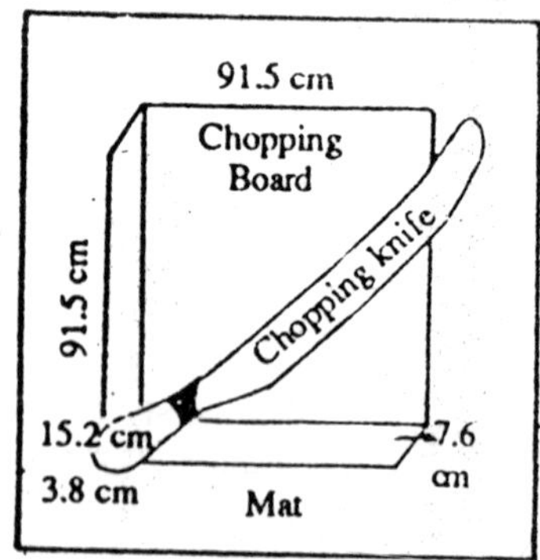

(c) Choping board and knife

Fig. 4.33

CHOPPING BOARD

This is made of soft wood and is used for cutting the leaf to the suitable sizes required for feeding the worms in the different instars. The size of the chopping board may be 0.9m × 0.9m and 5cm thick or of any other convenient size.

MATS

Mats usually 1.2 × 1.8m are used for collecting the leaves when chopping is done on the floor; they prevent the dust and dirt on the floor getting mixed in with the leaves.

LEAF CHAMBERS

Mulberry leaves harvested from the field are stored and preserved fresh for feeding the worms at set intervals during the day. The leaves are stored in cool rooms, or in underground masonry pits, or in rooms in heaps covered with cloth or polythene sheets. They can also be stored in convenient

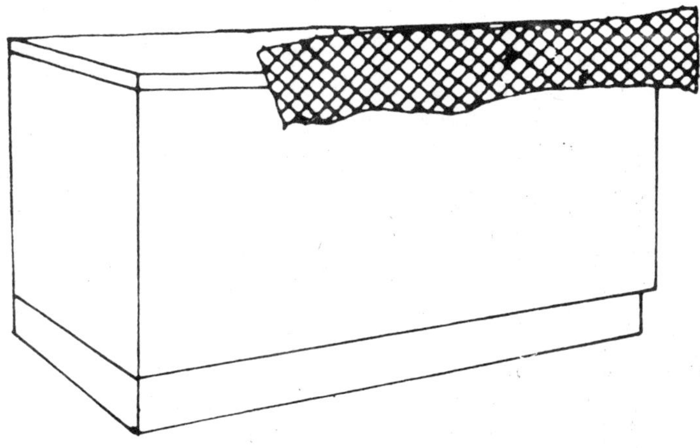

Fig. 4.34 : Leaf chamber.

leaf chambers, 1.5 m long, 0.9 m wide and 0.8 m deep. The side walls and bottom are made of wooden strips 7.5 cm wide and placed 7.5 cm apart. The chamber is covered on all the sides with gunny cloth which is kept wet. During the summer months and dry days the gunny cloth should be sprayed with water repeatedly to prevent the leaves withering.

CLEANING NETS

These are made of cotton thread or nylon, woven into nets of different meshes to suit the size variations of different stages of the silkworm. They are used for cleaning the rearing beds and at least two nets are required for each of the rearing trays.

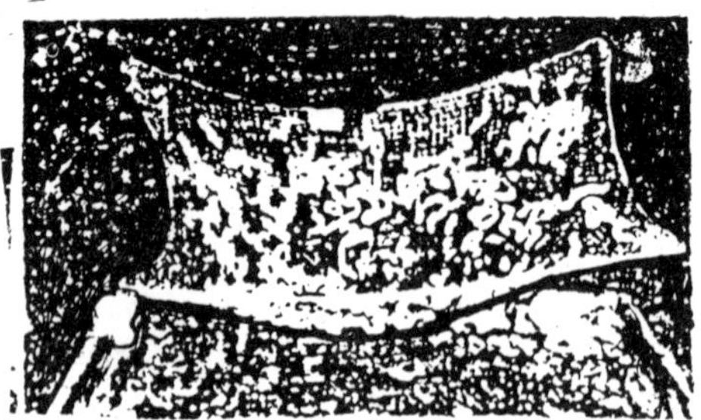

Mesh sizes suitable for I, II, III, IV and V instars are 2 mm^2, 10 mm^2 and 20 mm^2 respectively.

Fig. 4.34(a) : Cleaning Nets.

MOUNTAGES

These are used as supports for the silkworms to spin cocoons. In India they are made of bamboo usually 1.8 m long and 1.2 m wide. Over a mat base, tapes 5-6 cm wide woven out of bamboo are fixed in the form of spirals leaving a gap of 5-6 cm. They are also called chandrikas (Fig. 4.35). In Japan mountages called revolving mountages or turning cocooning frames are used. Other types of mountages such as centipede rope mountages, straw cocooning frames etc., are als.o used. In certain other countries small-sized plants are also used as mountages.

HYGROMETERS AND THERMOMETERS

These are used to record the temperature and humidity of the rearing rooms and help to maintain optimum room temperature and humidity for the healthy growth of the silkworms.

FEEDING STANDS

These are small wooden stands 0.9 m high used for holding the trays during feeding and cleaning (Fig. 4.36).

OTHER EQUIPMENT

Among the other items required are feeding basins, preferably plastic, to hold chopped leaves, and buckets and basin stands, etc., A sprayer is also required for disinfecting the rearing room, for spraying water on to the leaf chambers, etc. Leaf baskets made of bamboo are used for carrying the leaves from the garden to the rearing rooms.

Fig. 4.35 : Bomboo mountage (chandrika) with cocoons (India).

SANITATION AND PREPARATIONS FOR REARING SILKWORMS

The silkworms must be reared with the utmost care since they are very susceptible to disease. Bacteria, moulds and some protozoans attack the silkworms readily and any disease, once it breaks out, spreads quickly. To prevent disease, good sanitation methods and hygienic rearing techniques must be meticulously followed. The rearing room and the appliances used for rearing must be thoroughly cleaned. Dust, dirt and refuse of dead larvae found in the rearing house and on the rearing appliances should be removed and the house and appliances thoroughly washed with water and dried. Afterwards the equipment and the rearing rooms should be disinfected by spraying 2-4 per cent formaldehyde solution. "Formaldehyde" available commercially as a 40 per cent strong

Fig. 4.36 : Feeding Stand.

solution should be diluted to the required strength and used at the rate of 800ml to $10m^2$ (Fig. 4.37).

Fig. 4.37 : Disinfection of rearing house.

Before disinfecting the rearing room all crevices and holes should be filled in and the room rendered airtight as far as possible. Ideally the room temperature should be maintained as 25°C during disinfection for quick diffusion of the formaldehyde gas. All the rearing equipment as well as the walls, ceiling and roof of the rearing room should also be sprayed. After spraying the room is kept closed for 15-20 hours, then the doors and windows are opened and the room kept open for approximately 24 hours for all traces of formaldehyde to disappear. The proper time for carrying out disinfection is two to three days prior to the actual hatching of the eggs.

HATCHING

Hatching of silkworms generally starts early in the morning. The process of transferring the silkworms to rearing trays is called "brushing". A suitable time for brushing is about 10 a.m. as most of the larvae hatch by 8.00 and develop good appetites within two hours of hatching. In order to obtain uniform hatching, eggs at the blue egg stage are kept in black boxes on the day prior to hatching. In this way, the early maturing embryos are prevented from hatching and the late-maturing embryos given time to develop and catch up with the early maturing ones. The next day they are exposed suddenly to diffused light so that the larvae hatch uniformly in response to the phototropic stimulus. By this method hatching of 90 per cent and over can be obtained in one day. If, however, hatching is not uniform and only 50-60 per cent of the eggs hatch on the first day, brushing can be done on the following day as well. If necessary the hatched worms can be separated and kept in thin tissue paper, and stored in a refrigerator at 10°C till the next day to wait for the remaining larvae to hatch. Both the batches can then be mixed and brushed together. Newly hatched silkworms are black and hairy and look like black ants; in fact they are often referred to as "ants'.

METHODS OF BRUSHING

In the case of eggs prepared on egg cards, the cards with the newly hatched worms are placed in the rearing trays or boxes and tender mulberry leaves cut into small 0.5 cm squares are sprinkled over the egg cards. The hatched silkworms crawl onto the tender leaves and start feeding. Later the cards are removed and any worms still left on the card are tapped on to the rearing bed. After some time the bed is uniformly prepared in the rearing tray and first feeding of mulberry leaves is given.

In the case of loose eggs brushing is carried out as follows : a net with small holes (mosquito netting) is spread over the box containing the hatched larvae, and mulberry leaves cut to the size indicated above are scattered over the net. Paper with small slotted holes or perforations can be used in place of mosquito netting. When the worms have crawled over to the leaves, the net with worms and leaves is transferred to the rearing tray.

Sometimes the hatched worms are brushed with a feather from the egg cards directly on to the rearing tray. This is not a desirable method as it may injure the delicate newly hatched silkworms.

After brushing the bed is prepared by collecting the worms and the mulberry leaves together by using the feather. The bed is spread uniformly using chopsticks. After about two hours of brushing the first feeding is given to the silkworms.

Prior to the first feeding, as a precaution against muscardine, the larvae may be treated with a 5 per cent ceresan-lime powder.

Feeding Silkworms

Silkworms are fed to satisfy their appetites and to ensure their healthy and uniform growth. It is therefore necessary to preserve the quality of leaf during feeding and to keep the rearing beds clean. In order to save mulberry leaves and labour, the feeding must correspond to the eating habits and appetite of the larvae. Feeding with too many leaves is not economical. Generally early age silkworms eat leaves from the surface while late age worms from the edges.

Care should be taken to feed the silkworms a sufficient quantity of leaves. An insufficient supply of leaves results in irregular growth and irregularity at moult. In young age larvae this tendency to irregularity is high. Hybrids have larger appetities than pure races and consume more leaves.

The quantity of leaf required for rearing 50 layings or a box of 20,000 eggs of different stage of developments (instars) is given below. In univoltine and bivoltine species the quantity of leaves at 1st instar (1-2 kg); 2nd instar (5-6 kg); 3rd instar (20-25 kg); 4th instar (80-90kg); 5th instar (450-475 kg) and total mulberry leaves required is approximately 550-600 kg. In multivoltine and bivoltme species in tropical areas the requirement of mulberry leaves is as follows : 1st instar (1-2 kg); 2nd instar (2-3 kg); 3rd instar (15-20 kg); 4th instar (35-50kg); 5th instar (300-325 kg) and total requirement is about 350-400kg.

BED CLEANING

Removing the old mulberry leaves, faecal matter of silkworms, exuviae, any dead or unhealthy silkworms, etc. from the rearing bed is called bed cleaning. Accumulation of any such matter creates environmental conditions detrimental to the health of silkworms. Accumulated faecal matter and old leaves builds up humidity in the bed and favours rapid multiplication of micro-organisms. With fermentation setting in, the rearing bed temperature will rise and affect the growth of worms. Periodical bed cleaning to maintain hygienic conditions is therefore indispensable in silkworm rearing.

Too frequent bed cleaning during the early stages is harmful as there is every chance of losing worms since they are minute and may be overlooked. Usually the bed is cleaned once during the first age; twice during the second age *i.e.,* just after first moult and again before setting for second moult; and three times during the third age, *i.e.,* just after moult, in the middle of the age and again just before settling for next moult.

During the fourth and fifth ages, the beds should be cleaned once a day in the morning, particularly when the rearing is conducted on shelves where small shoots and plucked leaves are supplied to the worms. In the case of shoot and floor rearing where entire shoots and leaves are fed to the worms, normally only one cleaning is given to the fourth and the fifth age worms, *i.e.,* after two days of feeding after the third moult, and once just prior to mounting in the fifth age.

For cleaning purposes, nets, paddy husk or charred husk are used. In the case of net cleaning, the net is spread over the bed and one or two feedings are given ,before the nets are lifted and transferred to a fresh cleaned tray. The worms crawl through the meshes in the net and come up to feed on the leaves on the nets. If the worms are healthy, practically all the worms will come up leaving the old leaves and litter behind on the old trays which are cleaned later.

SPACING

Provision of adequate rearing seat space is of great importance for the vigorous and full growth of silkworms. As the worms grow in weight and size, the density in the rearing bed increases and conditions of overcrowding are faced. It is, therefore, essential that the density of population in the rearing bed should be regulated and ideal rearing bed conditions ensured. Most of the troubles in silkworm rearing arise from a lack of appreciation of the importance of spacing. There is always a tendency on the part of the rearers to rear as many laying as possible within a limited space with the result that overcrowding easily occurs and frequently leads to crop losses, particularly in tropical countries.

Overcrowding of silkworm means insufficient space for the free movement and free feeding of the worms and so the larvae crawl over one another. Crowded conditions also increase accumulation of gases, heat and fermentation of faecal matter particularly during the early age when temperature and humidity in the rearing room arc high. Under such unhygienic conditions, the worms do not feed on the leaves even though fresh and good quality leaves may be available in sufficient quantity. This results in unequal and unhealthy growth of larvae and often leads to crop losses.

However, if there is space beyond the optimum required, there is considerable wastage of mulberry leaves and also the labour costs for feeding increases. It will also involve avoidable unnecessary outlay on rearing equipment.

The spacing to be provided for different ages (instars) of silkworms of 50 layings or a box of 20,000 eggs is given below:

Univoltine & bivoltine species	**At beginning of each age (m^2)**	**At the end of each age (m^2)**
Isr instar	0.2	0.8
IInd instar	1.0	2.0
IIIrd instar	2.0	4.5
IVth instar	5.0	10.0
Vth instar	10.0	20.0
Multivoltine & bivoltine species (in tropical areas)	**At beginning of each age (m^2)**	**At the end of each age (m^2)**
1st instar	0.2	0.5
IInd instar	0.5	1.5
IIIrd instar	1.5	3.00
IVth instar	3.0	9.0
Vth instar	9.0	18.0

CARE AT MOULTING

Uniform growth of silkworms of any batch depends on the care given to rearing. Among the various steps involved in the rearing of silkworms the handling of silkworms during moult is of particular importance. Silkworms under proper rearing conditions would all settle uniformly for moult and also come out of moult uniformaly. When the worms are due to moult and again when they come out of moulting they should be handled with the almost care, as described below:

At the approach of moult, the silkworms attains its maximum body growth for that particular instar and as a result the body of the silkworm becomes stout and shiny. In relation to the size of the body, the head of the worm about the moult appears quite small and also darkish. At this time, a bed cleaning should be carried out. At the same time as the cleaning wider spacing is also provided. After the cleaning, the worms about to settle for moult are given one or two final feeds of the instar with leaves cut to a comparatively smaller size. Wider spacing and somewhat finely chopped leaves help to reduce the humidity in the bed which facilitates uniform moulting. If, however, the humidity is high, a very thin layer of lime of powder may be dusted over one bed after the last feeding to keep it dry. When all the larvae have settled for moult, feeding should be stopped.

The larvae normally take 15-30 hours to complete moulting during the different instars. After moulting the new instar larvae come out casting their old skin. At this point the head is comparatively bigger in relation to the body and it is always bigger than the head of the previous instar. The body of the worm is comparatively less shiny because of the loose skin of the new instar larvae. From these distinguishing features, the larvae still to moult or in moult and those already out of moult can be distinguished.

The first feeding of the new instar should start only after almost all the worms have come out of moult. In localities where muscardine is commonly prevalent, it is advisable to take anti-muscardine precautionary measures by dusting ceresan-lime powder on to the newly moulted worms prior to first feeding.

As pointed out earlier, it is very important that the rearing bed be as dry as possible when the worms are in moult. This enables the silkworm to crawl out of the old moult skin easily thus helping all the worms to come out uniformly. Since the newly formed skin is thin and delicate, under more humid conditions the worms become susceptible to fungal attack. Therefore, it is essential that the silkworm beds be kept as dry as possible during the moult.

ENVIRONMENTAL CONDITIONS

In the commercial production of cocoon crops optimum environmental requirements ensuring maximum productivity of good quality cocoons of very high silk content are aimed out. Among the various environmental factors that influence the cocoon crops, the most important are atmospheric temperature and humidity prevailing at the time of rearing, quality and quantity of leaf supply and the methods of rearing followed, such as feeding, cleaning, spacing etc.

TEMPERATURE

Plays a vital part in the growth of silkworms. Silkworms are cold blooded animals and as such temperature will have a direct effect on the various physiological activities of their system. The optimum temperature for normal growth in silkworm is between 20°C to 28°C. The temperature above 30°C directly affect the health of the worm. If the temperature is too low, *i.e.,* lower than 20°C, the growth of the silkworm owing to the very low rate of physiological functioning is considerably retarded, specially in the early stages, with the result that the worms become too week and susceptible to disease.

The room temperature is generally low during winter and rainy days, and it should be regulated by heating with electric heaters or charcoal fires. During the summer when the temperature is high, doors and ventilators must be opened to lower the temperature by free circulation of cool night air. The adverse effects of high temperature can be overcome through proper designing of the rearing house and by ensuring adequate ventilation and free ventilation of air inside the rearing house particularly during late age worm rearing.

HUMIDITY

Also plays an important role in silkworm rearing. The combined effect of both temperature and humidity largely determines the satisfactory growth of the .silkworms and the success of cocoon crops. Under too dry conditions the leaves wither very fast and become unsuitable on feed, on the other hand too much humidity in the case of late-age worms builds up bed humidity and creates conditions that favour outbreak of diseases. The optimum humidity during moulting should be maintained between 70-80%.

For regulating humidity in the rearing room paraffin paper is invariably used for the rearing beds curing the young-age worm rearing for raising humidity. Also, if necessary, wet foam rubber bands or paper bands may be used to increase humidity in the beds. In the late ages if the humidity is low, sprinkling water on the floor may be found to be helpful.

AIR

Like all other animals, silkworms require fresh air for their various physiological functions. Due to respiration of the silkworms, carbonic acid gas is released in the rearing beds and the rearing rooms, CO, ammonia, sulphur dioxide, etc. are also generated in the rearing rooms by the burning of charcoal to raise the room tumperalure. These gases are injurious to silkworms. Too many workers in a rearing room can also pollute the air. Therfore, care should he taken to allow for circulation of as much fresh air as possible through proper ventilation. so as to keep the toxic gases at a low level. Air plays an important role in regulating rearing room temperature and humidity free artificial circulation air is extremely useful for bringing down high temperature and also high humidity conditions particularly in tropical regions.

LIGHT

Silkworms are photosensitive and generally have a tendency to crawl towards dim light. Rearing in either complete darken or in bright light leads in irregularity in growth and moulting. Larval moult is uniform when silkworms are reared in 16 hrs light and the remaining period in darkness.

Given the above facts it is advisable to rear silkworms in the dim light during the daytime and in the dark at night.

LEAF QUALITY

The mulberry leaf is the exclusive food of the silkworm *B. mori.* It is essential that the mulberry leaves are not only in abundant supply but are also of good and suitable quality. Leaf quality has much to do with the success of silkworm rearing and the quality of the cocoons produced.

The maintenance of leaf quality, their harvesting and preservation is already discussed under cultivation and harvesting of mulberry leaves.

Rearing Early-age Silkworms

Different methods of rearing the early ages are practised. In all methods importance is given to maintenance of leaf quality and correct humidity and temperature in the rearing bed so that vigorous and healthy development is ensured. The two methods used at present are :

1. Paraffin-paper rearing
2. Box rearing.

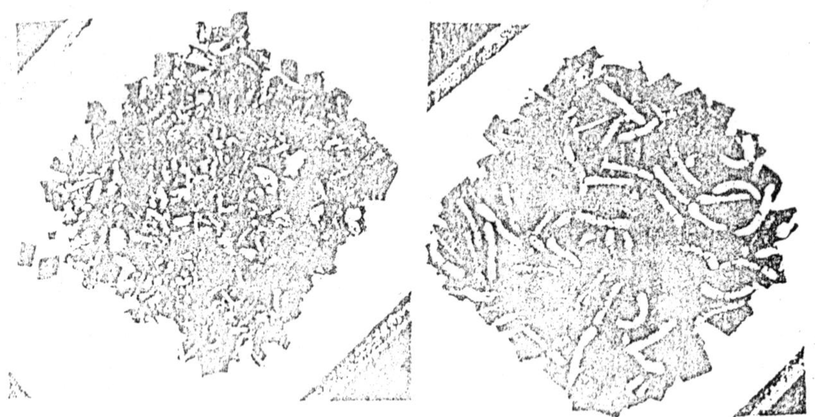

IInd instar silkworm **IIIrd instar silkworm**

Fig. 4.38 : Early-age Silkworms.

1. PARAFFIN—PAPER REARING

Paraffin paper is used as a bottom layer and as a cover for the rearing beds in the usual rearing trays (Fig. 4.39). The object is to maintain the optimum humidity in the bed and to keep the leaves fresh for a longer period by preventing moisture loss through evaporation. Feeding is restricted to two or three times a day and so the labour required for feeding is reduced.

For young-age rearing purposes superior quality paraffin paper preferably with a melting point of 55°C should be used. Although, the sheets can be used repeatedly they should not get torn or soiled through repeated use.

A sheet of paraffin paper is spread on the base of the rearing tray and the silkworms are reared on this sheet and under another. In between the sheets on all four sides of the rearing bed, strips of wet foam rubber or ordinary paper are placed to maintain the required humidity. If necessary, the edges of the top paraffin sheets may be weighted with light sticks to seal the edges for better maintenance or rearing bed humidity.

The top paraffin-paper sheet must be removed at least 30 minutes prior to feeding so that there is adequate aeration of the bed and the accumulated toxic gases dispelled. Once the larvae start settling for moult, it is no longer necessary to cover the bed with paraffin paper and so the bed is kept open at the top. When the worms are in moult the bed is kept quite dry. A thin layer of lime powder sprinkled over the bed will help to keep the bed dry.

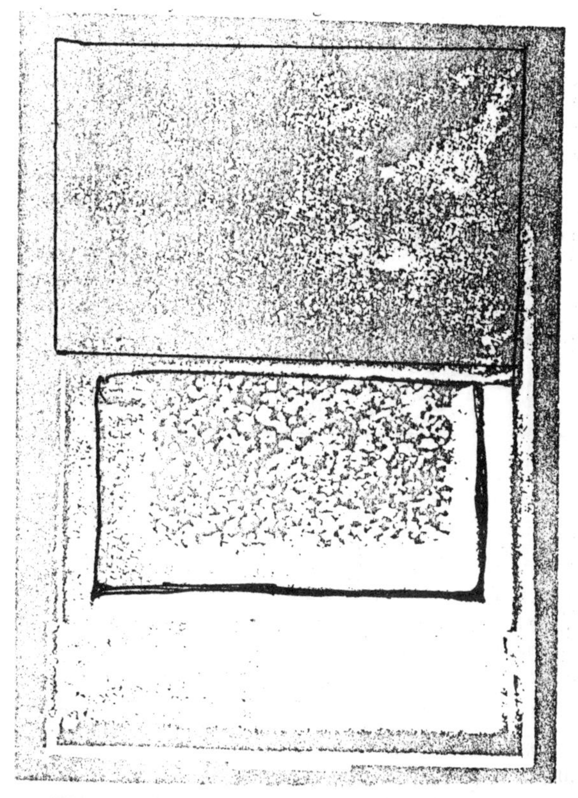

Fig. 4.39: Paraffin-paper rearing in trays.

With the paraffin method of rearing, particularly in humid places, there is danger of muscardine. To avoid this, a thin layer of anti-muscardine lime powder may be spread over the bed to protect the silkworm against the disease at the time of hatching, at each moult, and in the middle of each age.

2. BOX REARING

Boxes or trays made of wood or plastics or galvanized iron at least 10-15 cm deep are used for box rearing (Fig. 4.40). The boxes used may be with or without lids.

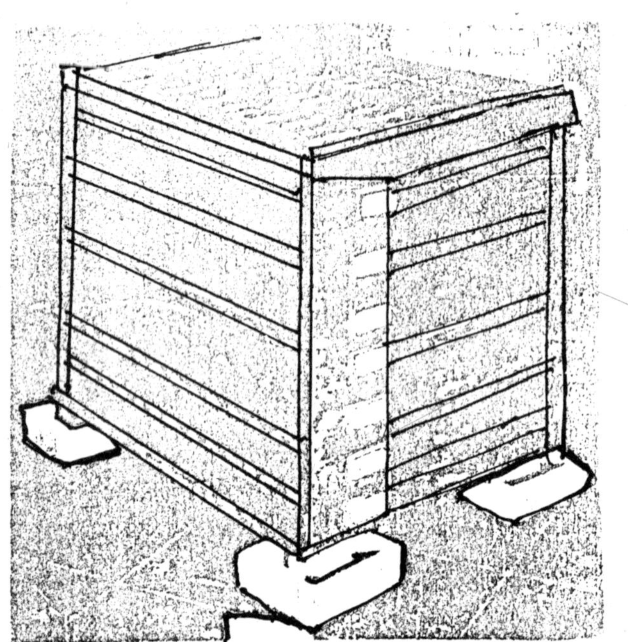

Fig. 4.40 : Box rearing.

REARING IN BOXES WITH LIDS

In this case a sheet of paraffin paper is spread over the bottom of the tray. The rearing bed is prepared over this and covered with another sheet of paraffin paper after putting in place the wet foam rubber pads. The lid is put on and the box is placed in the shelves. For rearing third instar larvae, the box is not covered with the lid. When larvae start settling for moult, the paraffin paper, wet foam pads and the box lids are removed to keep the bed dry.

REARING IN BOXES WITHOUT LIDS

Wooden boxes of uniform size and 10-15 cm deep are used. Each box is provided with a sheet of paraffin paper for the base, over which the rearing bed is prepared, wet foam rubber pads are placed on all the four sides and the bed is covered with another sheet of paraffin paper. The boxes are then piled one over the other for rearing first-age larvae. For rearing second-and third-age larvae, a space of 2-3 cm for ventilation is provided between the boxes by fixing in place pieces of wood. The boxes are kept open for at least 30 minutes prior to each feeding, and completely open when larvae start setting for moult. In using this method care must be taken to disinfect the silkworms by sprinkling anti-muscardine powder because due to the high humidity of the rearing beds the larvae are subject to muscardine.

CO-OPERATIVE REARING OF YOUNG SILKWORMS

Silkworms not reared properly in their ages are prone to disease at later stages, and crops may even fail. Maintenance of optimum temperature and humidity, supply of suitable leaves, and handling of the young-age rearing with utmost care, etc., demand highly technical skills, which are not often available to the ordinary silkworm rearers. Furthermore the rearers may not be able to afford the necessary equipment to rear silkworms under the ideal conditions. In order to overcome these difficulties, co-operative rearing centres have been organized where silkworms are reared under ideal conditions and technical supervision up to the second or third moult, and the larvae in the third or fourth ages are distributed to the rearers. The co-operative young age rearing centres are provided with ideal rearing houses with the necessary equipment and the rearings are conducted by technical experts.

Fig. 4.41 : Co-operative rearing of young silkworms.

Mulberry gardens suggesting the required quality of mulberry leaves for rearing the early age silkworms are generally attached to co-operative rearing centres.

Since thc larvae are reared carefully under ideal conditions and expert technical supervision the worms are vigorous and healthy and this ensures successful crops later in the rearers houses. Becausc of mass-scale rearing in co-operative rearing centres, the cost involved is reduced to the minimum and is generally charged to the rearer. The rearer is also saved much of the trouble involved in rearing young-age worms and he can handle the late-age worms for a fortnight or so without it seriously interfering with his other farm work. Generally a co-operative rearing centre can rear at a time 200-500 boxes (20,000 eggs each) up to the third moult or double that quantity up to the second moult. Today in Japan almost 90 per cent of the rearings are conducted at such co-operative young-age rearing centres and this is what has contributed mainly to the sustained bumper harvest of quality cocoons.

REARING LATE-AGE SILKWORMS

Late-age worms in the fourth and fifth instars need comparatively less humid rearing conditions and also preferably a lower temperature

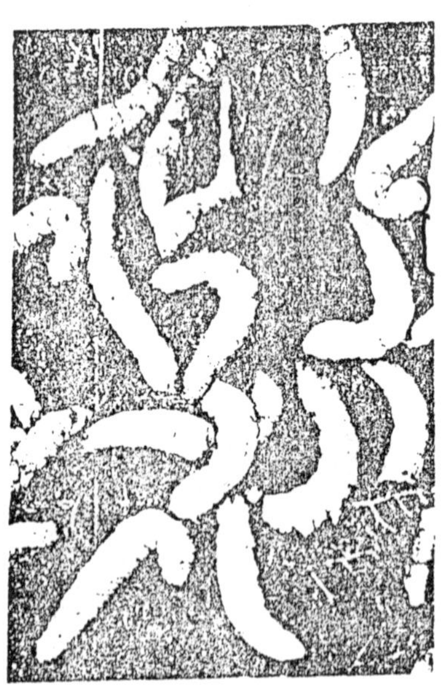

IVth instar silkworm

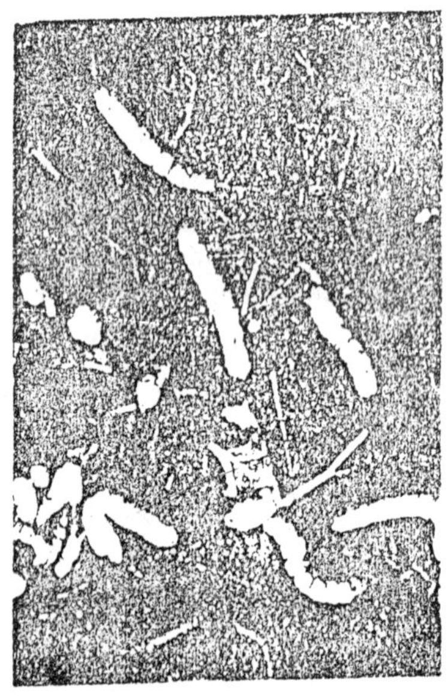

Vth instar silkworm

Fig. 4.42 : Late age worms.

(Fig. 4.42). These are real feeding stages when the worms consume about 90-95 per cent of the total feed, and therefore, adequate spacing and an adequate amount of feed should be given to these two ages. From the fourth age onward whole leaves may be given and even small chopped shoots fed to the worms. Because either whole leaves or even shoots are fed to the late-age worms and so the leaves stay fresh longer, only three to four feedings need be given during the 24 hours. It is, however, advisable to give large feed as the night feed.

METHODS OF REARING LATE AGE WORMS

There are three methods employed to rear late age worms :

(a) Shelf rearing

(b) Floor rearing

(c) Shoot rearing

(a) Shelf Rearing

Rearing of silkworms in rearing trays arranged one over the other in tiers on rearing stands is called shelf rearing. Generally rearing stands are arranged in two rows parallel to the wall, with adequate space in the centre removing the trays and attending to the clearing of the beds and feeding of the silkworms. Each rearing stand can accommodate up to 10 rearings trays. In India round bamboo trays 1.2-1.4 m in diameter

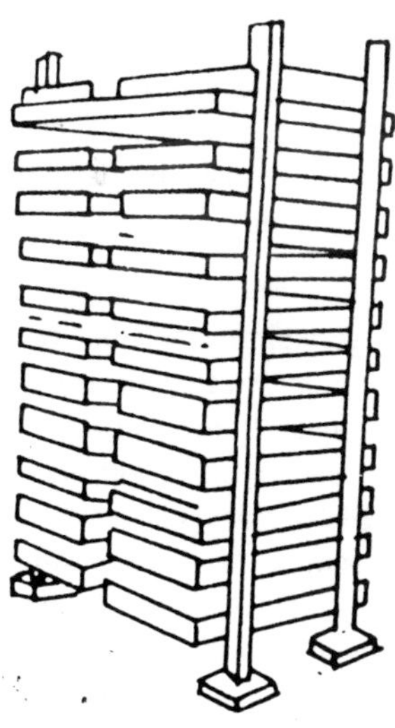

Fig. 4.43 : Shelf rearing stand with Ant wells.

are used for rearing late-age silkworms. Under this system of rearing leaves picked individually from the mulberry trees or branches and cut to a convenient 15 cm are fed to the silkworms during final instar. Usually 3 to 4 feedings are given a day, and nets are used for cleaning the rearing bed once a day. Shelf rearing has the advantage of accommodating more silkworms in a limited, area than the other two methods. Great care should be given to proper spacing of the worms. Since cleaning is more frequent and the trays are removed from the stands for every feeding, the amount of labour required is greater.

(b) Floor Rearing

This is another method of rearing silkworms on fixed rearing seats. The rearing seats are arranged in two to three tiers so as- to accommodate as many silkworms as possible. The rearing seat should measure 1-1.5m in width and 5-7m in length according to the rearing room. The space between the tiers should be 0.6-0.8m. Sufficient space must be provided all around the rearing seat for attending to feeding and other rearing operations. The rearing seat is usually made of wood or bamboo strips. As with shelf rearing the silkworms are fed with leaves or branches cut small. The number of feedings is three or four a day. Bed cleaning is also followed in the same way as for shelf rearing by using nets, but it is done less frequently; twice during the fourth age and three times during the fifth age. The main difference between the shelf rearing and floor rearing is that the labour required in shelf rearing to pull out the trays and put them back every time feeding and cleaning is done is eliminated. In floor rearing, which therefore, is less laborious and less labour consuming, as in the case of shelf rearing, care must be taken to provide sufficient spacing.

(c) Shoot Rearing

Shoot rearing is very similar to floor rearing. Silkworms are reared on big branches in one or two tiers. The big shoots harvested from the fields are fed straight away to the worms. The rearing seats are usually 1m wide and of any convenient length according to the measurements of the rearing house. Shoot rearing is usually on a single tier 20 cm above the ground or occasionally in two tiers with a gap between the tiers of about 1m. Shoot rearing can be carried out indoors or outdoors. Outdoor shoot rearing is possible only when the atmospheric conditions are favourable, the temperature approximately 25°C and no interference from rain anticipated.

As mentioned above, whole shoots are supplied for every feed so the larvae keep moving upward consuming the mulberry leaves. Since the leaf supply owing to shoot feeding is distributed in three dimensions there is better aeration of the rearing beds and, therefore, it is possible to have 50 per cent more worms per unit area of rearing seat as compared to the shelf or floor rearing. Cleaning is also reduced to a minimum; once in the fourth instar and once again in the fifth instar. To clean the bed, ropes of convenient length are spread parallel to each other lengthwise on the bed and under the branches and after two or three feeds when all the worms have crawled onto new branches, the bed held by ropes is rolled into loose bundles by cutting the ropes every 2 m. After cleaning the beds the rolled up branches and the worms are transferred back on to the rearing beds and spread out again. With this method both feeding and cleaning require less attention and are, therefore, far less laborious and labour consuming.

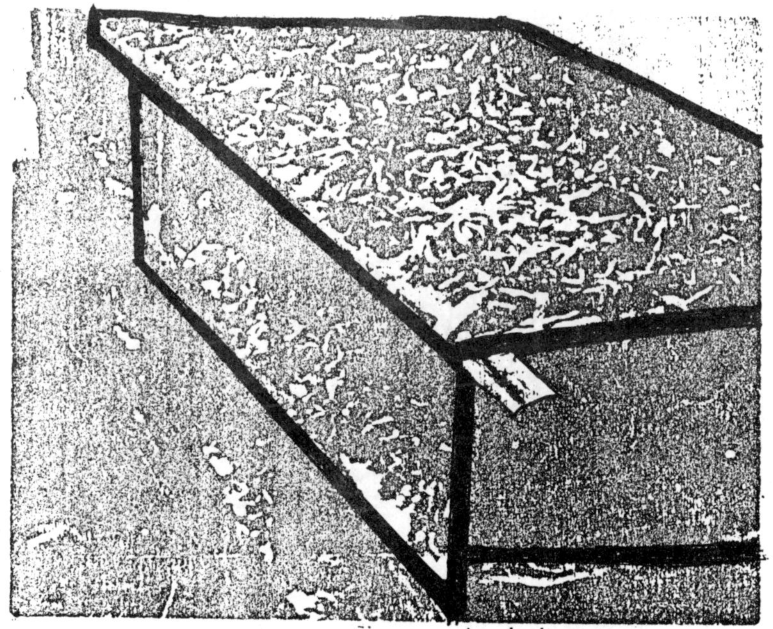

Fig. 4.44 : Shoot rearing bed.

The shoot rearing saves on labour costs which constitute a major expenditure in silkworm rearing at the fourth and fifth ages. With shoot rearing it is possible to reduce the labour force by approximately 60 per cent at the fourth and 50 per cent at the fifth age. The amount of leaves required is reduced by approximately 25 per cent at the fourth and 10 per cent at the fifth age. Since this method is the most economical it is becoming more popular in Japan.

(d) Mounting

Toward the close of the final larval instar the silkworm stops feeding and gets ready to build the cocoon prior to transforming itself into the pupa inside the cocoon. The feeding during the final instar may last from five to seven days in the case of multivoltine and bivoltine races, in the tropical areas, and seven to nine days in the case of bivoltine and univoltine races in subtropical areas. With the cessation of feeding and as the stomach contents empty, the fully mature larva becomes transluscent

Fig. 4.45 : Mounting of silkworms on bamboo mountages.

and yellowish in appearance, which is a clear indication that the worm is fully ripe and is ready to be mounted on to mountages or cocooning frames. Such ripening larvae normally crawl towards the periphery of the rearing trays in search of suitable supports for building their cocoons. Thus the process involved in picking the ripe worms and putting them on to the suitable mountages for the spinning of cocoons is called mounting of worms (Fig 4.45).

As the mature or ripe worms are mounted on the mountages they pass their last excreta in semi solid condition. During rains when the humidity is high, excess body moisture is also eliminated as urine. After defecating the worm starts constructing the cocoon. It anchors itself first to the mountage by oozing a very tiny droplet of silk fluid which immediately hardens and sticks to the mountage. Then by continuous movements of the head, silk fluid is excreted in minute quantities which goes on hardening to form a long continuous filament. The silkworm at first lays the foundation for the cocoon structure by weaving a preliminary web which provides the necessary foothold for the larva to spin the compact shell of the cocoon. Owing to the characteristic movements of the head of the larva the silk filament is deposited in a series of short waves or eyelets forming the figure eight and in this way layers of cocoon are built and added on to form the compact shell.

THE PRELIMINARY WEB

The floss of the cocoon—though formed of a continuous filament is tangled and is not reelable. The floss in uni- and bivoltine races of silkworms is about 2 per cent of the weight of the cocoon; in the case of multivoltine races, however it is higher, as much as 10 per cent or more.

After the compact shell of the cocoon is formed the shrinking larva wraps itself in a gossamer layer and detaches itself from the shell and becomes the pupa or chrysalis. This last layer is only a body sheath of the worm and does not form part of the main shell and, like the floss layer is not reelable.

The process of spinning the cocoon takes about two to three days in the case of multivoltine varieties of silkworms and three to four days in the case of uni and bivoltine races.

COLLECTING MATURE WORMS FOR MOUNTING

Collecting mature silkworms and mounting them on the mountages is a laborious job. Usually the mature larvae are picked by hand by skilled labour capable of identifying the mature worms, which are collected in trays and later put on the mountages. Since the mounting is done by hand the density of the mounted worms can be controlled and the incidence of double cocoons reduced. Diseased silkworms can easily be sorted out so as to ensure that the cocoons are all healthy and of uniform quality.

In order to save the labour involved in picking mature silkworms, some simpler devices are sometimes followed such as the use of green branches or nets for collecting the worms. With the branch method, branches of green leaves are placed over the rearing bed and when the worms crawl on to them they are taken out and shaken over a mat to dislodge the worms which are later collected and put on the mountages. Similarly with the net method, nets are spread on the bed after feeding and the mature worms, which are no longer feeding, then crawl up onto the nets which are taken out and shaken off over a mat and the worms are later mounted as with the branch method. With the shoot rearing method the early maturing larvae, which constitute 10-20% of the total larvae, can be picked by hand as they ripen.

Later on when the remaining larvae mature more uniformly and almost simultaneously, they can be collected by shaking them off the upper layers of the mulberry branches to mats. The larvae are then mounted on mounting, frames.

METHOD OF FREE MOUNTING

With this method instead of collecting the mature worms from the beds, the mountages are placed in the beds themselves and the worms crawl up onto them. Usually straw mounlages or revolving mountages are used (Fig 4.46). In the U.S.S.R. brush-type local weeds are used as mountages and are placed in the rearing beds for the worms to crawl up on and spin the cocoons.

In countries such as Japan where straw mountages are used, a thick layer of cut straw is spread over the rearing bed, the straw mountages are placed on it and the ripening worms crawl through the bed of cut straw onto the straw mountages. The mountages with worms in the process of spinning are later transferred to trays covered with sheets of newspaper which absorb the urine and faeces excreted by the larvae. Revolving mountages can also be placed over the rearing beds for free mounting of worms. These should be removed when a very thin layer of the cocoon shell has been formed and kept hanging undisturbed for completion of cocoon formation.

Fig 4.46 : Free Mounting (Japan).

ENVIRONMENTAL CONDITIONS REQUIRED FOR MOUNTING

Maintenance of the required environmental conditions during mounting is important as it determines the quality of the cocoons spun. Silkworms under mounting normally require a comparatively drier atmosphere. Ideal conditions for good spinning are temperatures not exceeding 26°C ana relative humidity between 60 and 70 per cent. If temperatures rise above 26°C the quality of the cocoon is affected. It is particularly important that during the first 50 hours after mounting ideal conditions are maintained for cocoon spinning.

Often the liquid excreta falling from the mountages collect on the floor and build up humidity in the room. To prevent this increase in humidity absorbent material or mats, which can be changed at intervals, can be spread out underneath the cocooning frames, or the rearing room can be adequately aerated to drive out the moist air.

HARVESTING OF COCOONS

After spinning the cocoon the larva undergoes metamorphosis inside the cocoon and becomes the pupa. Generally pupation takes place on the fourth or fifth day of spinning, in the case of bivoltines and univoltines in the temperate regions such as Japan, but in tropical areas the multivoltines pupate on the third, or fourth day of spinning. The pupa when formed has a thin cuticular skin which is soft to the touch and ruptures easily if disturbed; therefore early harvesting of cocoons should be strictly avoided. In course of time the pupal skin hardens and turns dark brown and it is at this stage that cocoon can be harvested. The proper time for harvesting cocoons is on the seven or eighth day of spinning in the case of bivoltine and univoltine races, and on the fifth day in the case of multivoltine races. Harvesting must not be unduly delayed beyond the period mentioned above as any parasitic insects infecting the silkworm larvae emerge from the pupa and damage the cocoons.

Fig. 4.47 : Harvesting of Cocoons.

Cocoons are normally harvested by hand (Fig 4.47). In the case of cardboard rotating mountages simple devices are used to separate the cocoons from the cavities in the blocks of the cocooning frames. After harvesting the cocoons are sorted; good, defective, double, pierced, stained etc. cocoons. The good cocoons are cleaned by removing any faecal matter found on the surface and marketed at once.

SILK GLAND

When a ripe silkworm is dissected a pair of large glands will be seen lying on either side of the intestines (Fig. 4.48). The ducts from these glands unite in the spinneret below the mouth of the silkworm and the entire structure is called silk gland. Each of the pair is very long and several times the body length of the larva and lies in a series of folds. The whole gland is made up of about 1,000 cells formed at the embryonic stage itself, the cells only enlarging with the growth of the larvae. The silk gland consists of three distinctive and functional parts - the posterior, middle and the anterior. The posterior ends of the glands are closed while the fore-ends converge to join together into a common excretory duct opening into the spinneret near the mouth.

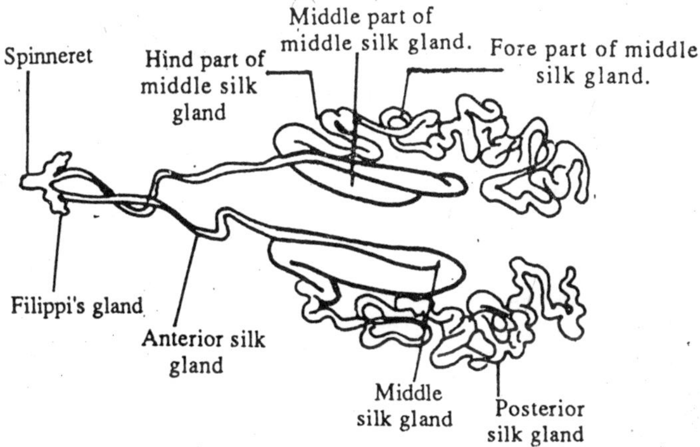

Fig. 4.48 : The silk gland.

The posterior part secretes the silk substance known as fibroin, while the middle part serves as a reservoir for receiving the fibroin from the posterior silk gland. It also secretes round the fibroin a natural gummy substance called sericin. The anterior part which is actually a silk duct or excretary canal moulds the sericin coated fibroin and conveys it to the spinneret. Here a pair of Fillippi's glands which are situated on either side of the common duct elaborate certain secretions which help to unite the moulded silk material from the two excretary canals and prevent the freshly moulded silk from hardening. The silk substance hardens into silk filament or have only on coming into contact with air. It is made up of two independent fibroins called Brins from the two glands and is covered completely with sericin. It is the continuous long silk bave that makes up the cocoon shell. The composition of the cocoon shell is given in the Table 4.3 below:

Table 4.3

Fibroin	72.0 - 81.0%
Sericin	19.0 - 28.0%
Fat & wax	0.5 - 1.0%
Colouring matter & ash	1.0 - 1.4%

Fibroin forms about 76 per cent of the whole weight of the composite bave and sericin about 22 per cent, the rest being wax and other substances.

SPINNING OF THE COCOON

Soon after the ripe larvae is left on the mountage it anchors itself to the mountage by oozing a very tiny droplet of the silk fluid which immediately hardens on the surface of the mountage and sticks to it. Then by continuous movements of the head, silk is excreted in minute quantities which goes on hardening to form long continuous filament or bave. The bave so formed is used for the construction or spinning of the whole cocoon.

The silkworm at first proceeds to lay the foundation for the cocoon structure by casting a loose tracery of silk from one point of anchorage to another drawn in long straight filaments of 2.5-3.5cm or even 5cm. This foundation layer is the preliminary web, made up of a continuous filament but the deposition of the filament is not orderly but tangled. This web provides the necessary foothold to the larva for leisurely and methodical spinning of a densely woven strong shell, which generally takes from three to four days for completion. This foundation layer which covers the compact shell is called floss or blaze.

After the preliminary web has been build, the larvae settles down to weave the compact shell depositing the bave in a more or less regular pattern. Due to the peculiar movements of the head of the larvae, the bave is deposited in a series of short waves or in eyelets like figure '8'. The former pattern is usually found in the outer layer of the shell and the latter in the inner layers. When 20-30 such patterns are made in one area the worm moves its head away to another zone to repeat the operation. Each group of such deposition of bave is called a bundle or group which commonly occupies an area 4 to 5mm^2. These groups, though formed in different areas of the shell, have a continuous filament running through them. A large number of such groups make one layer and five to six layers make the compact silk shell.

After the completion of the compact shell, the now shrunken larva wraps itself with a gossamer layer and finally it detaches itself from the shell to transform into an inert pupa or a chrysalis. This last layer is only a body sheath of the worm and does not form a part of the main shell. This body sheath is known as the pelade layer. The outer floss and the innermost pelade together made the waste known as frisons. Thus, the whole cocoon is made up of one continuous filament or bave. Breakages in the bave occur only when the larva is violently disturbed during cocoon spinning. A cocoon has three distinctive layers viz.; the outer floss or blaze, the middle compact shell which constitutes the reeled silk and the innermost pelade layer. The outer floss and innermost pelade which do not yield any reeled silk form waste.

RECOVERY OF SILK IN REELING

In terms of percentage of recovery of silk in reeling, it is not constant for all races. It is generally higher in uni-and bivoltine races than in multivoltine races. In the former 70-80 per cent of the silk is recovered as they are less flossy and have better reelability than the latter. The pelade layer is similarly between two and four per cent of the shell weight.

It is estimated that the larva moves its head about 70 to 80 times per minute in the process of spinning the cocoon and that the bave or filament is formed at a rate of 15 cm per minute. The total length of bave in a cocoon varies from 1,000 m to 2,000 m in uni and bivoltine races and from 300 to 800 m in multivoltine races. The thickness of the filament is usually measured gravimetrically in terms of the denier, which is the weight in grammes of 9,000 metres. The thickness is not uniform throughout but varies according to the position of the bave in the cocoon, being coarsest in the beginning and finest towards the end.

About 10 days after cocoon formation, the pupae enclosed in them are killed by soaking the cocoons in hot water, by steaming them, by giving them dry heat in an oven or in the sun, or by fumigation. Some cocoons are kept safe. Their pupae are allowed to develop into adults get seeds (eggs) for the next crop (brood).

REELING OF RAW SILK FROM COCOON

Reeling constitute an important aspect of sericulture because cocoon production is directly related to reeling industries. Before reeling the thread the cocoons, with dead pupae, are dipped in a container of hot water for more than 10 minutes (stifing). During this period they are continuously stirred with a rod. For tasar silk special cooking techniques is employed. The cocoons are dipped in 0.5%-Na_2CO_3, solution for 18 hours. Then cocoons are subjected to steam cooking for 2½hrs. After 24 hrs. the cocoons are treated with 0.5% formaldehyde solution for 15 minutes. Due to this process, their outer portion is loosened and removed in the form of long tapes and the end of the continuous filament is found. The filaments of several cocoon are picked up and passed through the 'glass eye' on to the reel. The thread thus reeled forms the raw silk of commerce. About one kilogram of raw silk is obtained nearly 55,000 cocoons and one cocoon yields about 300 metres of silk thread. The waste superficial threads and damaged cocoons are combed and teased. The resulting fibres are spun and this is spun silk.

SERICULTURE (ERI TYPE)

This type of silk is obtained from the worms of *Attacus ricini* which feeds on castor leaves. After mulberry silk worms it is the another type which can be artificially domesticated. But they differ from mulberry in two respects. Firstly, the silk filament is not continuous and secondly, the cocoon is made up of two layers.

Eri silk-worm is also a domestic insect as mulberry silk worm. Its main food plant is castor. Though, it can feed on other alternate host plants *e.g.*, Kaseru (*Heterophanex fragrans*), Cassava (*Manihot utilissama*), Papaya, Champa (*Plumeria acutifolia*), Gular (*Ficus glumerta*) and a few others. The quality of cocoon depends upon the food plants.

Cultivation of Food-plants

Cultivation of castor plants is done like the other crops of the year and this can be done on a larger scale. A fixed distance is maintained between two plants. Plants are not allowed to bear fruits and seeds, with the help of regular prunning.

Eri Silk-worm Rearing

Since the eri-worms are hardy, large in size and better able to withstand a high temperature (90°F) and moist atmosphere (85-95 per cent humidity), its rearing is simple than mulberry silk worms.

The eggs are obtained loose and placed on a tree. The incubation period is 8-10 days. The newly hatched worms are given very tender castor leaves split into 2-3 pieces. Food is supplied 4-5 times daily. There are 5 moultings. The worms grow larger after each moulting. In some places worms are not fed on trees in the last stage but transferred to bunches of leaves tied together by their stalks and hung saddle wise on horizontal bamboo support. The Eri-worms are multi-voltine and 5-6 crops in a year can be taken in the plains of U.P. except in seasons when the temperature is high. On an average 28gm eggs give rise to about 16000 live worms which consume about 370-450kg castor leaves for their full growth in 30-35 days.

Unlike the mulberry silk-worms the cocoons of the Eri-worms are not stifled but the moths are allowed to emerge from them. The pierced cocoons are then spun into thread either by means of hand spinning wheels called Taklies or an improved spinning machine operated by foot. The pierced cocoons are boiled in the solution of soda or potash and then carded.

For obtaining eggs a simple device is adopted by using small straw bunches called 'Kharika' to which the female moths are tied. The kharikas hooked at one end and hung vertically on strings. In some cases the female moths are confined in small bamboo basket. A vertical position is however, preferred for oviposition whichever mode of oviposition is adopted. The female moth should be kept tight to the receptacle on which the eggs are laid as otherwise the moths fly about and lay eggs all over the rearing house.

There are five broods in a year viz., *khatia* or Autumn crop (September-October); *Jarua* or winter crop (November-December): *Jethwa* or summer crop (May-June); *Aheyua* or early monsoon (July) and *Bhadia* or late monsoon crop (August). Of these the first three are main crops while the later two are only catch crops. The bulk of cocoons used for reeling comes from the autumn crop which is considered to be most suitable.

SERICULTURE (TASAR TYPE)

Although, Tasar and munga silkworms are wild in nature, but attempts have been made and are in progress to domesticate them too. Success has been achieved in this regard to some extent. A technique of "controlled rearing" has been evolved by the Central Tasar Research Station, Ranchi.

Cultivation of Food-plants

It is widely spread in all the forest areas in India on account of variety of forest trees on which the tasser-worm feed. The main areas where the industry is concentrated are Bihar, M.P., Orrissa. A.P. and W.B. This industry is predominantly carried out by the hill tribes and backward classes of people living in forests. The main food plants of the tasser worm are *Antheraea mylitta*. Asav (*Terminalia tomentosa*), Arjun (*T. arjuna*) and Sal (*Shorea rubusta*). A related species of tasser silk-worm has also been reported from the forests of Jammu feeding on the leaves of Ber (*Zizyphus Jujuba*) and certain Oak trees (*Querces sp.*).

Tasar Silk Worm Rearing

The rearers collect reproduced cocoons that are generally found in nature in selected areas in the reserve forest during summer months in April-May when the trees would have shed their leaves thereby making the cocoons visible for collection These cocoons are carefully preserved in bunches in the house and at the onset of rains in June-July. The cocoons are tied to the tip of pole fixed near the house ot the rearer.

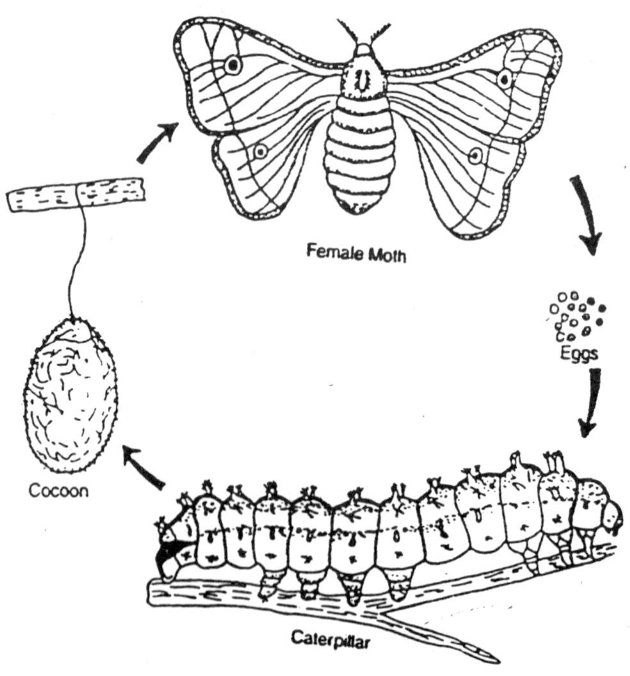

Fig. 4.48(a) : Life-cycle of Tasar silk-worm.

Soon after the rains set in the months emerges, male moth fly away and the female moths are tried in a thread so that other male moths may mate. The paired moths are picked up and confined to cup of leaves. After a day the male moths are separated from the female and discarded. The female lay eggs in the leafy cups. On an average each female moth lays 150-200 eggs. The tassar worms are generally multi-voltine but uni-volline varieties known as *modia* and a bi-voltine variety known as *ampatia* are also found to occur. In case of multi-voltine varieties three cups are taken in a year viz., winter crop (December-February). spring crop (March-April) and rainy season crop (July-August). The *uni-voltine variety* is reared in July-August while the bi-voitine is reared in August-September and December-January.

The rearing of tassar-worm is done out door on trees while the eggs hatch, the leaves cups are tied to the tender shoots of the food crop and the tiny worms crawl over the branches of the tree and start eating the leaves. The ground near the food plants are kept clean so that the ants lizards and other pests do not climb up the trees and attack the tassar-worms. When the leaves are exhausted in a branch or in a tree the worms are transferred gently to branches with fresh leaves or to separate tree. To keep off birds etc., an all day vigil is kept by the rearer with bows and arrows. Thus, the out door rearing requires a great deal of industry and patience as in case of muga silk-worms. The mature worms spin their cocoons on the branches.

Reeling of tassar cocoons is done on a crude country van called *Natwa* while the tasser silk waste is hand spun. Tassar is used for wearing fabric of dresses mostly in handlooms. The industry has come to lime light on account of tassar fabric demands in America and foreign market.

MUGA SILKWORMS

The muga silkworms (*Antheraea assama*) also belong to the same genus as tasar worms but produce an unusual lustrous golden- yellow silk thread which is very attractive and strong.

The primary food plants of A. assama Westwood are Som (*Machilus bombycina*) and soalu (*Litsaea polyantha*) leaves which are Assammese names.

Muga silkworm are wild in nature, *muga moths*, (muga is an Assamese word meaning brown or amber) are distributed from Western Himalayas to Nagaland, Cachar districts of Assam of south Tripura. But the sericulture practice is confined to the Brahmaputra valley of Assam and Foothills of East Garo hills of Meghalaya. Ideal temperature for muga silkworm growth is 24-30°C and humidity 75-85%.

The wings and body of the male moth are copper brown to dark brown while female is yellowish to brown in colour; both pairs of wings bearing eye spots. Males are smaller in size, slender abdomen, with bushy antennae. Assam and Meghalay account for more than 90 per cent of the production of muga silk. This silk is a traditional costume during the marriage ceremonies and festive occasions; the ladies garments 'Mekhala and Chadhar' made of muga silk are a priced commodity in Assam.

The Regional Muga Research station at Boko is developing appropriate technology to replace the traditional methods as this results in low productivity.

(C)

DISEASES AND PESTS OF SILKWORMS

As mulberry silkworm is a domesticated variety, it is susceptible to disease and attack by pests and parasites. Resistance or susceptibility to disease and pests is only comparative and there is no silkworm race totally resistant or immune to disease or pests. Many of these diseases and pests not only affect the physiological processes but the life and existence of the silkworm itself causing great trouble and loss to the silkworm rearers.

SILKWORM DISEASES

VIRUS DISEASES

Virus diseases found in silkworms are *grasserie* or *nuclear polyhedrosis*, cytoplasmic polyhedrosis, infectious flacherie and gattine.

GRASSERIE

This disease is also known as jaundice (yellowish colour of insect); or nuclear polyhedrosis. This disease is caused by a virus called *Borrelina*

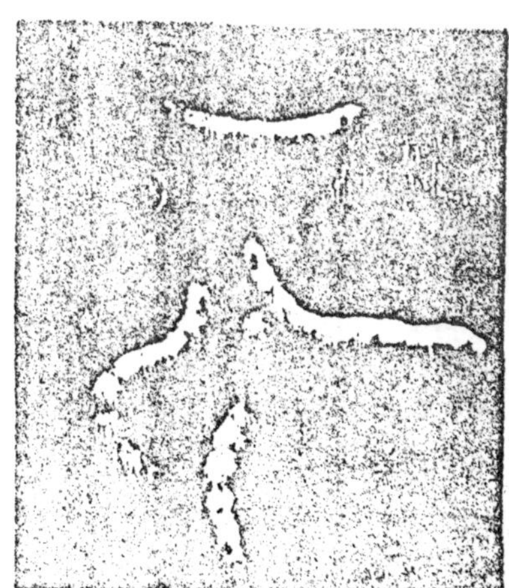

Fig. 4.49 : Left : Normal silkworm; Right : *Grasserie*-affected silkworm.

(85 × 330 nm size), which is a parasite principally in the nuclei of certain cells. The virus is contained in vast quantities in the supernatent haemolymph fluid in the nuclei of adipose tissue, tracheal membranes, dermal cells, and blood cells.

Preoral and wound infection are considered to constitute the main route of infection in early stage of infection, no symptoms are observed but with the progress of the disease, infected larvae loose appetite and tonus and are characterized by the swelling of intersegmental membrance and by restless behavior in locomotion. In infected tissues hexagonal polyhedra are generally observed which contain many virus lodes. A ripe larva attacked with grasserie may waste its silk or it may spin a flimsy cocoon and die before or after pupa formation.

CONTROL

Hygienic conditions, avoiding unsuitable leaves, proper ventilation and spacing, timely picking out of infected worms, handling worms carefully so as not cause injury or wounds are some of the precautionary measures that can be followed.

Polyhedral bodies retain their infectivity for a long time under dry condition but this infectivity is lost in 30 minutes at 70°C, and in three minutes at 100°C, therefore rearing tools and appliances should be sterilized with steam or hot water. The rearing room and rearing appliances can be disinfected with formalin or high power bleaching powder in order to inactivate the polyhedral bodies.

CYTOPLASMIC POLYHEDROSIS

This is also a virus disease and some times reported to be cause of some of the so-called flacherie.

CAUSAL AGENT

The causal agent is *smithia* virus. The polyhydra are formed in the cytoplasm of cylindrical cells of the mid-gut. The virus particles of the cytoplasmic polyhedra are spherical and measure 30-60 millimicrons. There are two types of cytoplasmic polyhydrosis in the silkworm one producing hexagonal polyhedra and the other tetragonal.

INFECTION

Infection can take place through the oral cavity and by induction. As polyhedra are freely released in excrements, infection in rearing trays is more common, in Japan the appearance of the cytoplasmic polyhedrosis is relatively low in the spring, and high in the summer and autumn. This is said to be due to difference in mulberry leaf quality. Outbreak of cytoplasmic polyhedrosis is more frequent in autumn.

SYMPTOMS

The symptoms of silkworms affected with nuclear and cytoplasmic polyhydrosis differ considerably because of different sites of virus multiplication. In the case of cytoplasmic polyhedrosis affected larvae appear like those affected with flacherie. They lose appetite and lag behind normal larvae in their development. The head is sometimes disproportionately large. Polyhedra developing in gut can frequently be observed through the dorsal integument as pale yellow or whitish area. The infected larvae passes soft whitish excrements. The most important symptoms is whitening of the midgut and this is clearly seen on dissection.

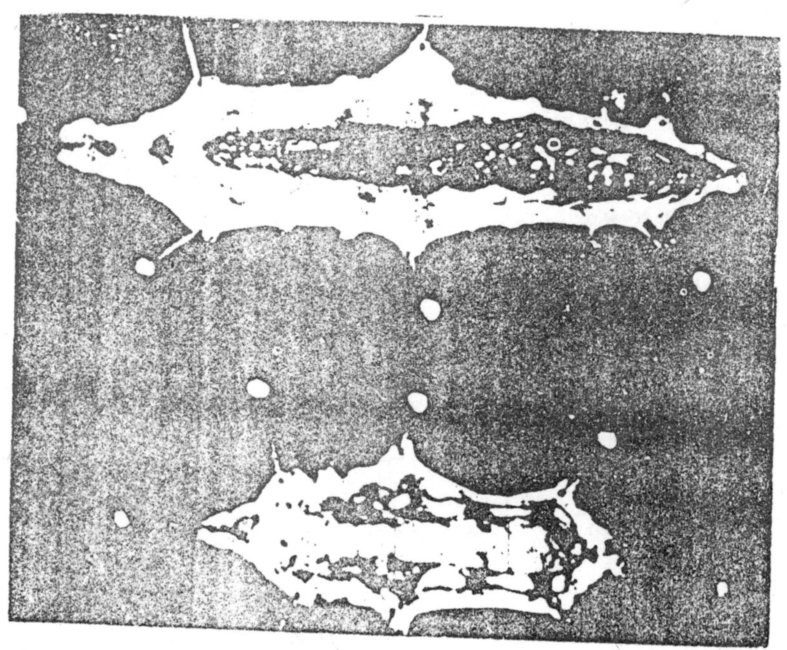

Fig. 4.50 : ***Above :*** **gut of normal silkworm;** ***Below :*** **gut of cytoplasmic polyhedrosis affected silkworm.**

CONTROL

The control measure are the same as for grasserie.

INFECTIOUS FLACHERIE

CAUSAL AGENT

A virus known as Morator virus has been isolated from the silkworm, Bombyx mori infected with flacherie. It is believed that this virus is free in the tissues and is not occluded in a crystal as in polyhedrosis.

INFECTION

Infection takes place mainly through the oral cavity. Soon after infection, the virus is liberated into the excrements, and secondary infection takes place in a rearing. Earlier instars are easily infected. Whether this disease can be caused by induction is not clear. Usually the latent period is seven to twelve days.

SYMPTOMS

The infected silkworm shows various symptom as in the so called flacherie. Diagnosis on the basis of outward appearance or autopsy is not possible. One of the methods of diagnosis is by histochemical techniques, With this method, the histochemical changes of mid-gut tissue are examinedly under a microscope. Following another method the virus is separated frorn the homogenate of the infected silkworms and given to healthy silkworms to see whether they are infected.

CONTROL

The control measures are the same as for grasserie.

GATTINE

It is also known as the disease of the clear heads.

CAUSAL AGENT AND INFECTION

Gattine appears to be caused by a submicroscopic virus to which the bacterium *Streptococus bombycia* is a secondary invader. At the beginning of the disease, when nuclear lesions are first discernible in the mid gut epithelium the intestinal contents of the insect are practically devoid of micro-organisms which, however multiply rapidly and cause the characteristic symptoms of the disease.

Streptococcus bombycis plays a significant part in the development of gattine, although it is not the principal cause. In the gattine virus, the characteristic tissue lesions are not found, not is the syndrome of gattine complete.

SYMPTOMS

The symptoms become apparent only when the combination of both the virus and the bacterium *Streptococus bombycis* occurs in the susceptible larvae. The characteristic symptom is that the anterior or cephalic end of the insect becomes swollen and almost translucent.

In addition to the clear heads, other external symptoms of gattine are lack of appetite, ejection from the mouth of a clear ropy liquid which is slightly more alkaline than the normal secretion, and somewhat accentuated diarrhoea.

CONTROL

Control of this disease depends largely upon the maintenance of sanitary conditions in the rearing rooms. If individual larvae have been affected, they should be immediately picked and destroyed. If heavy infection is noticed, repeated general disinfection during the suspension of all rearing operations is often required to control this disease completely.

FUNGAL DISEASES

MUSCARDINE OR CALCINO

The term muscardine originated from the Italian word "muscardina" meaning a muse comfit because the bodies of the insects infected with fungi are transformed into white mummified specimens resembling in appearance comfits or bonbons. The Italian name "Calcino" for muscardine refers to the chalky or lime like appearance of the dead silkworm infected with white, green, yellow or red muscardines.

WHITE MUSCARDINE

CAUSAL AGENT

The causal organism of white muscardine in silkworms is *Beuveria bassiana (Bals) vuill*. This fungus belongs to fungi imperfecti.. Conidia are round or globular, or rarely, oval, cylindrical conidia are also found. The conidia appear chalky white collectively but colourless individually. The conidium at the time of germination swells and gives out one or two germ tubes.

INFECTION AND LIFE CYCLES

Infection mainly takes place through the skin due to body contamination or infection through the tracheal opening may also occur but infrequently. Infection begins soon after the insect's body is contaminated with the conidia of the fungus. Usually infection occurs by a direct penetration of the integument by the infecting germ tube of the fungal conidium. Cylindrical conidia form and spread throughout the body fluid. The fungus continues its development once it is within the body cavity. Blood becomes scanty, blood cells are destroyed and the acidity of the blood reaches neutrality. When the infection has reached an advanced stage, circulation slows down and the consistency of blood becomes pasty. Finally, circulation stops and the insect dies.

The dead insect at this stage also shows a very characteristic production of white crystals on the body surface which have been reported as double oxalate of magnesium and ammonium. A dead caterpillar is a dangerous source of infection.

SYMPTOMS

The progress of the disease in the infected larva is very rapid. The larvae loses appetite and becomes inactive. Specks of an oozing only substance may appear on the skin when the disease advances (Fig. 4.51).

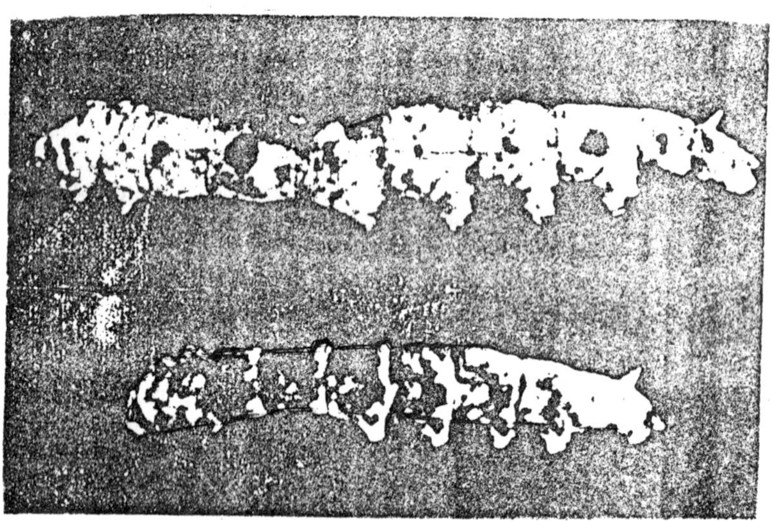

Fig. 4.51 : White-muscardine affected silkworms showing spots resembling oily specks.

The larva may also have diarrhoea and vomit juice. The body becomes limp and loses its elasticity. The larvae cease to move and generally dies within three to five days of infection. Appearance of a pinkish or reddish tinge has also been reported after death, due to the presence of a red pigmented bacterium called *Serratia marcescens.*

PREVENTION AND CONTROL

There are two types of control measures :

(a) Control measures taken during the course of rearing,

(b) Control measures taken in between two rearings or seasons.

(A) CONTROL MEASURES TAKEN DURING THE COURSE OF REARING

Infected larvae must be removed and burnt prior to the appearance of the conidia. A convenient method for separating infected larvae from healthy larvae is to use a net. The healthy larvae will crawl to the fresh leaves in the net while the infected ones, which will be inactive, will remain with the litter. Litter should be removed as frequently as possible, and burnt. Muscardine develops rapidly in hot and humid weather. At such times it is better to have good ventilation and low moisture in the rearing rooms.

BACTERIAL DISEASES

FLACHERIE

The word "flacherie" which is one of the most general and common of the descriptive terms, is used to described the flaccid condition seen in silkworm troubling from all sorts of dysentries.

CAUSE OF FLACHERIE

There are number of theories put forward to explain cause of disease, this may he due to rapid multiplication of a large number of certain kind of bacteria in the intestine, which adversely affect digestive functions, causing disease.

Other attributed causes are:

1. high temperature, high humidity and bad ventilation,
2. bad leaves,
3. over feeding,
4. decreased alkalinity of the gut,
5. over-crowding etc.

On the basis of the latest information, flacherie is now classified into 5 types depending upon causal agent and symptoms : *bacterial* diseases of digestive organs; *septicemia; sotto disease; cytoplasmic polyhedrosis*; and *infectious flacherie.* Cymplasmic polyhedrosis, infectious flacherie, gattine, and grasserie and discussed under *virus diseases*.

BACTERIAL DISEASES OF THE DIGESTIVE ORGANS

Intermediate theory which is accepted to explain cause of infection, according to this theory, silkworms become weak after hatching, their metabolism becomes inactive, causing imbalance of metabolism. As the sterilising power of the digestive fluid weakens, the bacteria devoured together with mulberry leaf multiply in the digestive tract and take nutrition from the body of the silkworm, destroying the membranous tissue of the intestine.

SYMPTOMS

The general symptoms are : the larvae loses appetite, becomes sluggish and grows slowly. Inelasticity of the skin and softening of the body. In later stages diarrhoea and vomiting of fluid are observed.

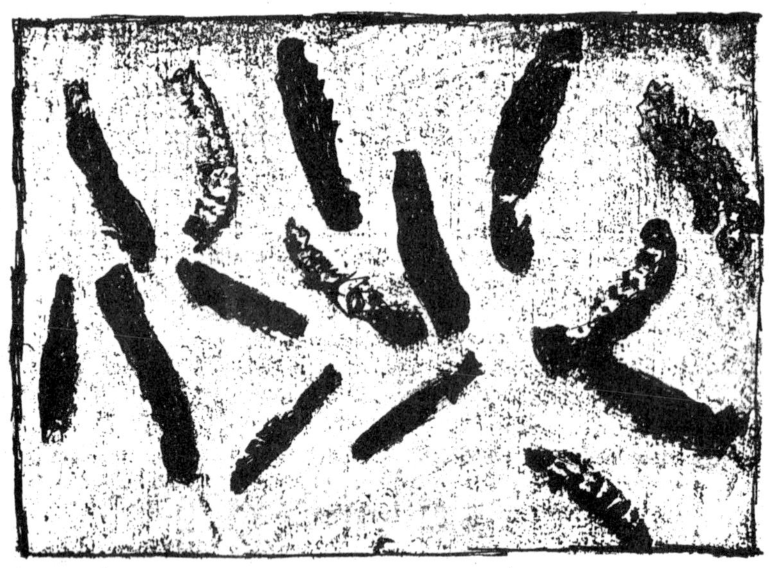

Fig. 4.52 : Flacherie affected silkworm.

CAUSAL AGENT

As initial infection stage Streptococci at final stage (near death) *coli* such as *bacillus* are infectious causative agents.

In most cases body becomes black but sometimes turns red owing to the presence of *Bacillus prodigiosus* or green because of Bacillus pyocyaneus.

CONTROL

(1) The first and foremost requisite is to raise healthy and strong silkworms since the primary cause is the weakness of the silkworms.

(2) *Proper incubation of eggs :* The silkworms become weak if the eggs are incubated at higher temperatures or greater dryness than approximately 60 per cent of relative humidity. Temperature and humidity should be maintained at 22°-25°C and 80-85 per cent respectively.

(3) *Selection of suitable race :* It is essential to select a race that is sturdy and resistant to adverse conditions for summer and autumn rearing.

(4) Feeding of good quality leaf; the health and development of the silkworm is closely related to the quality and quantity of the leaf fed. It is essential to feed plenty of high-quality mulberry leaves to avoid flacherie. Leaves grown in too little sunshine, dry weather, shade, or with lack of fertilizer, or muddy, dusty, withered leaves, etc. are poor in food value, and detrimental to the health of the larvae, particularly in the early larval stages.

SEPTICEMIA

CAUSAL AGENT

Penetration and multiplication of certain kinds of bacteria in the haemolymph cause septicemia. The principal pathogenic bacteria are large and small bacilli, streptococci, staphylococci, etc. Infection is through an injury or wound in the skin. Pupae or moths which do not feed, are also affected with septicemia. Even with bacterial diseases of the digestive organs, bacteria may sometimes penetrate into the body fluid through the membranous cells of the alimentary canal towards the final stage of the disease and cause septicemia.

SYMPTOMS

They vary depending on the kind of bacteria. When a silkworm is infected with more than one kind of bacteria, the symptoms are determined by the predominantly propagated one. As the disease advances, the prolegs lose the capacity to clasp and the worm dies. The body of a diseased larvae does not differ much in appearance to that of a healthy larvae until it dies, but when the larvae vomit fluid, the body shrinks. Soft and liquid-like excrements irregular in shape may be found. Towards death the excrements become brown or dark brown. Depending on the kind of bacteria present, the colour of the dead body also varies. In many cases black or greyish black colours may be observed. But colours such as dark brown (*Proteus* such as *Bacillus*) red (*Bacillus produgiosus*) green (*Bacillus pyocyaneus*) light yellow, etc., are also to be observed. Occasionally, many small brown or dark brown spots appear on the skin. Depending on the kind of bacterial infections, the corpse may or may not emit a foul odour. In general, the fore-intestine part of the body may be swollen and the posterior part shrunken in dead larvae. The disease occurs in a similar manner as regards colour, progress and odour in pupae and moths.

CONTROL

Irrespective of the health of the silkworm the disease is transmitted mainly through an injury or wound. Hence the infected silkworms should be treated as for any other infectious disease. The affected silkworms should be isolated from the healthy ones as soon as possible and destroyed by burning or burying deep in the soil. General disinfection of rearing rooms and rearing appliances with a 2 per cent formalin must be carried out after rearing is over.

SOTTO DISEASE

PATHOGEN AND INFECTION

Bacillus sotto Ishiwata (*Bacillus thuringiensis* var. sotto) or a large bacillus is involved in this disease. It produces a toxic substance and the disease is a toxicosis. The toxin produced is dissolved in the alkaline digestive fluid and absorbed through the gastric wall. It affects the nerve centres, causing spasm and paralysis.

SYMPTOMS

The symptoms are lack of appetite, sluggishness, lack of skin tension followed by shrinkage of the body, diarrhoea, complication of both shrinkage and diarrhoea, constipation, loss of clasping power of prolegs followed by death. The body of the insect become black shortly after death. The time from initial infection to death is shorter when temperature and humidity are high. The corpse becomes dark brown and the inner organs of the body are liquified. If the skin is cracked, a black foul-smelling liquid oozes out.

CONTROL

Bacillus sotto is found in silkworms killed by this bacteria, in diseased larvae in mulberry leaves and even in the air and water. Since this disease is a toxicosis caused by swallowing bacterial toxin, regardless of the health of larva, prevention of swallowing of the toxic substance by the larva is the primary step in controlling this disease. Diseased silkworms must be removed from the healthy ones and destroyed. As usual the rearing rooms and rearing equipment must be thoroughly disinfected before starting the next rearing.

COURT

This disease is known as court in Europe and rangi in India. It is a minor bacterial disease. The causal agent is the bacterium Serratia marcescens Bizio (*Bacillus prodigious* Fluigge). These bacteria are short, spherical rods, measuring 0.5 × 0.6 to 1 micron and occur singly or in small chains. They are motile with peritrichous flagella, and gram negative. The genus *Serratia* usually produces a characteristic red or pink pigment, although white to rose strains also occur.

INFECTION

The bacterium almost always produces a fatal infection when inoculated into the silkworm, but when introduced with the insects food, it frequently does not kill the host. When ingested, it is pathogenic to varying degrees according to the stage of the larvae. The percentage of mortality is higher when the insect is nearer to the pupal stage.

SYMPTOMS

The disease is identified by the crimson-red colour of the affected silkworm larvae and pupae at the time of death or after death. The affected larva also turns flaccid. Normally only the larva and pupa are affected.

CONTROL

This bacterium occurs commonly in nature, and individual silkworms are sometimes found infected with it. However, clean, hygienic rearing, careful handling of the silkworms without causing injury, isolation and destruction of diseases individuals will help to control this disease. Extensive epizootics, however, are apparently very rare. Silkworms showing symptoms of this disease should not be used for seed preparation. General disinfection of the rearing rooms and rearing appliances also must be carried out with a 2 per cent formalin solution immediately after rearing or before the next rearing.

PROTOZOAN DISEASES

PEBRINE

This is one of dreadful disease, also known as pepper disease or corpuscle disease. The name pubrine was given to the disease in 1860 by *De Quartrefages* because of the black spots that appear on the diseased silkworm look like pepper grains (Marathi-miri).

CAUSAL AGENT AND INFECTION

The disease causing organism is the spore of *Nosema bombycis* belonging to the family Nosematidae. It is observed that the disease may be transmitted through the egg, by contact with the diseases silkworms and through ingestion of contaminated food. There are two stages in the life cycle of this organism, the spore stage and the vegetative stage.

SPORE STAGE

The mature spore is oval and measure 3-4 microns by 1.5-2 microns (Fig. 4.53) The spore consists of :

(i) Spore membrane which encloses the sporoplasm,

(ii) Sporoplasm in the form of a girdle across the width of the spore,

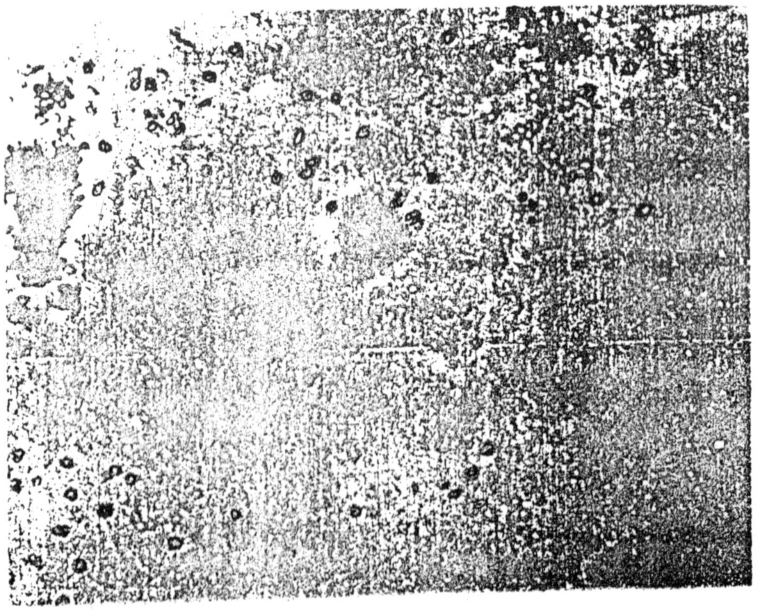

Fig 4.53 : Photomicrograph of pebrine spores (*Nosema bombycis*).

(iii) Anterior and posterior vacuoles

(iv) Two nuclei in the sporoplasm

(v) Polar capsule : The polar capsule is a sac-like structure that bulges out into the spore cavity from the anterior end. It is surrounded by the sporoplasm and connected at one end to the outer membrane of the spore and communicates with the outside through a small opening.

(vi) Polar filament : The polar capsule encloses a spiral polar filament which is projected through the small opening mentioned above. The polar filament is more than 30 times the length of the spore (Fig. 4.54).

Polar capsule

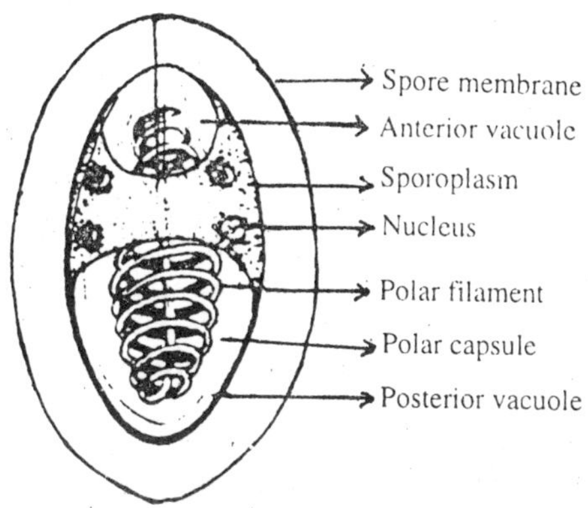

Fig. 4.54 : Structure of pebrine spores.

Schizogony (vegetative stage or asexual cycle) occurs in the caterpillars and sporogony (sexual cycle) in the imagines. Spores are passed on to the eggs by the female and newlyhatched caterpillars get diseased because of the eggs from which they hatch. Spores germinate in the gut of the newly hatched caterpillar, enter the epithelial lining of gut, the cells break and parasites pass out with the faeces. *Nosema* grows best at 80°F and growth gets inhibited out 90°F hence infection is low during summer and high during winter and spring.

PREVENTION AND CONTROL

Provide running water and full sunlight. Dryness should be preferred to shade and accompanying dampness. Keep colonies packed and strong in winter and avoid winter brood rearing.

PRODUCTION OF PATHOGEN FREE EGGS

Since pebrine is spread by trans-ovum transmission, mother moths must be inspected. If they are found to contain spores; all their eggs must be burned.

DISINFECTION OF REARING ROOM AND TOOLS

Rearing tools are immersed in 2% formaldehyde or in 0.50% chlorinated lime for more than 30 min. When fumigating chemicals are used in the rearing room, the room temperature must be maintained above 20°C for 30 min.

PREVENTION OF INFECTION BY WILD INSECTS

The pathogen of pebrine disease of the silkworm is also pathogenic to *Bombyx mandarina, samia cynthia pryeri* etc. Pest control in the mulberry fields and near silkworm rearing establishment must be practised.

SILKWORM PESTS

Among the insects that attack silkworms, the most important ones are the parasites which are commonly known as 'Uzi' or 'Uji' files. The other are dermestid beetles, ants, wasps, earwigs, crickets and preying mantis also attack silkworms but they are generally of minor importance as pests. Pests and parasites from group other than insects are mite, spiders, some nematodes, Rats and squirrels also can do considerable damage to silkworm rearing.

UZI FLY (TRICHOLYGA BOMBYCIS)

It is dipteran fly of family Tachinidae, is a serious pest of silkworm larvae and pupae. It is distributed in Burma, China, Thailand and some states of India. This fly can cause considerable damage to silkworm rearing and as a result the sericulture industry frequently suffers in the cocoon crop production.

LIFE CYCLE, HABIT AND NATURE OF DAMAGE

T. bombycis is a large fly with prominent black and grey stripes and dark ebony in colour. The female fly when it gets to the silkworm larve of later instars lays white, glossy, small, oval eggs on the body or at the inter segments. In the first and second instars silkworms are seldom host to this fly pest as they are too small and are not sufficiently attractive to the fly pest and the young stages of silkworms are practically free of infestation. Ordinarily the fly pest prefers the later instars (third, fourth and fifth) to the earlier ones of the silkworm for oviposition.

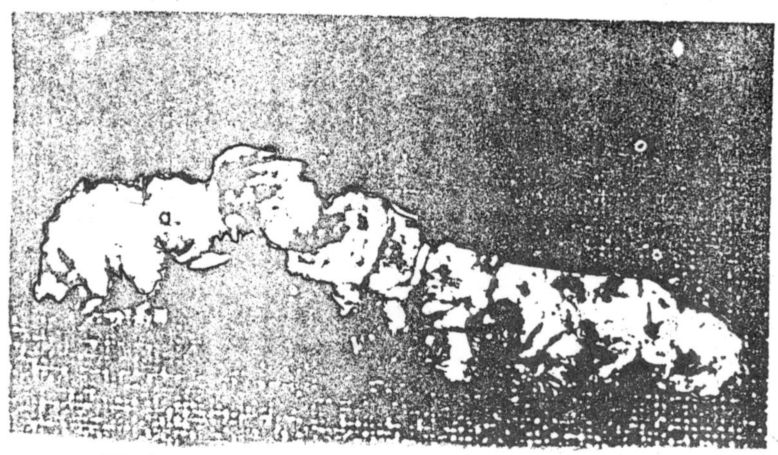

Fig. 4.55 : Larva affected with Tricholyga bombycis.

Each female fly lays about 300 eggs. The egg hatches within one to three days of oviposition and the young maggot bores its way into the body of the silkworm larvae. A black spot or scar appears on the body of the silkworm at the point of entry after penetration of the maggot (Fig. 4.55). The maggot lives on the tissues of the silkworm especially the fat bodies, for about a week and gets its air supply from the aperture through which it entered. The maggot lives in the body cavity of the silkworm for about a week and having finished feeding, the fully grown maggot wriggles its way out killing its host. If the infestation takes place in the late fifth instar of the silkworm, the larva may be able to spin its cocoon, but the cocoons cannot be used as seed cocoons and are also unreliable since the fully grown maggot makes its way out of the silkworm pupa and cocoon, for pupation in the soil, by killing its host and cutting the cocoon layer. In any case the damage is serious because either the silkworm larva itself is killed or the spun cocoons are rendered useless for reeling or for seed purposes.

The maggot selects soft soil and pupates within a day. The adult fly (Fig. 4.56) emerges in about 10-14 days and starts a fresh life cycle. The entire life cycle takes about 17-23 days under Indian climatic conditions.

The larvae affected in the third and fourth instar die invariably before they reach the spinning stage. If they are affected in the early fifth. instar, most of them die prior to spinning, whereas others die after spinning their cocoons normally. In such cases, since the worms die inside their cocoons and subsequently rot, the cocoons are stained and unfit for reeling. This fly infestation has a significant negative effect on the length, width and also the shell weight of the cocoon.

Fig. 4.56 : Female uzifly of silkworm (Araki).

PREVENTION AND CONTROL

The most effective method of control is to prevent the entry of the fly into the rearing rooms by providing suitable wire mesh (fly-proof nets) for doors, windows and ventilators, wherever rearing is done in separate rearing houses or rooms. Pest - trapping rooms may also be provided at the entrance, but this may not be financially feasible for poor rearers. The maggot infected larvae and cocoons must be destroyed by burning or by treating with chloropicrin. All crevices on the floor must be closed up to prevent the maggots from creeping under the floor to pupate.

DERMESTID BEETLES

This group of insects belongs to the order Coleoptera, family Dermestidae. This family contains many genera and a number of very destructive species. Most of the damage is done by the larvae when cocoons are stilled and stored for a long time; holes are bored and the pupae in them eaten. In addition to this, they damage animal and plant products including leather, furs, dried fish, carpet, woolen and silk materials.

The following are some of the dermestid beetles that are important from a sericulture point of view :

Dermestes cadverinus, D. valpinus, D. vorax, Trogoderma versicolar, Anthrenus verbasi, Attagenus japonicus etc.

MORPHOLOGY OF DERMESTES CADVERINUS

The adult is oval, elongated, dark brown with club shaped antennae, and is about 1 cm in length. The larvae have their upper surfaces covered with hairs of various lengths. The larvae are reddish -brown and spindle shaped (Fig4.57). They moult five to seven times and attain a length of about 15 cms. The elongated oval eggs are 2 mm length, and are milky white in colour.

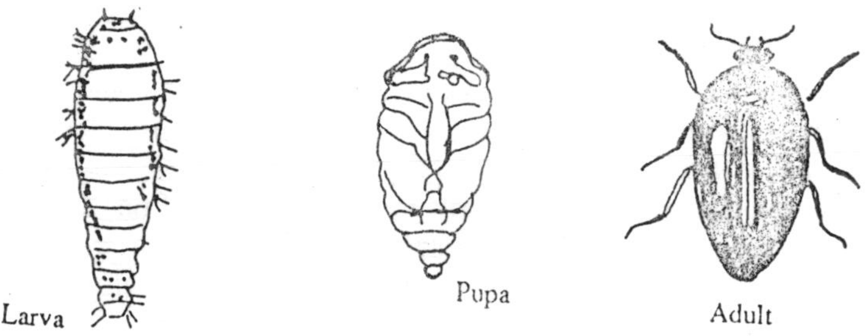

Fig. 4.57 : Dermestes cadverinus Larva, Pupa and adult.

LIFE CYCLE, HABIT AND NATURE OF DAMAGE

Generally the insect passes the winter in the adult stage, begins mating and starts laying eggs in May. The adult lives on animal matter for one year after oviposition. The female beetle moves around in dark areas of the cocoon storage and deposits its eggs in the crevices of dried organic matter, slits in wood, etc. The egg hatches in a week and the young larva starts feeding on dried animal matter. The insect prefers to live in darker places. It moults five to six times in one or two months and becomes a pupa. The adult, after emergence, mates and lays eggs which develop to become the adults of the second generation. Although the insect generally passes the winter in the adult stage, since the time of metamorphosis is not fixed, both the larval stage and the pupal stage may be encountered in winter.

The larvae and adults are attracted by the smell of stifled cocoons and the dried pupae inside. They bore into the cocoons and eat the dried pupae and sometimes the eggs and the damaged cocoons become unfit for reeling. Rarely, the young larvae attack living silkworms.

CONTROL

Rooms where cocoons are handled must be provided with concrete floors, and an insecticide CVMP (2-chloro-l-(2, 4, 5 trichlorophenyl) vinyldimethyl phosphate), sprayed on the ground. Also, an insecticide, Dichlororos, can be sprayed in the mulberry field for pest control.

ANTS (ORDER : HYMONAPTERA. FAMILY : FORMICIDAE)

The insect pest attacks silkworms in the rearing trays and this can be prevented by placing the legs of rearing stands in ant wells. When the ants attack is at the time of spinning, a little kerosene can be poured on the floor around the legs of the chandrikes or bamboo mountages, two of which are usually kept slanting in the form of inverted "V"

NEMATODE

Nematodes belong the phylum, Nemathelminthes and class Nematoda. The nematode *Hexmermis microamphidis* is mostly found in silkworms of late- autumn rearing. This organism attacks the young-stage larvae and penetrates into their bodies. The head of this affected silkworm becomes transparent and the body milky white.

LIZARDS

These commonly frequent the rearing houses and often pick up the young-age worms and eat them, thus causing considerable damage.

RATS AND SQUIRRELS

These mammalian predators eat the silkworms avoiding only the silk glands and the pupa after biting open the cocoons. Their entry into the rearing house can he prevented by providing suitable wire mesh (fly-proof nets) for doors, windows and ventilators. Traps and poisonous baits can also be used.

BIRDS

Birds such as crows and sparrows pick up the silkworms when the mountages are kept out of doors at the time of spinning. Damage by these predators can be avoided by keeping the mountages indoors at the time of spinning.

(D)

ECONOMICS OF SERICULTURE

The suitability of the climatic condition for mulberry cultivation alone cannot sustain continued culture of silkworms on industrial scale in a country. The deciding factor is the economy of the industry *i.e.,* the cost of production which in turn reflects the efficiency of leaf production, utilization of the leaf by the silkworm race, the spinning quality of the silk etc. Added to these, the availability and cost of labour are also important factors contributing to the cost of production.

Various scientific methods of mulberry cultivation have been adopted to maximize production in unit area of land and the cost analysis of input-output ratios have been worked out. The new innovations have been reported not only to increase leaf production but also to reduce the cost of production of unit weight of leaf. The another interesting feature is the net profit earned from unit land through a sericulture is more than any other agricultural or commercial crop, especially in developing & underdeveloping countries where average income is low. The product of the land in mulberry leaf which is converted into raw silk by rearing silkworms and extraction of silk, both of which require large labour force the quantity of silk produced varies from 20 to 120 kg per hectar while even approximate monetary comparisons are difficult to make, the value of silk is obviously greater than the value of rice, sugarcane or any other crop produced per unit area of land.

From the point of view of national economy one of the important advantages of sericulture is that it does not require much land. The area needed to produce one ton of raw silk depend on the intensity of mulberry cultivation and quantity of silkworm that is with rain fed mulberry cultivation without the application of artificial fertilizers and rearing inferior quality of silkworm. 20 kg raw silk is produced from one hectare of mulberry plantation as other extreme cultivation the best varieties of mulberry on irrigated land with the application of fertilizers and rearing the most suitable silkworm races up to 120 kg of the raw silk is produced from one hectare of mulberry plantation for the country independent to start silk production assuming that from the very beginning the most suitable varieties of mulberry is cultivated and high quality silkworm races are reared a yield of 40 to 60 kg of raw silk could easily be produced. On this basis a foreign exchange in the range of Rs. 5000 to 9000 per hectare of mulberry can be expected.

It may be seen that cocoon production in 1970 was 23%, below that of 1938 (pre-war level), but showed a steady and gradual increase between 1968 and 1970. The main declines, between 1938 and 1970 have been in Japan, where production is less than half the 1938 level; and in Italy and Greece where cocoon production has fallen to less than 10%. There has also been a substantial decrease in Yugoslavia while the cocoon production seems to have almost disappeared in France.

On the other hand there have been significant increases in cocoon production in China, the USSR, India, Rumania, Syria, Spain. In China, production almost doubled between 1938 and 1970. India has silk-producing centre in Assam, Bengal, Madras, Mysore, Punjab and Kashmir. Mulberry silk is restricted to Purnea district along the border of West Bengal. The total production of silk is of the value of about 3.5 crores in Bihar. Muga silk is produced only in Assam.

India has the unique distinction in the world as it produces all the four major varieties of silk viz. Mulberry, Tasar, Eri and Muga. However, India occupies fifth position in the world for silk production. The first four in order of production of silk are Japan, China, South Korea and U.S.S.R., Brazil, Italy come behind India.

In India, of the 6, 29, 143 villages, sericulture is being practised in about 59, 528 villages (9.5%) and this provides work (employs) to about 60 lakhs people; mostly tribals. Indian sericulture has shown a dynamic growth during last few years. Its silk production during the last 14 years has gone up from 3475 MT in 1978 to 14,140 metric tonnes in 1992-93.

The major silk producing states are Karnataka, A. P. Tamilnadu, W. Bengal and Jammu & Kashmir for mulberry silk. Sericulture is an export oriented industry exporting silk to over 50 countries of the world. (European, U.S.A etc.) and bringing in foreign exchange worth about Rs. 250 crores.

India produces 5% mulberry silk, 10% tasar silk and 100% muga silk.

Table 4.4 : Production of raw silk (metric tonnes) by India

Mulberry	Tasar	Eri	Muga	Total
13,007	361	700	72	14,140

SERICULTURE RESEARCH AND TRAINING CENTRES

In India, there are 4-research centres working under the central silk board under the Ministry of Trade and commerce.

1. Central Sericulture Research Station, Behrumpura, West Bengal (Mulberry).
2. Central Sericulture Research and Training Institute, Mysore (Mulberry).
3. Central Sericulture Research and Training Institute, Mysore (Mulberry).
4. Central Tasar Research Station (Ranchi, Bihar).

These centres have extension centres in different parts of the country trying to reach to almost every silk producing region to give advise, guidance and undertake short and long term training courses.

5

LAC INSECT AND LAC CULTURE

In general most of the scale insects are harmful because they suck the sap from many of the our cultivated plants. But some of them however, are beneficial and lac-insect is one. Lac as a resinous protective secretion of the tiny lac insect, which is of great economic importance. In order to obtain and collect the lac, these insects are cultured and the technique for proper care of host plants, regular prunning of host plants, propagation, collection and processing of lac is called as *lac culture.*

The name lac seems to be derived from the word Laksha meaning *one lac* in Sanskrit which is suggestive of the large number of insects settled on young succulent shoots of the host plant involved in the production of lac. The minute red coloured larvae of the insects, or crawlers, settle on young succulent shoots of the host plant in groups of thousands, derive their long proboscis into the bark and plant tissue, grow and secrete a resinous material, which ultimately covers them. The thousands of crawlers settle side by side and the resinous secretion builds up around them and completely encases the twig. After the completion of the life cycle, and just about the time the larvae of the next generation begin to emerge, the twigs are harvested and the encrustations scraped off, dried and processed to yield the *lac* or *shellac* of commerce.

HISTORY

The references of lac insect are found in 'Atharv ved' regarding its habits and usefulness. In Mahabharta references are there that kauravas had build "Laksha graha" or palace of lac for pandavas. There are also records to show that ancient Greek and Roman also know the use of lac. The scientific study of lac was studied by *father Tuchard* in 1709. *Kerr* (1782) gave the name *coccus lacca.* Later Green (1922) and chatterjee (1915) called the lac-insect as *Tachardia lacca* (Kerr); finally, the name was given as *Laccifer lacca.*

Classification

Phylum	—	Arthropoda
Class	—	Insecta

Order	—	Hemiptera
Sub-order	—	Homoptera
Super-family	—	Coccidae
Family	—	Leciferidae.
Genus	—	Tachardia (Laccifer)
Species	—	*Lacca.*

India held a vertual monopoly of lac until about 1950, accounting for nearly 85% of the world's production of sticklac. Since 1950, however, Thialand has become the main competitor of India in export of lac; and it now supplies 25-30% of the world's requirement. Other countries like Africa, Australia, Brazil, Burma, Shri Lanka, Chiana, Formosa, France, West Germany, Japan, Malaya, Nepal, Spain, Turkey, U.S.A. and some others also produce lac. In India, major lac producing places are Assam (Kashi hills), Bengal (Calcutta, Jangipur, Murshidabad, Mathrapur, Malda), Bihar(Manbhum, Palamau, Ranchi, Santhal Pragana), Delhi, Gujarat, Hyderabad, Kashmir, Madhya predesh (Damoh, Champa, Bilaspur, Rewa, Umaria), chennai, caimbatore, Mysore, Orissa (Cuttack, Mayarbhanj), Punjab (Hoshiarpur, Shahpur), Rajasthan (Indergarh, Kota, Jaipur, Jhallawar, Karauli), and Utter Pradesh (Ghazipur, Mirzapur, Agra) etc. The distribution of lac production in different states of India is Bihar (41.3%), M.P. (26.5%), W.B. (20.6%), Maharashtra (5.9%) and U.P., Orissa, and Assam (less than 4%).

Host Plants

In India, the host plants of the lac insect belong to family, papillionaceae, among which the most preferable trees are::

1. *Zizyphus mauritiana* (Ber)
2. *Z. Jejuba* (Ber)
3. *Z. Xylopyraya* (Ghont)
4. *Cajanus cajan* (Arhar)
5. *Grewia* sp. (Grewia)
6. *Acacia arabica* (Babool)
7. *Acacia catechu* (Khair)
8. *Butea monospeme* (Palas)
9. *B. frondosa* (Palas)
10. *Schleicheia oleosa* (Kusum)
11. *S. trijuga* (kusum)
12. *Shorea talura* (Jallari)

13. *Ficus religiosa* (Peepal)

14. *Leela* sp. Leea.

For commercial lac production the plants of kusum, palas and ber are widely used for rearing of lac insects.

MORPHOLOGY

The lac insects live as a parasite, feeding on the sap of certain trees and shrubs. Lac insect is a minute, resinous, crawling scale-insect which inserts its beak into plant tissues, sucks juices and grows, and secretes lac from the hind end of the body. Its own body ultimately gets covered with lac in the so called *'CELL'*. Lac is secreted by the insect for its own protection and not for the food of the insects. The commercial lac is produced by only the female insect and not the male.

Male

It is larger as compared with the female and can be identified from its red colour. The body is divided into as usual cephalothorax and abdomen. Head consists of a pair of antennae and eyes but without mouth parts, so unable to feed. Thorax bears 3 pairs of legs, and may be with wings or without wings. Abdomen is the largest part bearing a pair of caudal cerci and male accessory reproductive organs in the form of penis present in a sheath.

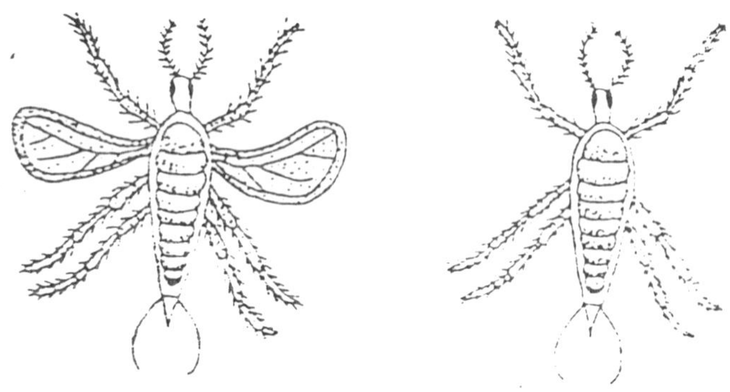

(a) Male with wings **(b) Male without wings**

Fig 5.1 : Adult male lac-insect.

Female

It is smaller, crimson coloured about 1.5 mm long, cephalothrax is oval while abdomen is short tabular structure bearing fringed anal end.

Head bears a pair of antennae and a single proboscis. Eyes are absent. Thorax is without wings and legs, so motionless. Abdomen bears a pair of anal cerci. It is female lac insects which secretes the bulk of lac for commerce.

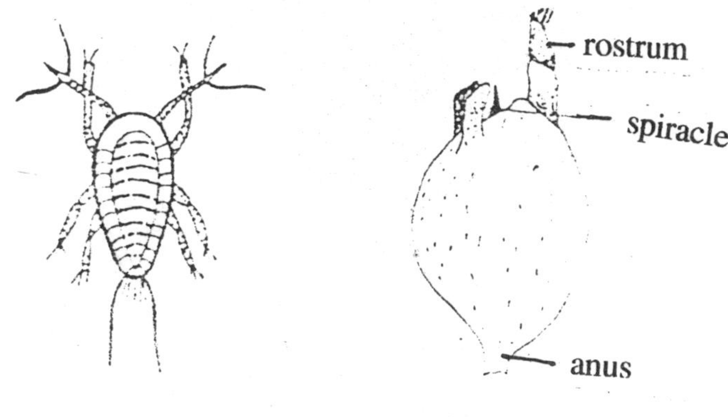

(a) Immatured stage **(b) Adult female**

Fig 5.2 : Showing loss of some vegetative parts in an adult female lac-insect.

FERTILIZATION

During fertilization males emerge from their cells and walk over the lac incrustation. The male enters the female cell through anal tubular opening and inside the female cell it fertilizes the female. After copulation, the male dies. One male is capable of fertilizing several females. Females develop very rapidly after fertilization. They take more sap from plants and exude more resin and wax.

Life-cycle

The female lays about 200 to 1000 eggs. She contracts her body repeatedly so that large number of small extensions (brood cells) are formed at the periphery of the female cell. The eggs are lodged in the brood cells. The eggs are laid in the month of October and November. Egg laying particularly ceases if the temperature falls below 17°C in summer and below 15°C in winter. The eggs hatch within 2-8 hours and develop in to first in star larvae (nymph) in the months of November and December. When larvae (nymph) emerge they are in quite large number, these mass emergence is called as a swarming. This first instar larvae are about 0.5 mm in length, red coloured and boat shaped. The head bears paired antennae, ocelli and ventrally situated piercing and sucking type of mouth parts. The three segmented well developed thorax contains two pairs of spiracles and only one pair of legs and terminates into a pair of long caudal setae or cerci. As soon as first instar larvae hatch out, they bore brood cells and swarm to very tender succulent shoots or branches at about 150-200 larvae per linear inch area. They start sucking the plant sap voraciously on one hand and secreting resinous secretion over the body on the other form special dermal glands located all over the body except mouth parts, two breathing pores and anus. As the resinous secretion comes in contact with air, it soon becomes hard and forms a coating over the body of nymph and is called as '*Cell' Within* this cell various life processes like growth of the nymph, morpohological changes and lac secretion takes place.

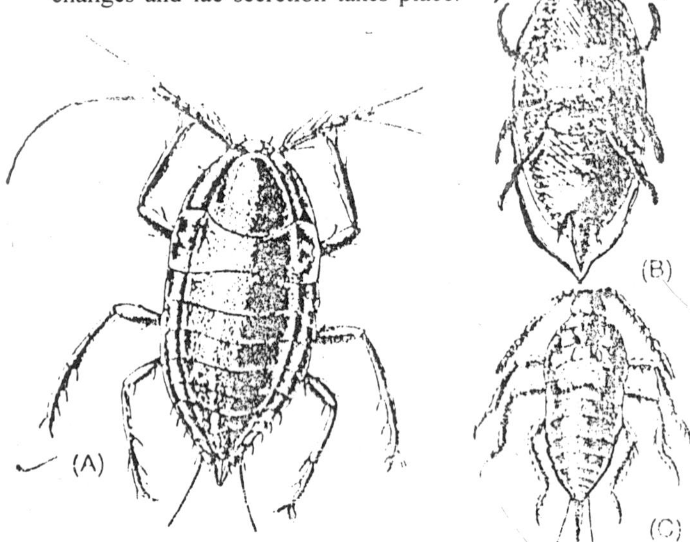

Fig: 5.3 : Life cycle of lac insect (A) Larva Nymph (B) Nymph I (C) Nymph II.

The 'male cell' is elongated and cigar-shaped having two hole *i.e.*, anterior and posterior. From the posterior hole which is covered by a flap or operculum and the male insect comes out through this opening. After 6 to 8 weeks of stationary life nymphs moult thricely and develops into the adult. Out of total population 30% becomes winged males and 70% females which are wingless. The females get fixed on the host plant in resinous mass. The males walk over the encrustations of females and fertilize them within their oval cells through anal opening. The males leave the parent cell after fertilizing the female. The female nymph once settled never moves but undergoes 3 moults inside her cell loosing its eyes and legs and with sedimentary antennae only. The fertilization of female is followed by a rapid growth of the female body till it begins to lay eggs in October and November. From these eggs male and female emerge in February to march. The male fertilizes the females of this generation and the fertilized female lays eggs in months of June to July and lies secreting lac all the time. Thus the life cycle repeated twice in a year on the same host plants.

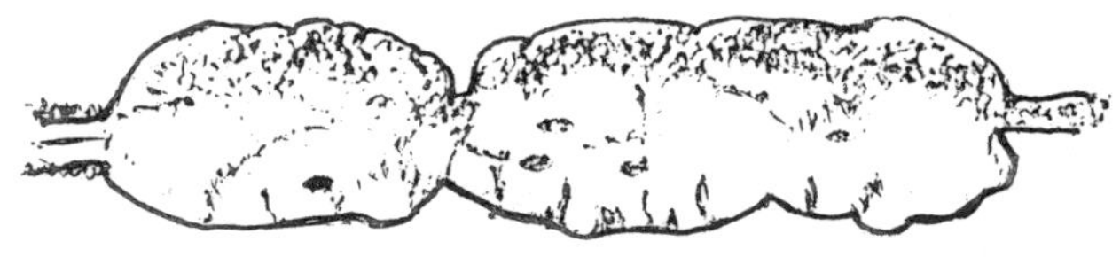

Fig: 5.4 : Kusum lac (encrustation).

Strain of Lac Insect and Lac Crops

There are two strains of lac insects grown in India. The *Kusum* grown on kusum plants and *Rangeeni* on Butea, zizipus and other plants. Each strain completes its life cycle twice a year but the duration of life cycle and season of maturity differs consideratelу. Thus, in case of kusum strain the *Aghani* crop is harvested in January and February from plants inoculated in June and July, the *Jethwi* crop is obtained in June and July from January-February inoculation. Similarly, in case of *Rangeeni* strain the katki crop is harvested in October and November from inoculated made in June-July. The Baisakhi crop is obtained in June-July from plants innoculated in October-November. Some times the *Baisakhi* crop is harvested in April or May.

On like the kusumi strains the crop in the case of *Rangeeni* strains are unequal in duration. The katki crop has a duration of about 120 days while the *Baisakhi* crop is 240 days duration. The Rangeeni strain is raised practically in on all the known lac hosts except the *Kusumi*. Nearly 90% of four production of stick lac is from this strain through the remaining 10% of the production from kusumi strain is considered superior in quality because of lighter colour of resin.

The four lac crops derived their names from Hindi months in which the main harvest is collected. Of the four crops the *Baisakhi* and *Aghani* are the two main commercial crops. From the two strains as they together contribute about 72% of the annul production.

Lac-Culture or Lac-Cultivation

Lac cultivation is initiated from brood lac, which is a twig of host tree carrying lac encrustation, with larvae about to emerge from mother cells. The twigs are cut, bundled and tied at convenient places on a fresh host plant, so that the emerging larvae swarm and settle on nearby succulent shoots. In recent years, much information of value to lac culture has emerged as a result of researches at the *Indian Lac Research Institute* at Nankun, Ranchi; Bihar.

Important steps involved in lac cultivation are:

(i) Prunning

In prunning, the overall consideration should be that the general health of the tree must be maintained, and for this purpose, the nature and extent of prunning of host plants and the time of the year, have been found to have an important effect on the production of succulent branches for feeding of lac insects and therefore on the lac crop.

It has been found that branches more than 25 mm in diameter should not be cut . Those 12.5 mm or less in diameter should be cut flush with the main branch from where they arrive whereas those 12.5mm to 25mm in diameter should be so cut as to leave behind a stalk of 450 mm in length. The proper time for prunning of palas trees meant for *katki* innoculation in June-July in about mid February but for Ber; trees prunning in may of the previous year has given best results. The proper time for prunning of *Palas* and Ber trees meant for Baisakhi innoculation in October-November is about mid April. Prunning for kusum trees has to be carried out only once when the tree is first taken up for lac cultivation for which July prunning has given better results than January prunning. There after, the harvesting operation itself serve as pruning also.

(ii) Coupe System

If the same tree is continuously inoculated or attack of lac insects, its vitality suffers and the yield of lac progressively diminishes.

It is therefore, important that host plants should be given periodic rest. So the plants of a farm are numbered into 5 groups or blocks or of plants. This artificial division or marking of trees is called coupe system of crop cultivation. In this system when one group of host plants is under the process of cultivations of lac, other groups of host plants would be under rest. Each coupe is inoculated according to a schedule.

(iii) Inoculation

The method by which the lac insects are introduced to the new lac host plant is called as inoculation. This may be of two types, namely:

1. Natural Infection
2. Artificial Infection.

1. NATURAL INFECTION

When infection from one plant to other occurs through natural movements of insects. This may be due to over-crowding of insect population and non-availability of tender shoots on a particular host tree.

Drawbacks

(a) **Incomplete Nutrition:** Lac insects with their piercing and sucking mouth parts, suck the cell sap of the same host plant for nutrition. If the cell sap of the same host plant is further sucked out by the swarmed nymphs of the second crop continuously, the growth of the host plant would be retarted. In this way lac insect may not be able to get enough nutrient from the same host plant.

(b) **Unfavourable climatic conditions:** At the time of swarming a number of environmental factors like high intensity of sunlight, heavy rainfall etc. affect the proper inoculation of nymphs. These factors may also affect the host plant at the same time and may cause a gap of inoculation resulting in irregularity of lac crop.

(c) **Irregular Inoculation:** During natural inoculation it is not sure that uniform sequence of inoculation taken place. If inoculation is not of continuous fashion, a regular crop of lac may not be obtained.

(d) **Multiplication of Parasites and Predators:** Like other animals lac insects have enemies in the form of parasites and predators. It the crop is not harvested in time and lac is allowed to remain on the same twig, the multiplication of parasites and predators takes place which suppress the population growth of the lac insects.

Due to above draw backs, this natural method is avoided and a artificial method is developed with certain devices.

2. ARTIFICIAL INFECTION (INNOCULATION)

Artificial Infection takes place through the agencies other than those of nature. Prior to about two weeks of hatching, lac bearing sticks are cut to the size of six inches. They are called *Brood lac*. Brood lac are then kept for about two weeks in some cool place. When the nymphs start emerging from these brood lac, they are supposed to be ready for tieding the brood lac with the host plant. It may be of different types. In longitudinal infection the brood lac is tied in close contact with host branches. In lateral infection the brood lac is tied across the gaps between two branches. In interlaced method, brood lac are tied among the branches of several new shoots.

The following precautions should be taken in artificial inoculation:

(i) One must ensure that the twigs, which are going to be tied on fresh host plant, are having good number of nymphs or eggs. It is also possible that from many of the twigs nymphs have swarmed out, thus inoculation would prove unsuccessful.

(ii) The eggs or nymphs present on the twigs should be healthy and about to swarm so that one has not to wait for longer period and thus save time.

(iii) For the uniformity of inoculation, 3 to 4 twigs should be utilised.

(iv) The twigs provided with eggs or nymphs should be without any parasite and predator.

(v) Host plants should be changed from time to time for the proper nutrition of the nymphs.

(a) Longitudinal infection, (b) Lateral infection, (c) Interlaced infection

Fig. 5.5 : Three different ways of artificial inoculation of lac.

3. PRESERVATION OF BROOD SAC

Successful lac cultivation depends on the continuous availability of brood lac in the required quantities. To ensure a steady supply of brood lac, it is advisable to have a large variety of trees under lac wherever possible rather than restrict cultivation to one host. One way of preserving adequate brood during the hot summer months is to select only those hosts which normally have green leaves during summer.

4. ALTERATION OF LAC HOST OR BROOD LAC

If brood lac from a particular host is used year after year in the same locality and on the same host species, the quality of the lac crop is likely to deteriorate. It would be advantageous, therefore, to exploit more than one kind of host for cultivating lac. *Khair*, which is more or less widely distributed, is host for both *Rangeeni* and *Kusumi* strains and can be successfully alternated with kusum. It has been established also that where *Palas* and *Ber* occur together, the former should be utilised both the *Baisakhi* and *Katki* crops and the latter for raising only *Katki* crop.

5. HARVESTING

Period

The common practice among cultivators is to cut the crop partially as *Ari* (Immature) April-May and fully as mature in *Katki* crop in October and November. On the basis of earlier investigations a view was held that *Ari* cutting was harmful and should be discouraged. Recent experiments on *Palas* in Bihar have however, indicated that obtaining the maximum yield from stick lac it is best to cut the crop in April-May as *Ari* and for brood purposes it should be harvested fully in October-November. This is however, true only for hot areas. In cooler regions the yield is better if the lac is harvested at maturity *i.e.,* in July.

Processing of Lac

The largest yield of lac and dye are obtained by harvesting the infested twigs while the female are still living. Encrustations of lac from excess crop are separated from the twigs by scrapping. The scraped material is called as *stick lac* or *green lac*. Lac can be scraped from the twigs before or after the emergence of larvae. If it is used for manufacturing before the emergence of larvae, the type of lac produced is called as *Ari lac* and if it is used for manufacturing purpose after swarming of larvae has occurred, the lac is called as *phunki lac*.

The scraping of lac from twig is done by knife, after which it is spread thinly (10-15cm-deep) in a covered and well ventilated place and periodically raked until dry. It should not be exposed to sun. The scraped lac is grinded in hard stone mills. The unnecessary materials are sorted out. In order to remove the finer particles of dirt and colour, this lac is washed repeatedly with cold water. Now at this stage it is called as *seed lac* and is exposed to sun for drying *seed lac* is now subjected to the melting process. Colour or different chemicals are mixed during melting process for particular need. The melted lac is sieved through cloth and is given through the final shape by moulding. The final form of lac is called *shellac*.

The quality and colour of the lac is variable according to the presence of gum and resins in the host plants. *Kusumi* lac is said to be the best while Dhak is supposed to be the worst and cheapest quality one.

Composition of Lac

The lac is secreted in the form liquid from the dermal lac glands, which open on the upper surface of the body through many pores; which becomes solid on drying or when exposed to air. Lac is a resinous matter which is secreted by the female lac insect around its body for its own protection. Besides the lac glands, there are three wax glands situated on mouth, mid-abdominal respiratory hole and sides of anus; which do not allow the lac to deposit on the mouth, respiratory hole and anus.

The major constituent of stick-lac is the resin (70-90%); other constituents present are sugar, proteins and soluble salts (2-4%); colouring matter (1-2%); wax (4-6%); sand, woody matter, insect bodies and other extraneous matter (8-12%), a volitile oil is present in traces. Colouring matter includes a water soluble red dye, *lacçaic* acid, and an alkali and spirit soluble yellow dye, *erythrolaccin.*

Properties

(a) Lac is not soluble in water but easily soluble in alcohol and alkalies. This property of lac has great value for insulation of electrical connections.
(b) Lac is easily fusible on heating.
(c) Lac has adhesive quality.
(d) Lac is bad conductor of heat.

Uses

Lac is used for various purposes, the chief among these are as follows:

(1) Lac dye was formerly used for dying silk wool and feathers, but has now been largely replaced by the more versatile synthetic dyestuffs.

(2) The former chief use was in the manufacture of gramophone records but in recent years, it has been replaced by synthetic substitutes.

(3) It is used in the electric industry in the form of insulating varnishes and moulded insulators.

(4) It has the biggest consumption in the surface coating industry; like laminated paper boards, photographic and engraving industry; plastic moulded articles etc.

(5) It is used as an ingredient of varnishes, polishes, finishes, lacquer (wave toys) etc. For protective and decorative purposes.

(6) It is used in sealing wax, Lac coated earthen ware, latch primer, Hessian (Gunny bags) coated with shellac.

(7) It possesses very good adhesion to mica.

(8) Lac is also used as coat for preserving archaeological and zoological specimens.

(9) Lac is also used for various purposes such as bangles and jwellery and leather dressings, water proof inks, nail polish, glazing confectionary, stiffen for shoes and heals of shoes, felt and furr heads, playing card finishing, finishing for oil clothes, printing inks, coating pills, backing of mirrors, coating of silvers, cosmetics and hair dyes, dental plates etc.

PARASITES AND PREDETORS OF LAC INSECT

Parasites

Several species of parasitic chalcidoid wasps (*coccophagus tschirchii; Erencyrtus dewitzi; Eupelmus tachardiae; Marietta javensis; Parechthrodryinus elavicornis; Tachardia phagus tachardiae; T. somervilli; Tetrastichus purpureus)*. These parasites lays their eggs in or on the body of the lac insect. The grubs on hatching feed on the lac insect and mature adults emerge from the lac cell by chewing out circular holes. The damage resulting from parasites is usually from 5-10% of the lac crop.

Predators

Predators of lac insect which are of serious consequence are two moths, *Eublemma anabilis* and *Holocera pulvera* and three species of *chrysopa* (lacewings). The caterpillars of the two moths tunnel through and eat away lac encrustations as well as the insects, whereas larvae of chrysopa sp. suck the body fluids of lac insects. The damage caused by predatory insects amounts to about 40% of the lac crop.

Control Measures

(a) Immersion of freshly harvested stick lac, not wanted for brood, in running or deep stagnant water.

(b) Scrapping of lac from twigs immediately after the harvesting and killing larvae and pupae of the pests by burning, crushing, or fumigating with carbon disulfide before storage.

(c) Avoiding cultivation of early and late maturing varieties of lac in the same locality to prevent the spread of pests.

(d) Brood should not be kept on trees for more than 21 days.

(e) Encasing of brood lac for inoculation in 60-80 mesh wiregauze baskets. The baskets permit free exit of lac larvae but not of their insect enemies.

(f) Biological control through the release of microbracon green which is parasitic on E. *anabilic and Apanteles tachardiae parasites H. pulverea.*

Non- insect enemies—Squirrels, monkeys, rat, bat, wood peckers, and man itself are enemies which also destroy lac crop in different ways. Damage may be due to adverse environmental conditions (excessive heat, cold, heavy rains, storms) as well as faulty cultivation methods.

6

AGRICULTURAL PESTS AND THEIR CONTROL

Terminologies used for types of damage to crop plants **:**

Damaged roots and *tubers* :

(a) *Tubers Tunnelled :* Usually infected with secondary rots; Potato tuber moth larvae bore down the stem into the tubers.

(*b*) *Tubers with narrow tunnels :* Bored by wireworms and some small species of slugs.

(*c*) *Tubers with wide, sometimes shallow tunnels :* Bored by cutworms, chafer grubs and slugs.

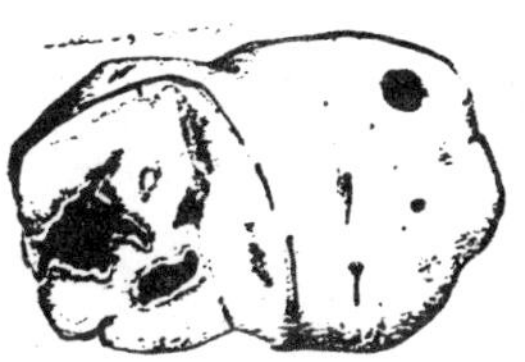

Fig. 6.1

(*d*) *Sweet potato tubers bored :* Infected with secondary rots; larvae and adults of sweet potato weevils.

Fig. 6.2

(*e*) *Roots with large swellings* : Root-knot nematodes (eellworms), cause extensive root swellings on many crop plants.

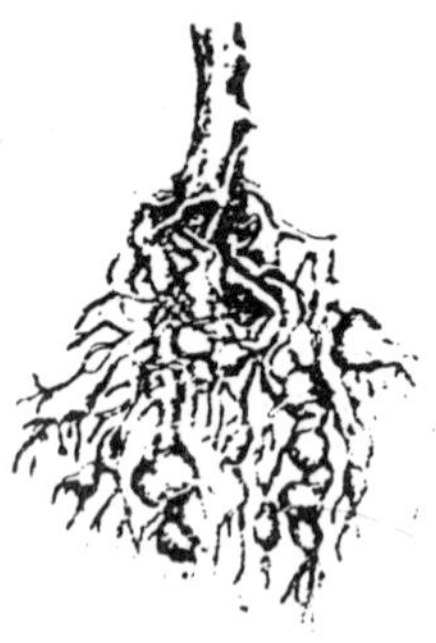

Fig. 6.3

(*f*) *Roots stunted but bushily prolific* : Citrus root eelworm a free living nematode.

Fig. 6.4

(*g*) *Roots with small globular cysts* : Small globular cysts are produced by a nematode on a wide range of host plants.

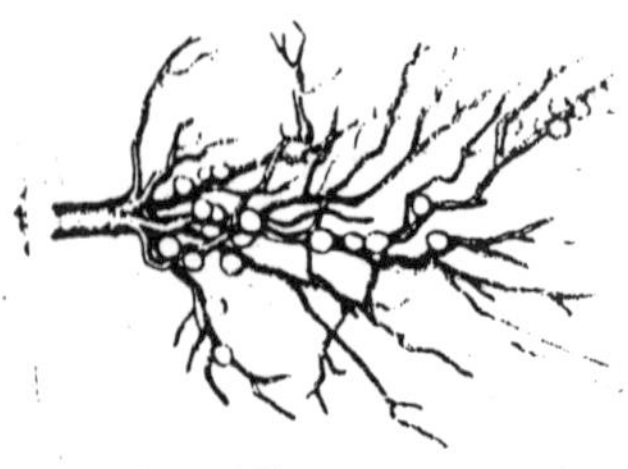

Fig. 6.5

(*h*) *Taproot eaten or hallowed* : Cutworms and chefer grubs will eat both fine roots and the main taproot of herbaceous plants, especially vegetables.

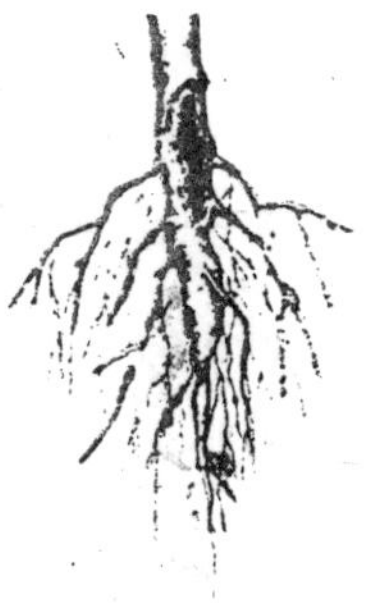

Fig. 6.6

(*i*) *Taproot tunnelled and eaten* : Symptoms are wilting and plant collapse, larvae of root maggots, and other diptera such as carrot fly maggots; with root crops, plants wilting does not normality occur.

Fig. 6.7

(*j*) *Rhizome bored* : Banana weevil larvae bore the rhizome extensively, and both pupac and adults may also be found in the tunnels; various fly maggots belonging to the chloropidac bore rhizomes of ginger.

Fig. 6.8

(*k*) *Yam tubers with wide tunnels* : Bored by adults and larvae of yam beetles.

Fig. 6.9

Damaged stems

(1) *Shoot and distal part of stem wilted and during :* The shoot may be bored by a caterpillar (spiny bollworm on cotton) or a long horn beetle larvae (coeloptera).

Fig. 6.10

(2) *Sweet potato stem (vine) bored :* Caterpillar of some clearwing moths and plume moths

Fig. 6.11

(3) *Tree trunk and branches bored, sometimes bark eaten externally:* Timber borers belong to two orders of insects, the Lepidoptera and coeloptera, metarbelidae feed externally at night on the bark under a silken tube coated with frass but retire during the day to a deep tunnel in the heart wood.

Fig. 6.12

(4) *Tree trunk or branch with single round emergence hole from 4-25 mm. diameter (longhorn beetle)* : With a pupal exuvium protruding from the hole (clearing moths).

(5) *Herbaceous stems and woody shoots bored,* swollen often with a single emergence hole, example, Sisal weevil and temparate cabbage stem weevil and turnip weevil.

Fig. 6.13

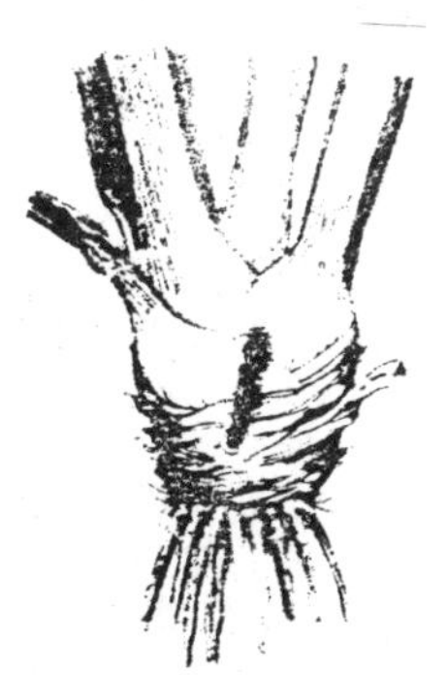

Fig. 6.14

(6) *Cereal shoots with dead-hearts :* Caused by boring larvae of shoot flies and caterpillars of pyralidae.

Fig. 6.15

(7) Buprestidae are called jewel beetles and the larvae known as flat- headed borrers, and the emerging adults leave an oval exit hole in the tree bark.

(8) Scolytidae (shot-hole and twig borers) belong to the group of ambrosia or fungus beetles; the adults bore into tree and make extensive breeding galleries under the bark; they carry fungus spores on their bodies with which to inoculate the new fungus galleries where the larvae feed,; some species bore down the centre of twigs on bushes such as tea. (Fig. 6.16)

Fig. 6.16

(9) *Twigs galled :* Made by feeding larvae of gall midges (Diptera), some gall wasps, old gall have multiple exit holes; a few twigs galls are made by weevil larvae and some woody aphids.

Fig. 6.17

(10) *Twigs-covered bags dangling from twigs and thin branches :* Bag worms (Lepidoptera) usually feed on the leaves but pupate with the bag firmly attached to twigs by a silken thong.

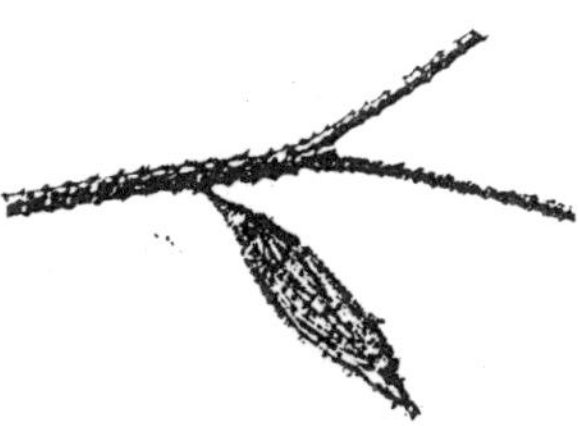

Fig. 6.18

(11) *Cereal and grass stems bored :* Caterpillars of the family pyralidae generally bore rice and grass stems, while the larger caterpillars of the noctuidae bore stalks of maize, sorghum and other larger species of Gramineae; tunnels in sugarcane are usually very short because the stem is solid and pithless. Stem sawflies bore apically in temperate cereals, especially wheat.

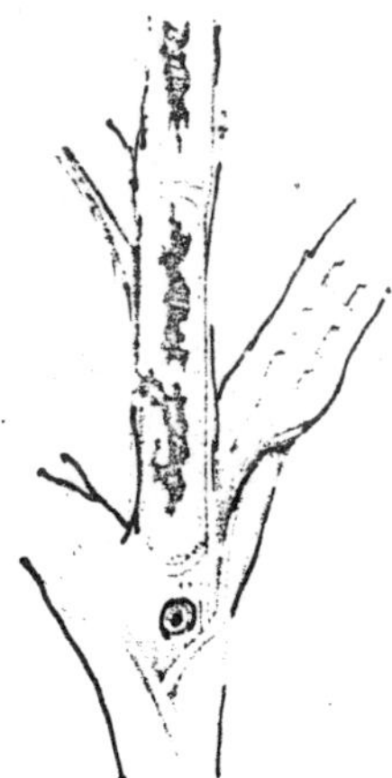

Fig. 6.19

(12) Bostrychidae are black, cylindrical beetles completely circular in cross- section and the tunnels are bored by the adult beetles called back-borers.

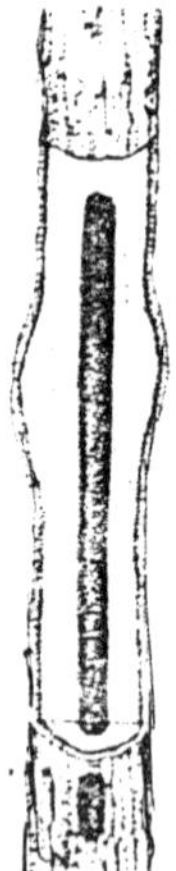

Fig. 6.20

(13) *Banana pseudosterm bored :* Banana stem weevil larvae make extensive tunnel galleries in which they pupate and adults may also be found.

Fig. 6.21

(14) *Woody stem with apical shoots killed and wilting :* Some Heteroptera feed on young shoots of woody shrubs and their toxic saliva enters the vascular system and kills the stem distally; shoots on trees are killed terminally in some regions by ovipositing cicadas, and also by some long-horned grasshoppers.

Fig. 6.22

Damaged Leaves

(*a*) *Margin with regular notches :* Many adult brood nosed weevils (Coeloptera) feed on the leaf margins of many different plants throughout the world, producing characteristics notches.

Fig. 6.23

(*b*) *With bubble -froth :* Either on leaves or in leaf axils; spittle mass built by nymphs of cercopidae for protection; spittle bugs.

Fig. 6.24

(*c*) *Lamina scarified :* Adults and nymphs of some thrips and some spider mites make tiny epidermal feeding lesions which give the lamina a silvery, bronzed or scarrified appearance; soft leaves will be caused to wilt. This is also seen in onions.

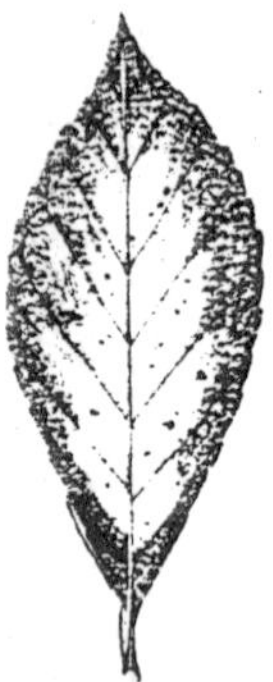

Fig. 6.25

(*d*) *Leaf curled under* or generally destorted, sometimes completely destorted into bunchy lump of tissue: the former damage is done by Aphididae, psyllidue and Homoptera the latter damage is by thrips.

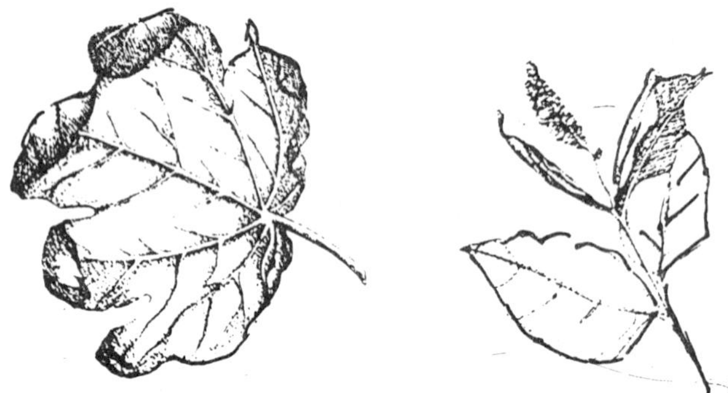

Fig. 6.26

(*e*) *Lamina cut and rolled* : Caterpillars at skippers, and other cut the lamina and make a leaf-roll, binding the edges with strands of silk; the leaf lamina is eaten within the protection of the roll; leaf rolling can be induced by some Aphididae, nesting spiders sawfly larvae.

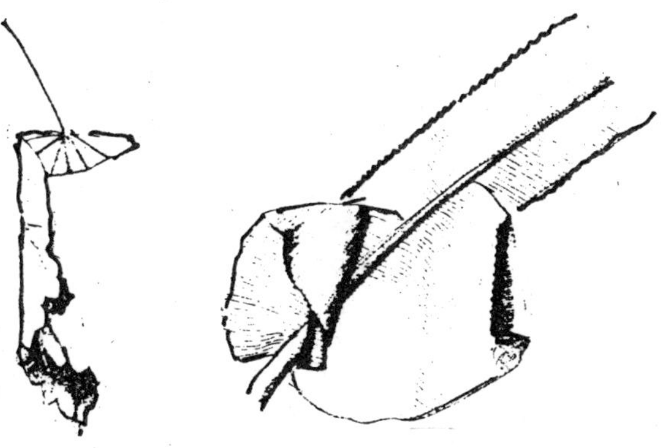

Fig. 6.27

(*f*) *Lamina pitted* : Psyllids cause ventral leaf - pits at the sites where the nymphs sit and feed. Young leaves some times may be considerably deformed.

Fig. 6.28

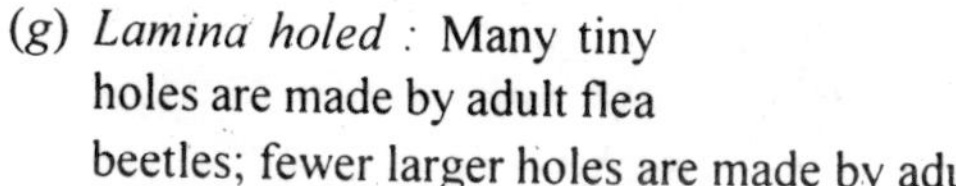

(g) *Lamina holed* : Many tiny holes are made by adult flea beetles; fewer larger holes are made by adults of tortoise beetles and caterpillars.

Fig. 6.29

(*h*) *Central leaves with longitudinal feeding scars* : Made by adult leaf beetles; lispine larvae mine inside the leaves.

Fig. 6.30

(*i*) *Lamina with erinia* : On the lower surface of some leaves are found wart-like out growths (erinia) inhabited by microscopic gall mites.

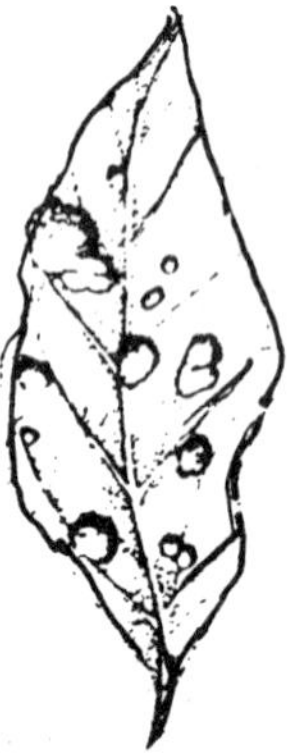

Fig. 6.31

(*j*) *Lamina skeletonized* : Caterpillars of several families like epiplemidue, Bombycidae espicially when young eat part of the way through the leaf lamina but leave the veins and one epidermids intact.

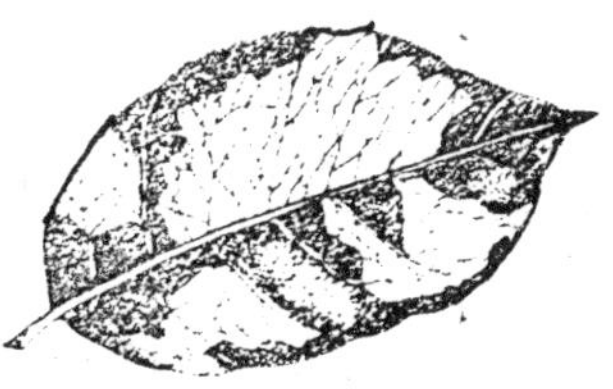

Fig. 6.32

(*k*) *Lamina windowed :* A window is a small hole in the lamina with one epidermis left intact, but after a while the thin epidermis dries and ruptures leaving a small hole; such damage is characterstic of larvae of the diamond-back moth.

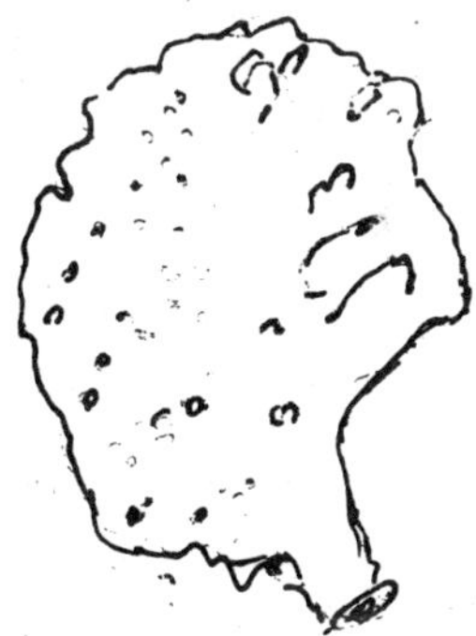

Fig. 6.33

(*l*) *Margin irrgularly eaten :* The commonest form of leaf damage by defoliating pests caused by grasshoppers, locusts many caterpillars, leaf beetles, sawfly larvae, some slugs and snails, birds, in a severe attack the entire leaf lamina may be eaten away, sometimes leaving only the main veins intact.

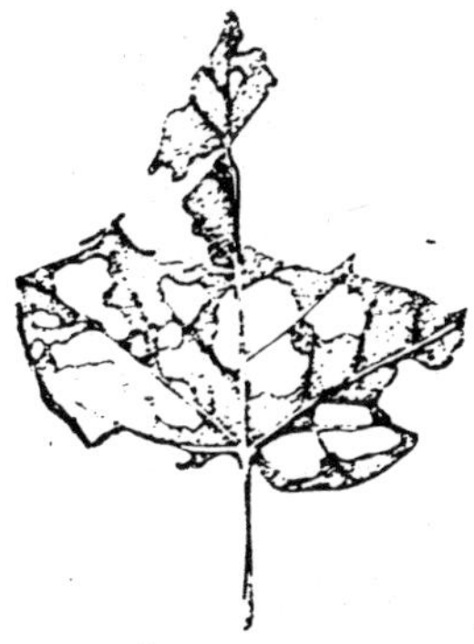

Fig. 6.34

(*j*) *Leaf galls :* Round or elongate galls small or large, few or numerous found on either upper or lower leaf surface, sometimes arranged randomly, some times distributed peripherally or alongside veins; made by feeding larvae of gall midges, gall mites gall wasps.

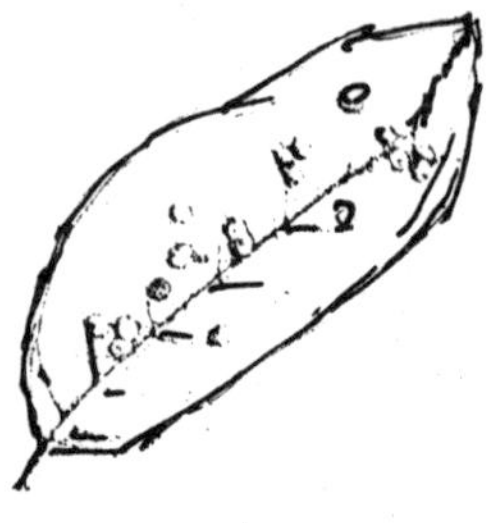

Fig. 6.35

(*k*) *Lamina tattered with irregular holes and tears :* Margin usually intact; money miridae and other Heteroptera feed on young leaves and their totic saliva results in small necrotic spots. As the leaf grows and expands the dead areas enlarge and tear, resulting in the characterstic tattering of the expanded leaf.

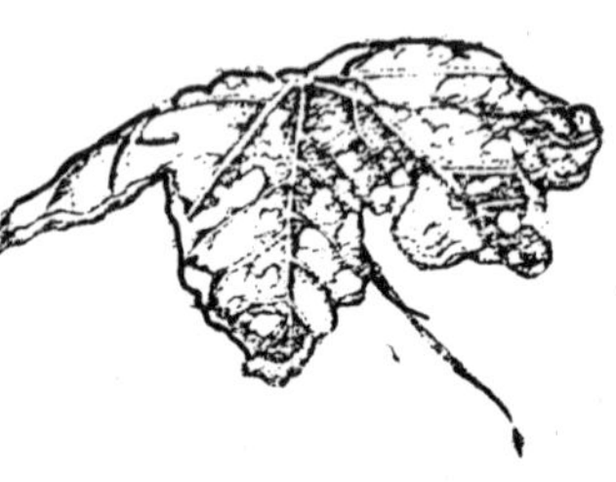

Fig. 6.36

(*l*) *Lamina mined :* The mines may be either tunnel mines or blotches, sometimes starting as a tunnel and ending as a blotch mine; tunnel mines with a central line of faecal pellets usually belong to caterpillars, tunnels without evident faecal pellets are made by maggots and beetle larvae.

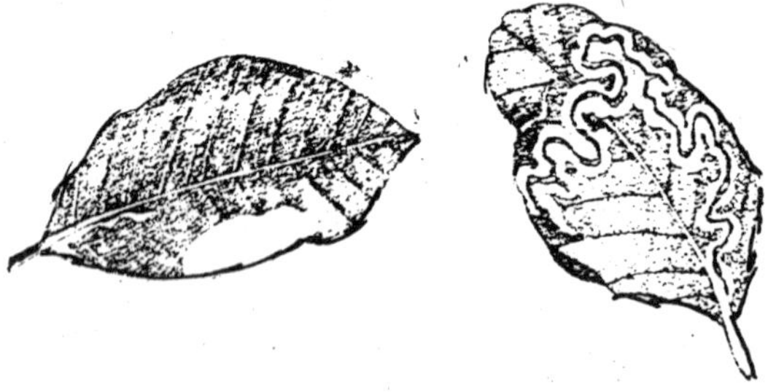

Fig. 6.37

(*m*) *Cereal leaves with serial holes :* These were the feeding sites of young cereal stem borers made when the young leaf folded prior to their penetrating the stem; as the leaves grow and expand, the tiny feeding sites become expanded into a series of matched holes.

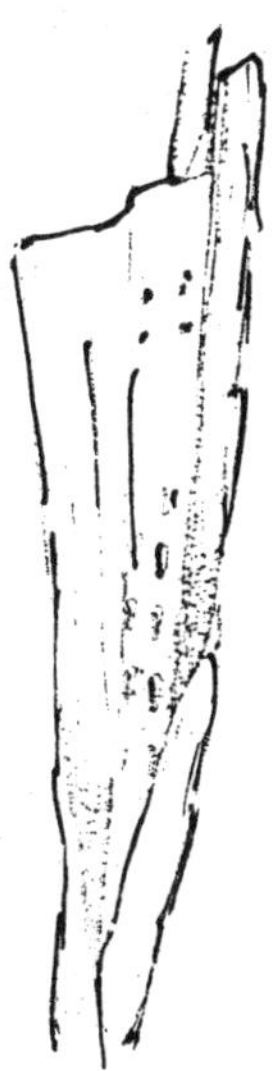

Fig. 6.38

Damaged Flowers and Buds

(*a*) *Petals grawed :* Adult blister beetles chew petals of many plants, often common on malvaceae, adult flower beetles make small holes in petals, *Popillia* being especially injurious.

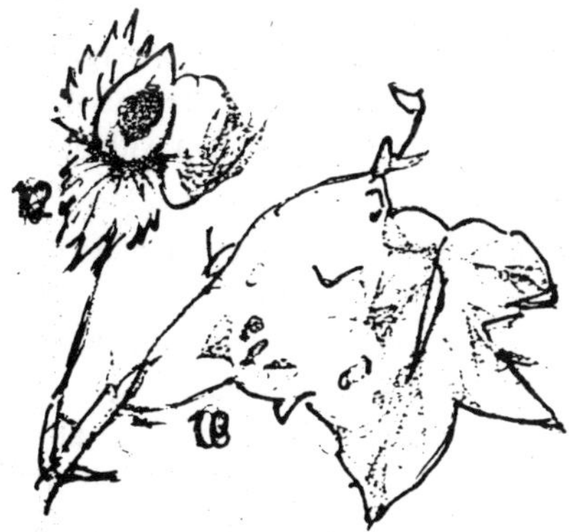

Fig. 6.39

(*b*) *Buds bored :* Caterpillars of some tortricidae bore into large buds of shrubs and trees.

(*c*) *Buds grawed with large holes :* Eaten by large caterpillars (*e.g.*, cotton semi- looper).

(*d*) *Buds and shoots stunted/wilting :* Heavy infestations of aphids, scales, mealybugs, can stunt young shoots during active growth; Hibiscus mealybug is unusual in causing severe shoot telescoping and often shoot death.

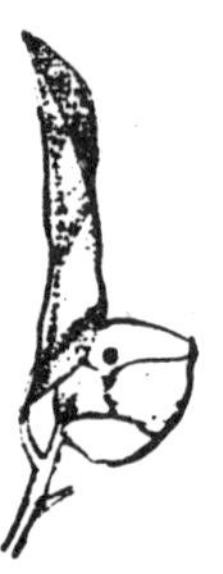

Fig. 6.40

Fig. 6.41

Damaged Fruits and Seeds

(*a*) *Cereal panicle with few grains developed or grains small and distorted :* Larvae of sorghum gall midge small distorted grains results from feeding of sap-sucking Heteroptera with toxic saliva.

(*b*) *Fruits nuts and seeds bored :* Bored internally by weevil larvae and hole made by emerging adult examples are mango weevil, Hazel-nut weevil, cotton boll weevil, maize weevil in small ovaries one weevil larvae will eat all the endosperm and thus each ovary contains only one weevil pupa (ex. clover seed weevils); several maize weevils can develop in one maize kernel.

Fig. 6.42

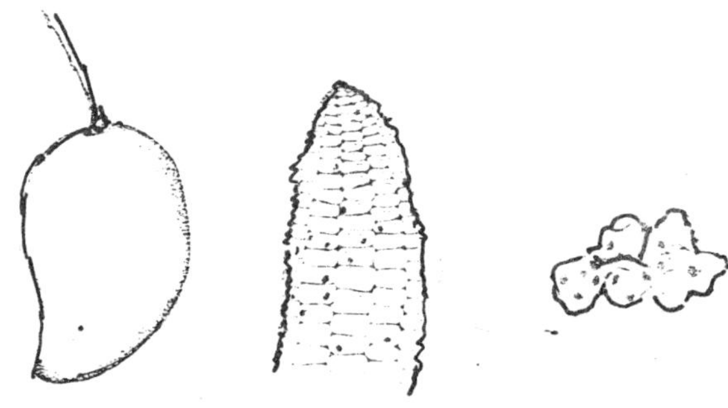

Fig. 6.43

(*c*) *Fruits with numerous small necrotic patches :* Heteropteran bugs have toxic saliva, and when feeding on fruits they cause necrosis at the feeding sites which usually becomes infected with fungi and bacteria, resulting in rotting, death and premature fruit-fall *e.g.*, cocoa capsid cotton stink bugs and coconut bug.

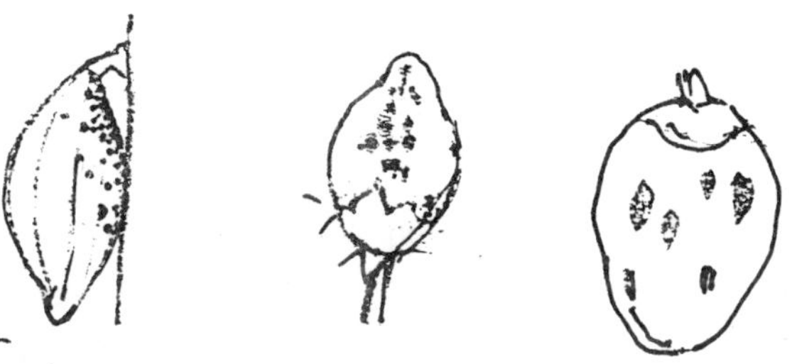

Fig. 6.44

(*d*) *Fruits with large internal tunnels sometimes large holes to the exterior :* Sometimes infected with fungal and bacterial rots sometimes fross expelled; made by caterpillars of the families tortricidae, pyralidae and noctuidae, eggs are laid externally, and the tiny caterpillars (first instar) bore into the fruit sometimes infested fruits fall prematurely, cotton bolls bored by many bollworms; top fruit bored by many different larvae; pulse pods attacked by two basically different groups of Lepidoptera: small caterpillars of tortricidae Lycaenidae by Lycoenidae and pyralidue which live inside the pods and feed on the developing seeds and the larger caterpillars of the noctuidae (Heliothis etc,) which are two large to live inside the pulse pods, so bore holes into the pods in the vicinity of the seeds which they eat by pushing the anterior part of their body into the pod cavity. One large caterpillar may bore a number of holes in several pods.

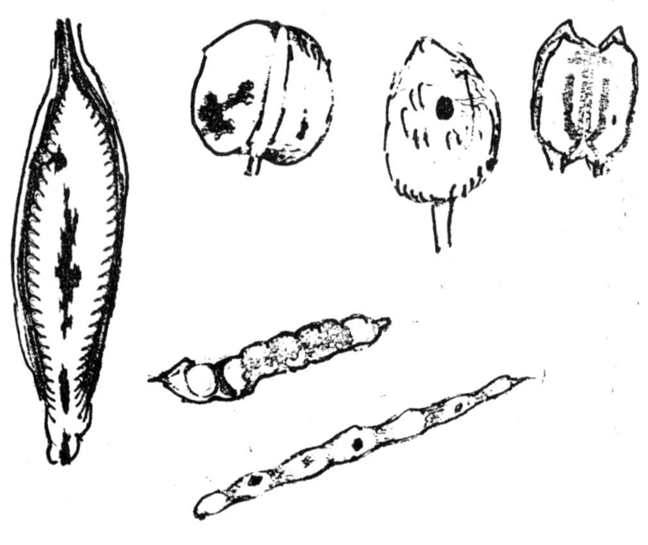

Fig. 6.45

(*e*) *Fruits webbed and qnawed* : Some caterpillars will spin silk over young fruits and feed on the fruits (coffee Berry moth); sesame webworm spins silk over the pod and shoot before boring the pod.

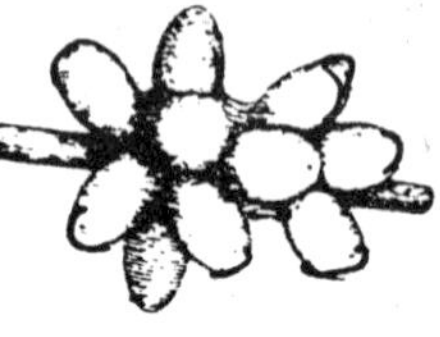

Fig. 6.46

(*f*) *Fruits with maggots inside a necrotic area and sometimes small holes* : Fruits flies attack almost all larger types of fruit; eggs are laid under the skin of the young fruit and the fly larvae develop internally as the fruit ripens; the fruit often falls prematurely, pupation usually takes place in the soil; secondary infection of the tunnels and oviposition site by bacteria and fungi is common; typically there is neither frass hole, entrance nor exit hole evident in the skin of the fruit while still on the tree but sometimes there may be sap exudation.

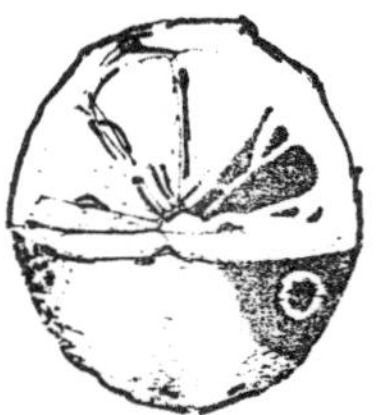

Fig. 6.47

(*g*) *Fruits tunnelled and bored* : Adults of scolytidae bore host plants to make breeding tunnels and galleries in which the larvae develop; coffee berry borer bore coffee berries. Bruchidae (adults) enter ripened pulse pods which have split open; the larvae develop inside the ripe seeds, finally pupating beneath a thin tramslucent 'widow' before the emerging adult makes the final exit holes.

Fig. 6.48

(*h*) *Fruit deformed* : Various phytophagus mites feeding on flowers and young fruits cause distortion and deformation of fruits.

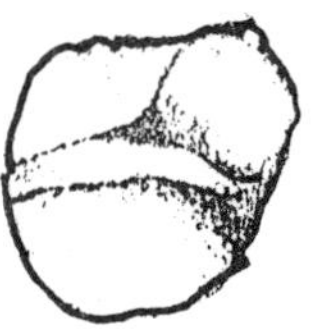

Fig. 6.49

Damage to Sown Seeds and Seedlings

(*1*) *Cereal seedling stem with dead heart* : Maggots of various diptera bore into the young stem, usually killing the growing point and making the apical leaf turn brown and die; some caterpillars also bore seeding stems of graminaceous plants, but typically they attack older plants.

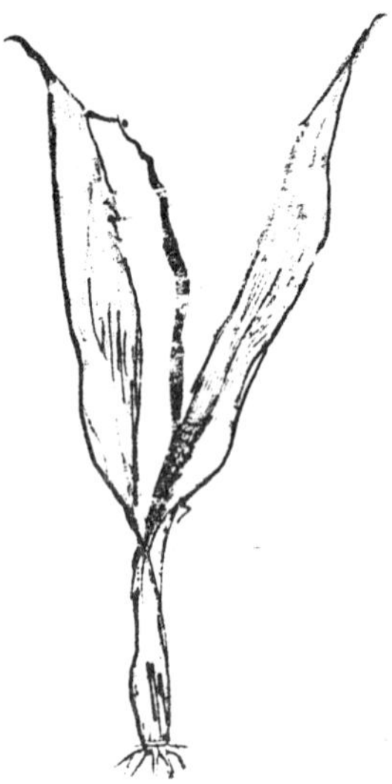

Fig. 6.50

(2) *Stem gnawed at about ground level or below ground :* A few beetles belonging to small families and a few adult scarabs.

(3) *Stem bored usually swollen :* Several species of Diptera have larvae that bore in the stem of various seedings, best known is probably the beam fly widespread and abundant on legumes, it will also bore into pitioles on beans.

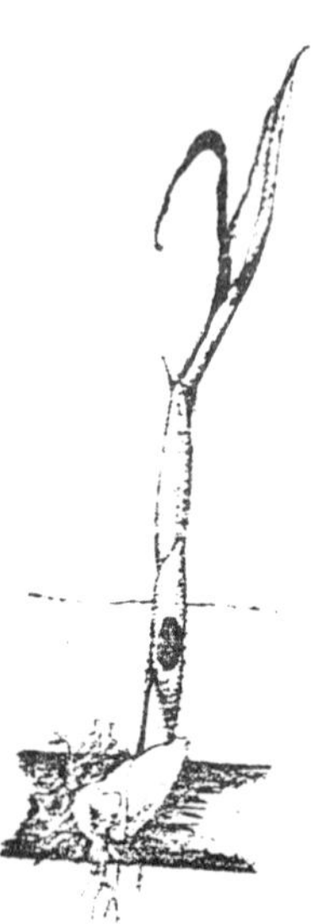

Fig. 6.51

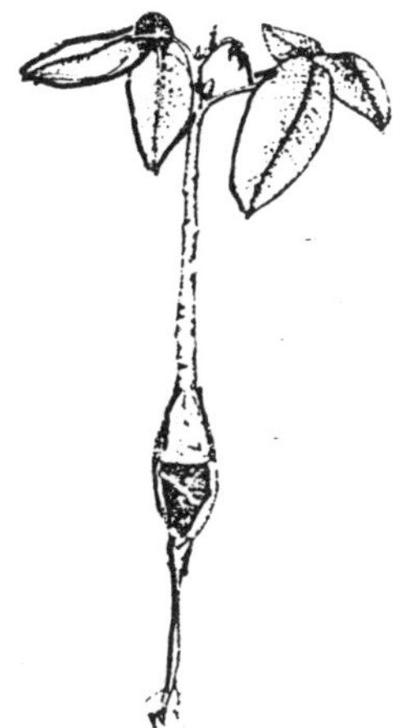

Fig. 6.52

(4) *Woody seedling or young plant wilting with earthtube up part of the stem; under the tube the bark is eaten away by termites or ants.*

Fig. 6.53

(5) *Cotyledons of large seeds bored and eaten* : Larvae of bean seed fly, bored into cotyledons, epicotyl and hypocotyl and prevent germination; similar damage done by small species of slugs.

Fig. 6.54

(6) *Seed dug up and eaten :* Various species of birds (sparrows etc.) and mice.

(7) *Cotyledones or first leaves pitted and eaten :* Adult flea beetles make a shot-hole effect on seedlings of cruciferae, cotton and other crops, frequently stunting and killing the seedling. Stem severd with shoot lying alongside, which is seen in typical cutworm and cricket damage, often several cousecutive seedlings are cut through; with crickets the cut plant is left for a day to wither and the next night is pulled down into nest; such damage may be done by surface slugs.

Fig. 6.55

(A)

AGRICULTURAL PESTS

Jowar Stem Borer

Scientific Name : *Chilo zonellus* (Spotted stalk borer)

Family : Pyralide

Host (main) : Sorghum, maize, bulrush, millet, sugarcane and rice. Alternative; Various species of wild grasses.

Type of Damage

Larvae which feed on leaves cause many small holes on the lamina of leaves, later catterpiller bores into the stem and damage the central pith of the stem which results in the death of upper part of stem which is called *dead heart* disease.

In India though it is primarily pest of sorghum, it occurs on rice in Assam, West Bengal, Bihar and Orissa.

It is one of the major pests on full-grown crop of sorghum which causes infections even after the formation of ear head in the plant. Usually the infection is found to be occurring in young plants is far more serious infection becomes more prominent in rabi crop and hot weather crops but found to be active throughout the year. Hybrid sorghum varieties are found to be more susceptible than the local varieties.

It causes damage upto 80 per cent.

Life History

The eggs are flattened ovoid and scale-like, about 0.8 mm long. They are usually laid on the underside of the leaf near the midrib in 3-8 imbricated rows in groups of 50-100. Hatching takes 7-10 days.

The young larvae migrate to the top of the plant where they mine the sheaths and tunnels in the midrib for several days producing characteristic leaf windowing. They then either bore down inside of the stem and bore into it just above an internode. In older plant the larvae may live in the developing heads. Larval development takes 28-35 days. The mature, catterpiller is 25 mm long, buff coloured with four longitudinal strips and a brown head capsule and thoracic shield.

Pupation takes place in the stem in a small chamber and takes 7-10 days.

The adult moths are not large being 20-30 mm across the wings. The male is smaller and darker than the female. The male has fore wings pale brown with dark brown scale forming a streak along the costa; the hind wings are pale straw colour. The female has much paler fore wings and hindwings almost white. The adults are shortlived.

The life-cycle takes about 29-33 days and there are at least six generations per year.

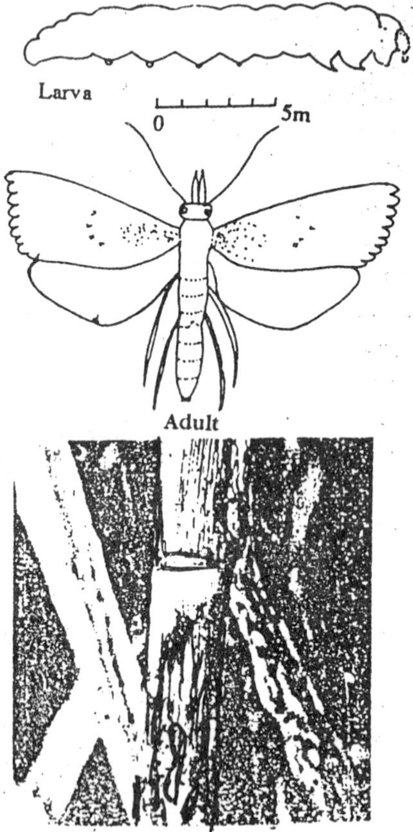

Fig. 6.56 : Jowar stem borer.

Distribution

East Africa, Sudan, Malawi, Afghanistan, India, Sri Lanka, Nepal, Bangladesh, Sikkim & Thailand.

Fig. 6.57 : Distribution of *C. Zonellus.*

Control

As borer will be present in the stem it is too difficult to control this disease. Following preventive measures are recommended for controlling the disease.

1. Removal of dead shoots and destruction of caterpillars.
2. After harvest removal and burning of stubbles of crop will destroy the hibernating larvae.

Chemical control of *Chilo* is not generally very successful. The insecticides used generally are DDT or endrin as dust or sprays, foliar sprays of parathion andparathion methyl are successful but their mammalian toxicity is too high for safe use on large scale. Endrin, dieldrin, trichlorophon, dichlorvos and diazinon have also been successful but several applications at weekly intervals are required. Application of two or three rounds of sprays of carbaryl 0.1% at 15 days intervals from a month after sowing minimises the incidence of the pest. The application of sevidol (4 : 4) or Thiodan 4% granule on 25th and 35th day of sowing at 8 and 10 kg respectively in leafwhorls controls the pest.

Biological Method : Natural enemies

(i) *Trichogramma minutum*

(ii) *Brecon chinensis* (egg parasite)

(iii) *Microbracon chilocida*

(iv) *Xanthopimpha* sps. (larval parasite)

(iv) *Apanteles flavipes*.

Blister Beetles (Flower Beetles)

Family : Meloidae

Hosts (main) : Pulse crops.

Alternative : Cotton, and many flowers and ornamentals (*e.g.*, Hibiscus)

Mylabris phalerata, M. pustulata and M. balteata are common species in India.

Type of Damage

The flowers are eaten by the adult beetles, causing a loss of pods in leguminous crops and conspicuous damage to various flowering ornamentals.

Life History

Eggs are laid in the soil in batches, the first instar larva that hatches out from the egg is active with long legs and is referred to as triungulin. It moves in search of eggs of grasshoppers and bees. In the species that

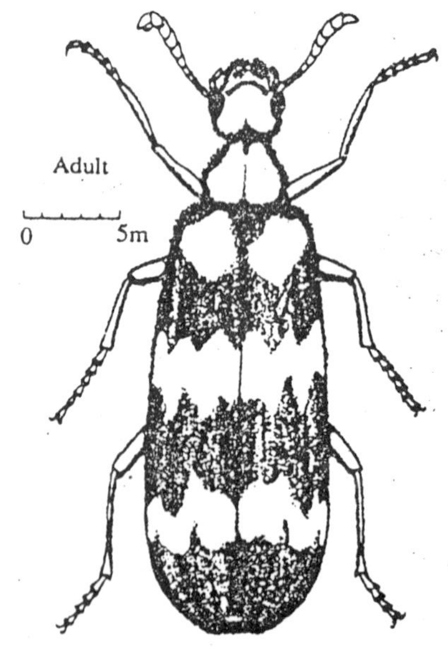

Fig. 6.58 : Blister beetle (adult).

develops in bee's nest, the triangulin climbs upon a flower and from there gets attached to the visiting bee and transported to the nest. The triungulin feeds on the eggs of its host and moults. In the second instar it bears short legs. In the third, fourth and fifth instars it is somewhat scarabaeiform. The sixth instar is coarctate forming a pseudopupa devoid of functional appendages and usually hibernation takes place in this instar. This is followed by small, white and apodous larva of seventh instar, which transforms to a pupa soon.

Pupation takes place in the soil.

The adult beetles are medium sized, 1.25 to 2.5 cm long, and conspicuous covered with bright metallic blue, green, black or brown. The head is hypognathous and is joined to the prothorax by a distinct neck. The wings are well developed and loosely cover the body; in some they are vestigial while others are apterous. The legs are long, appendiculate and usually with serrate tarsal claws. When disturbed the beetles emit a fluid Containing the oil principle, cantharidine which has irritant properties, through the openings in the apices of the femur; the effect of which upon the skin accounts for the common name of 'blister beetle'. The beetles are usually found in flowers and leaves and cause injury by feeding on them.

Distribution

Africa, India, Bangladesh, Sri-Lanka and S. E. Asia.

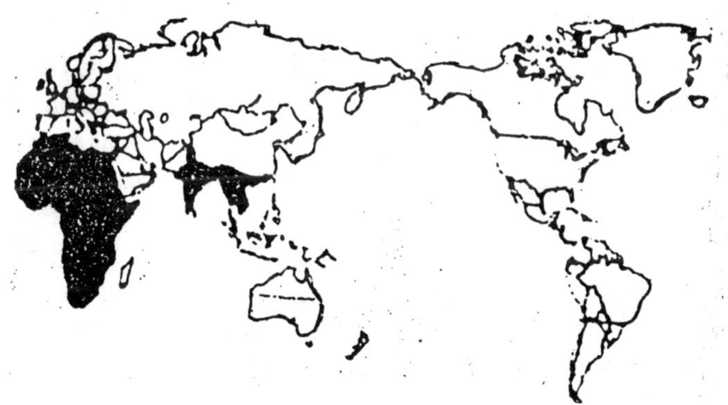

Fig. 6.59 : Distribution of Blister beetles.

Control

The adult beetles are: difficult to control because of their mobility; also the damage done is usually only slight and so chemical control is seldom warranted commercially, but it is expected that DDT sprays might be effective.

Pyrilla Perpusilla Wlk

Common name : Indian Sugarcane Leafhopper.

Family : Lophopidae

Order : Hemiptera

Hosts (main) : Sugarcane.

Alternative : Pearl millet (bajra), sorghum, maize, wheat and barley.

Pyrilla is a serious pest of sugarcane all over India.

Type of Damage

Adults and nymphs suck the sap from the leaves of sugarcane, which become yellowish and dry. Plant growth is arrested and even the canes dry up. The infested leaves appear white in colour while the leaves below are black due to sooty mould growing on the honeydew secreted by the nymphs. The yield of canes, sugar recovery and also quality of jaggery (gur) suffer badly; in severe cases, a loss of 30-40 % of sucrose can occur.

Life-cycle

The eggs are white, ovoid, 1.0 × 0.5 mm, and are laid in clusters of 20-25 on the undersurface of the leaves and are covered with a white waxy material. A single female lays 600 -800 eggs during its life-time. They hatch in 7-12 days. The newly hatched, nymph is pale-brown with a pair of long wax-covered and processes. The nymph jumps from leaf to leaf, grows by sucking sap from the leaves passing through five instars in 20-30 days. The average adult life is 6-8 weeks for the female and 4-6 weeks for the male. Three to four generations are completed in one year.

The adult is a straw-coloured bug, about 1 cm long with the head projecting forward in the form of a rostrum. Forewings show conspicuous veins and scattered tiny dark spots.

P. perpusilla appears in March-April and continues till December; the maximum activity of the pest being during July- September. The factors which favour the multiplication of the insect are easterly winds, heavy manuring, copious irrigation resulting in luxuriant growth and high humidity.

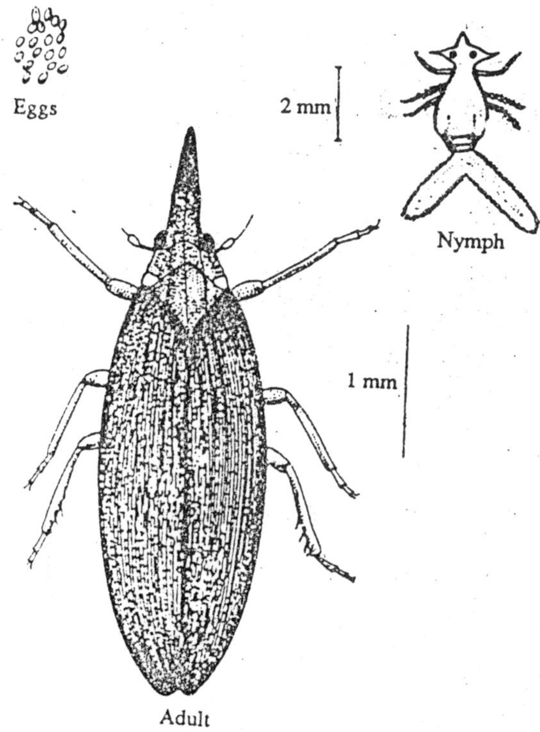

Fig. 6.60 : Life-stages of *Pyrilla perpusilla*.

Distribution

Pakistan, India, Sri Lanka, Afghanistan, Burma, and Thailand.

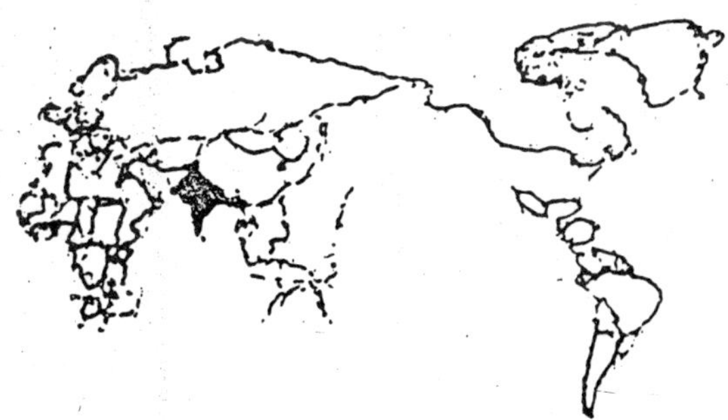

Fig. 6.61 : Distribution of *Pyrilla*.

Control

Resident varieties of sugarcane which are hard, non succulent and narrow-leaved should preferably be grown. All sprouts, stubbles and trash should be destroyed immediately after harvest. Conserve and introduce eggs parasites and destroy egg masses during April-May. Dust 5% BHC or spray 0.03% endrin, monocrotophos, dimethoate, phosphamidon or methyl demeton, especially on the undersurface of the leaves.

Biological Method : Natural enemies

(i) *Tetrastichus pyrillae*

(ii) *Cheiloneurus pyrillae*

(iii) *Ageniapis pyrillae* (egg parasite)

(iii) *Epipyrops melanolecca*

(iv) *Cestaryinus Prillae* (nymphal parasite).

Red Cotton Bug (Cotton Strainer)

Scientific name : Dysdercus cingulatus

Family : Pyrrhocoridae

Hosts (Main) : Cotton.

Alternative : Many plants and trees including sorghum, Hibiscus spp., Abutilon, and other malvaceae plants.

Type of damage

There are conspicuous red bugs on the cotton bush, which fall to the ground if the bush is shaken. Small green bolls may and go brown, due to death of the seeds, but they are not shed. No damage is visible externally on large green balls but if the inner boll wall is examined; warty growths or water soaked spots can be seen corresponding to patches of yellow staining on the developing lint. In very severe attacks the whole lock may be brown and shrunken.

Life History

The eggs are ovoid, 1.5 × 0.9mm, yellow when laid but turning orange. Hatching takes about 5-8 days. They are laid in batches of about 100 in moist Soil or plant debris; moisture is essential for development and the eggs die if the soil dries out.

There are five nymphal instars. The first instar nymphs do not feed but require moisture and usually congregate near the empty shells. The second and third instars feed gregariously on seeds on or near the ground. Later instars wander freely over the plant seeking suitable fruits and seed. Nymphs are often found in large numbers on tree trunks, etc. where they prefer to moult. The full grown nymph is a bright red bug with black wing pads, and is about 10-13mm long, according to species. The total nymphal period lasts about 21-35 days.

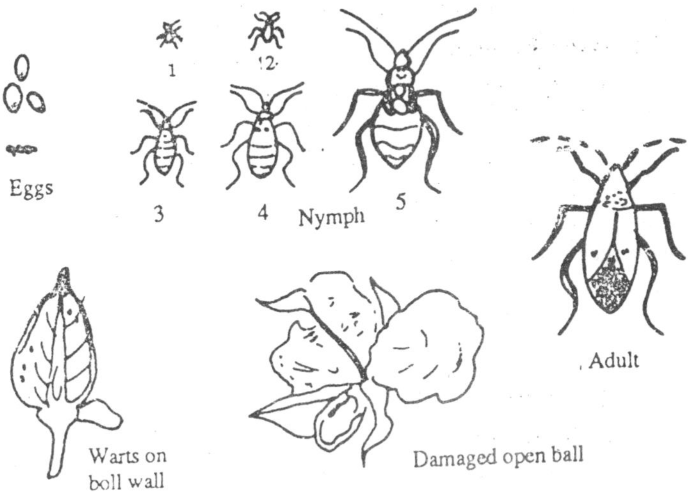

Fig. 6.62 : Life-cycle of Red cotton bug.

The adult male strainer is 12-15mm long, according to species; the female is slightly larger. The wings are reddish and each has a black spot or bar near the middle. Stainers are able to fly strongly but usually drop to the ground and crawl if disturbed on the bush. The bugs feed by sucking sap from the seeds; piercing and sucking proboscis is pushed through the boll wall and into the seeds. More eggs are laid by the female if she feeds on mature exposed seeds. After a pre-ovipositive period of 5-14 days adult females may live a further 60 days and lay a total of 800-900 eggs.

Distribution

Found in the cotton-growing areas of tropical Africa, tropical Asia, Australia, USA, Central and South America.

Fig. 6.63 : Geographical distribution of Red cotton bug.

Control

Cotton stainers may be controlled by caging chickens in cotton plots using chicken wire; about 15 birds will keep about 0.1 ha free of stainers. This method should not be combined with a chemical treatment.

The usual insecticides are either carbaryl as a spray or a dust mixture of BHC and DDT (5%).

Cultural Method

(i) Plough the cotton field to expose the insect eggs.

(ii) Hand picking of insects is also very effective.

Biological Method

(i) *Harpactor costalis*

Pest of Castor

Scientific Name : *Achaea janata*

Comnon Name : Castor semilooper

Family : Noctuidae

Order : Lepidoptera

Host (main) : Castor

Alternative : Citrus plants.

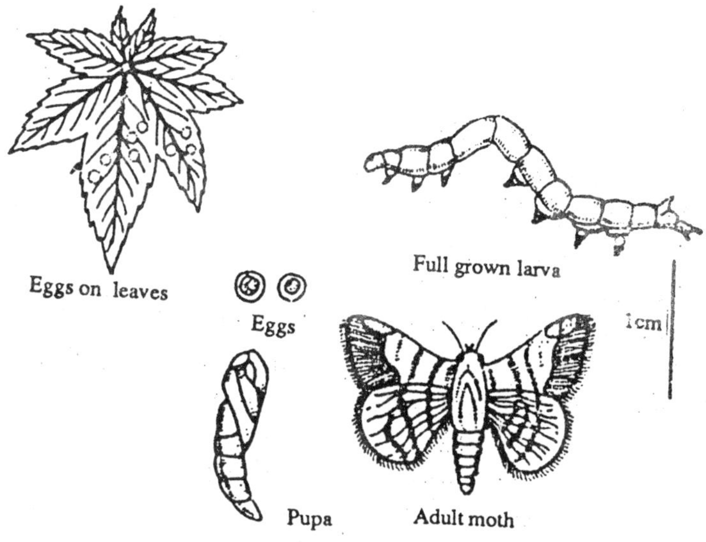

Fig. 6.64 : Life-stages of *Achaea janata.*

Type of Damage

The caterpillar is a voracious leaf-eater of castor, starting from the margins, eating inwards behind only the mid-ribs and the stalks. Maximum damage is caused by the second and third instar caterpillars; with the excessive loss of foliage; the seed yield is drastically reduced.

Life Cycle

The adult is a greyish brown moth with wavy lines on the forewings and black and white blotches on the hind wings. After a preoviposition period of a week or so, eggs are laid singly on tender shoots and leaves. The eggs are elongate, and bluish-green. They hatch in 3-4 days. The larva feeds on the leaves of the castor and is full grown-in about 2 weeks. Newly emerged larvae are white or green with a light brown posterior region. There are 4 instars, each taking 2-4 days. The full-grown caterpillar is about 7cm long and brownish black in colour with pale-white stripes. It pupates in soil or in leaf folds. The pupal stage lasts about 10 days. After, emergence, moths feed on citrus fruits by piercing the skin and feeding on the juices. Five to six generations are completed in one year.

Control

Control measures against this pest include the collection and destruction of the larvae. Dusting 5 per cent BHC or spraying 0.05 per cent endosulfan or 0.1 per cent carbaryl can control the pest. In biological control the larval parasites *Apanteles sudanus* Wlk., *A. radius* Wlk. and *Tetrastichus ophiusae* Craw and egg parasites *Trichogramma achaeae* and *Telenomus* sp. are successful.

Brinjal Shoot and Fruit Borer

Scientific name : Leucinodes orbonalis

Family : Pyralidae

Order : Lepidoptera

Hosts (main) : Brinjal

Alternative : Other solanaceous plants and peas.

Types of Damage

The pest starts damaging the brinjal plant a few weeks after its transplantation. When the shoot is attacked by the caterpillar it droops and withers (shrivel), finally drying up. When the petioles of the leaves are bored into by the larva the leaves wither and drop. The attacked fruits show holes on them, plugged with excreta. Upto 70% loss of crop is caused by this pest.

Life Cycle

Larva

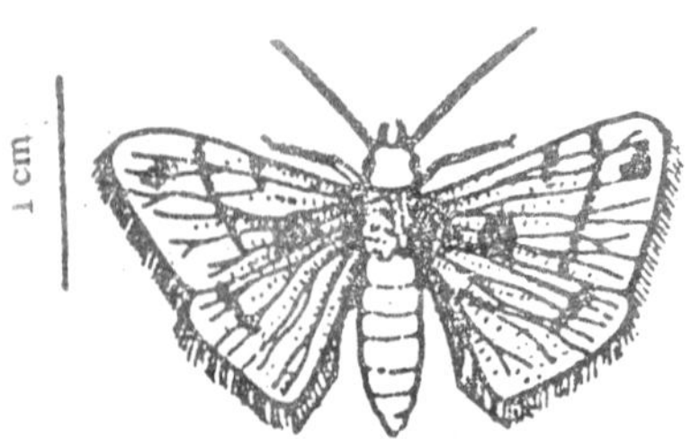

Adult

Fig. 6.65 : Life-stages of *Leucinodes orbonalis*.

The adult is a grey-brown moth, with blackish spots on the dorsum of the thorax and abdomen. The wings are white with pinkish-brown markings and are fringed with small hairs along the apical and anal margins. The moth lays elongate eggs singly or in small batches, on the leaves, shoots and fruits. They hatch in 3-5 days. The larva is a borer within the shoot, leaf midrib, petiole and fruit and feeds on the internal tissues. The larva undergoes 5 moults in 10-15 days. The fifth instar caterpillar is stout, pink, and measures about 1.6 cm in length. Pupation takes place in a cocoon on the plant and lasts for 6-8 days. Moth lives 2-5 days and the female lays upto 250 eggs. The larva is parasitised by *Pristormerus testaceus*. Morl., *Cremastus flavoorbitalis* and *Bracon* sp.

Control

Prompt collection and destruction of the plant parts harbouring the larvae help in reducing the infestation. Persistent insecticides such as carbaryl, quinalphos, endosulfan, Lindane and diazinon, when applied at regular intervals, give relief from heavy-infestations. However, precautions need to be taken to prevent toxic residues of the insecticides being left on the fruits. Therefore, remove all fruits before spraying and do not harvest any fruits for the next 3-4 days.

Aphids or Plant Lice

Family : Aphididae

Hosts : All vegetables including those of the cabbage family cucumbers, melons, beans, peas, potatoes, tomatoes, lettuce, turnips, spinach, and other garden crops have serious aphid pests.

Type of Damage

Apid is a large group of small, soft-bodied, delicate, pear-shaped insects that infest the stem or leaves of a variety of plants in large numbers and suck up the sap causing serious damage. A large number of them are economically important as pests of crops. The cotton aphid *Aphis gossypii*, the lab-lab or bean aphid *Aphis craccivora*, the wooldy aphis *Eriosoma lanigera*, the banana aphid *Pentalonia nigronervosa*, the peach leaf curl aphid *Brachycaudus helichrysi*, coconut aphid *Cerataphis variabilis* etc. are some common pests to mention here. They all feed by thrusting sharp hollow stylets, from their beaks, in among the cells of the plant and sucking out the sap. This causes the blighting of buds, dimpling of fruits, curling of leaves or appearance of discoloured spots on the foliage. As the aphids become more abundant, the plants gradually wilt from loss of sap and possibly safe by being poisoned by the saliva of the aphids, the leaves may become yellowish or brown and the plant usually dies.

The presence of aphids makes vegetables unattractive and detracts from their flavour and market value; and when they occur even in small numbers much work is to be done to remove them in preparing the vegetables for use. They excrete a honeydew from their intestines that gums up the plants and often serves as a medium on which a fungus may grow, that further spoils vegetables. Recently it has been shown that aphids are important natural agencies in spreading certain viral plant diseases, such as mosaics, and blights; and this phase of damage may exceed in seriousness through direct injury by feeding.

Distribution

As a group, world-wide and many species are nearly cosmopolitan.

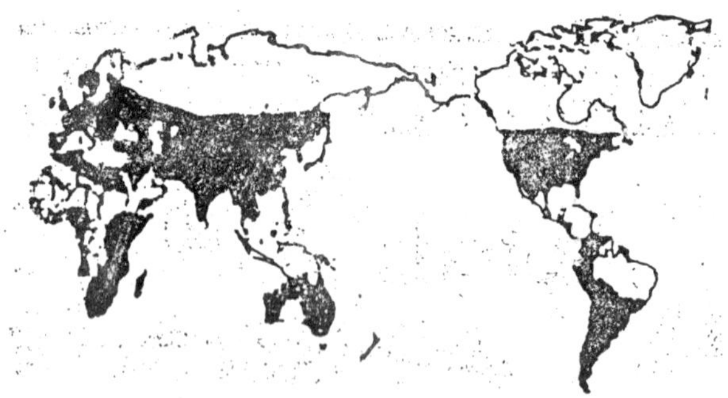

Fig. 6.66 : Geographical distribution.

Life History

Appearance and Habits—Aphids typically winter as fertilized eggs on some perennial plant; some winter on the dead remnants of annual vegetables; while the overwintering condition of some species is not known. The eggs are small, ovate blackish objects, glued on their sides generally to the stems of plants or in crevices about the buds. When the weather becomes warm enough, small nymphs hatch from the eggs, which grow quickly to full size but never get wings. Since each of these is the start for a great colony of aphids that may be produced during the season, they are called *stem-mothers*. They are all females which have the remarkable ability to reproduce young like themselves without mating. These young are born ovoviviparously, *i.e.*, already hatched from the egg; and differ from their stem-mothers, in having-only one parent and in not passing through an exposed egg state. They are like the stem-mothers in being wingless and in producing young

Fig. 6.67 : Garden aphids.

ovoviviparously, in beginning when they themselves are only a week or so old and producing from a dozen to 50 or 100 active nymphs within the next week or two. In this way a succession of generations is produced, the young clustering about their mothers until patches on the plant-may be crowded with them. At some time during this period, either all or a part of certain generations of these females may develop wings. These may fly to other plants of the same kind, or in some species they habitually fly to a different kind of plant (usually an annual), known as the *summer host*. Such winged ones are called *spring migrants*. They settle down on the new host plant and start a succession of generations there; all are produced as before from unfertilized eggs that hatch in the body of the mother.

As shortening days forecast the end of the season, and before the summer-host plant dies, a generation is usually produced that is all winged but is often of two kinds. Some of them are winged males, the first appearance of males in the aphid colonies being at the approach of cold weather. The others are winged females which are called *fall migrants* and which may serve to return the species to the kind of perennial plant from which their distant ancestors flew away in the spring. These fall migrants give birth to nymphs in the normal manner, but the nymphs when grown, are wingless true females that cannot reproduce unless they mate with the males which are of the preceding generation. After mating, the true female lays from one to four or more large fertilised eggs in a sheltered place about the plant, and dies; or sometimes, simply dries up about the single egg she is capable of maturing. From these eggs arise the stem-mothers of the next spring, which differ from all the hosts of other aphids produced during the year in having both male and female parents. In some species the males and true females have no mouth parts.

This is a kind of standard life-cycle. Variations from it will be noted in the discussion of particular species, but the essential features of this life-cycle should be fixed in mind because the details are not repeated for the aphids discussed under particular crops.

Control Measures

Aphids may be satisfactorily controlled by dusting with:

(a) 1 to 4 per cent actual nicotine;

(b) 0.5 to 1 per cent parathion;

(c) freshly prepared 0.5 per cent actual tetraethyl pyrophosphate; or by spraying with;

(d) 0.05 to 0.06 per cent actual nicotine sulfate is soapy water (see Fig.);

(e) freshly diluted 0.02 to 0.05 per cent actual tetraethyl pyrophosphate in water. Other materials which have proved effective against many species of aphids are—dusts of 0.5 to 0.75 per cent rotenone, 1 to 2 per cent γ-benzene hexachloride, and sprays and dusts containing the pyrethrins and the organic thiocyanates.

Heavy infestations of aphids will usually require several applications of these materials at weekly intervals for control. For certain low-growing dense crops such as melons, turnips, and the dusting is superior to spraying because the dust will circulate among the plants and even penetrate into curled leaves and reach the aphids in positions where a spray could not be driven. Benzene hexachloride and parathion should not be applied to edible portions of crops later than 30 days before harvest.

Biological Method

(i) *Minochiles sexmaculata.*

(ii) *Ishiodon javana.*

Mango Stem Borer

Scientific Name : *Batocera rubus (L)*

Family : Cerambycidae

Hosts (Main) : Mango, fig, jackfruit

Alternative : Various other tree species.

Type of Damage

The larvae burrow through the sapwood under the bark of the trees, either on the trunk on the main branches. Frass expulsion holes are made at intervals and sometimes sap oozes out of the holes, making obvious symptoms. On damaged branches the foliage may die and fruitset will be impaired.

Life History

Eggs are laid singly in the bark of the tree, upto 200 in total, into small cuts made by the female's mandibles. The young larvae bore straight into the wood, feeding in the vascular tissues of the tree, eventually making long irregular tunnels in the sapwood. In multiple infestations the flow of sap and water is interrupted, which has obvious adverse effects on the tree. Pupation takes place within the larval tunnel system, usually just under the bark.

The adult is a greyish brown beetle, about 3-4 cm in body length, with a series of conspicuous white spots on the elytra. The adults live for several weeks and feed on the bark of the tree to some extent.

The complete life-cycle could be as short as one year, but might be two years.

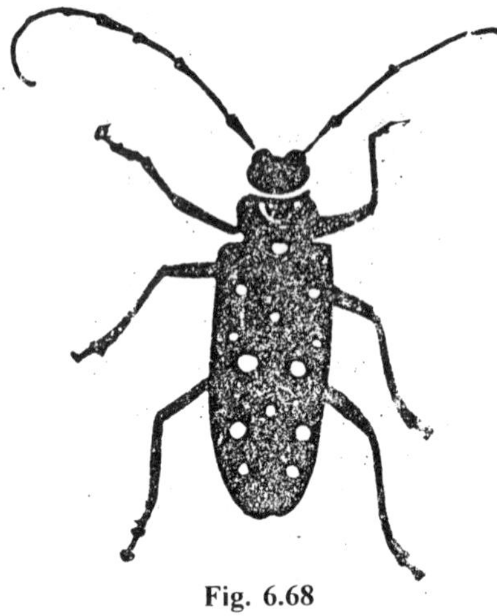

Fig. 6.68

Distribution

From India through South East Asia to South China.

Fig. 6.69 : Geographlcal distribution.

Control

Heavily infested branches should be cut and burned to destroy the larvae, and pupae within. Spraying the trunks and branches with dieldrin has in the past proved to be of some value. In some cases a mixture of dieldrin and kerosene has been injected into the frass holes to kill the boring larvae. A combination of these three methods usually gives adequate control. Sometimes it is recommended that a marker dye (Methylene blue for example) be added to the dieldrin solution so that it will indicate the extent of the solution penetration within the larval tunnel system.

Cultural Method

(i) Removal and destruction of affected branches.

(ii) Killing of grubs by packing a stiff wire inside the hole.

Lemon Butterfly (Citrus Swallowtail)

Scientific Name : Papilio demoleus L.

Family : Papilionidae

Order : Lepidoptera

Hosts (Main) : Citrus plants

Alternative : Other members of the Rutaceae; *Zizyphus*, and bael fruit.

Type of Damage

Defoliation caused by the caterpillars to citrus plants is often serious. They feed on the leaves from the margin inwards to the midrib. Seedlings and young plants suffer the most.

Life History

The adult is a black and yellow swallow-tail butterfly. Its antennae are black and club-shaped. Eggs are laid singly or in small groups of 2-3 on the undersurface of leaves and shoots; upto 180 eggs are laid by each female. The eggs are minute, spherical and white turning grey on maturity. They hatch in 3-6 days. The larva has five instars; the first three are brownish-black with white patches; the last two greenish with brown and grey markings. When disturbed the caterpillar pushes out from the top of its prothorax a bifid, purple structure called the osmeterium which emits a distinct smell. The larva becomes full-grown in 13-26 days and is stout and about 4 cm long. It pupates on the plant, the chrysalis being attached to the plant by its tail and held by strands of silk. The pupa is yellowish-green or brown and is 3cm long. The pupation period is 7-24 days. The peak population of the pest is in April and again from July to October.

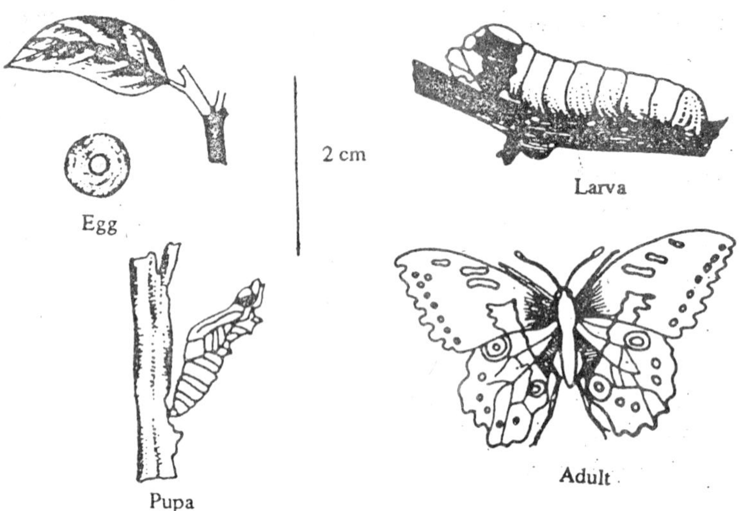

Fig. 6.70 : Life-stages of *Papilio* demoleus.

Distribution

Saudi Arabia, through Pakistan, India, S. E. Asia upto S. China and Taiwan, and parts of Australia, Papua, New Guinea and West Iran.

Fig. 6.71 : Geographical Distribution.

Control

Handpicking of the caterpillars and pupae from small orchards in the initial stages of infestation and removing alternative host plants are mechanical control measures. The pest can be controlled by treating the plants with a short term contact insecticide such as 0.04% monocrotophos or phosphamidon.

Stored Grain Pests (Pulse Beetle)

Scientific Name : Callosobruchus chinensis

Family : Bruchidae

Order : Coleoptera

Hosts (Main) : Gram, peas, cowpeas, lentil, arhar.

Alternative : Chick peas, maize, soybean, other pulses.

Type of Damage

The pest attacks leguminous pods in the field from where they are carried to storage godowns. The larvae bore into the pulses and grains and feed and develop inside. The infestation in case of grains in early

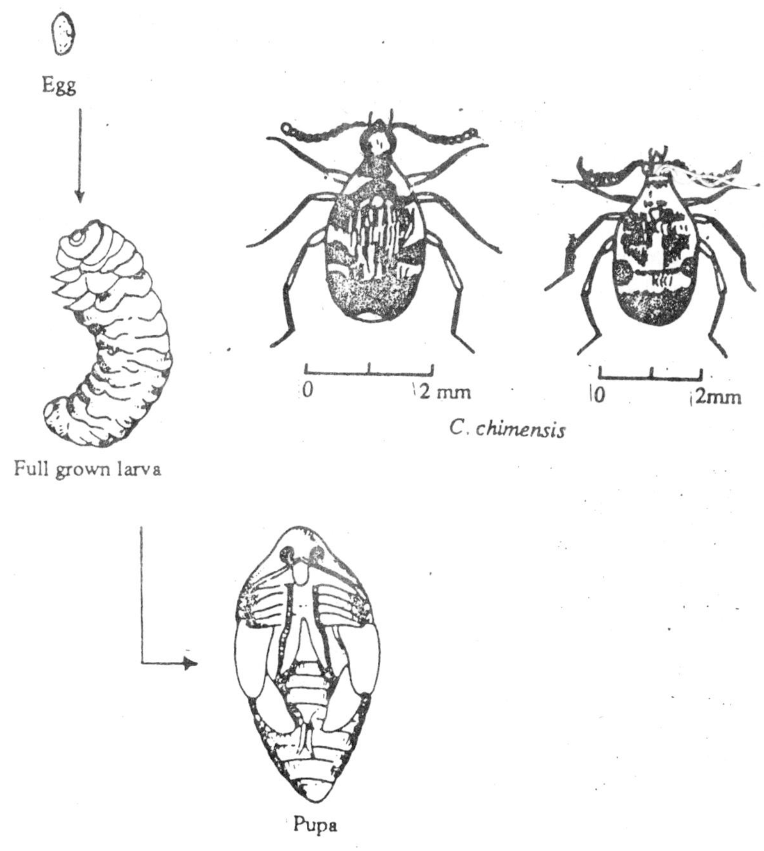

Fig. 6.72 : Life-stages of *Callosobruchus chinensis.*

stages cannot be detected since the hole through which the larva enters is very minute. The damaged grains are hollow inside, bearing small holes and are unfit for human consumption.

Life Cycle

The adults are small, roundish, brown beetles, 4 mm long. There are dark markings on the elytra and a white, raised spot on the middle of the body. The female lays upto 90 eggs singly on the seeds or on the pods in the field. The incubation period is 4-6 days. The young larvae burrow into the grain and feed inside for 2-3 weeks, undergoing 4-5 moults. The full-grown larva is about 1 cm long, white and curved in appearance. Pupation occurs within the grain or grain dust and takes about 7 days to complete. The adult emerges from the grain after cutting a small, round hole. Six to seven generations a year of the pest are common in India.

Distribution

Cosmopolitan throughout most of the tropics and subtropics.

Fig. 6.73 : Geographical Distribution.

Control

Cultural control can be achieved by growing susceptible crops at least a kilometer away from storage godowns which are the main source of infestation. Fumigation with methyl bromide in the stores is very effective but proper precautions must be taken because of the high toxicity of this compound.

Rice Weevil or 'Sursuri'

Scientific Name: Sitophilus oryzae L.
Family : Curculionidae
Order : Coleoptera

Hosts (Main) : Rice

Alternative : Maize and other cereals in storage.

Type of Damage

Generally infestation starts in grains only during storage, which may lead to heat spots in the grain. The grains are hollowed and weight is reduced. Adults which are winged are even known to fly from the godown to the fields in the vicinity where they begin to infest the grain in the field.

Life Cycle

The adults are small, reddish brown weevils, about 4 mm in length, with a large rostrum and thorax. The head is prolonged into a slender snout or rostrum with chewing mouthparts located at its tip. The wings hidden beneath the dark brown elytra are functional. The life-span of the adult weevil is 4-5 months. The female starts egg laying 5 days after emergence and lays a total of 300-400 eggs. The eggs are translucent white, oval and narrow, at the tip. The female lays the eggs inside the

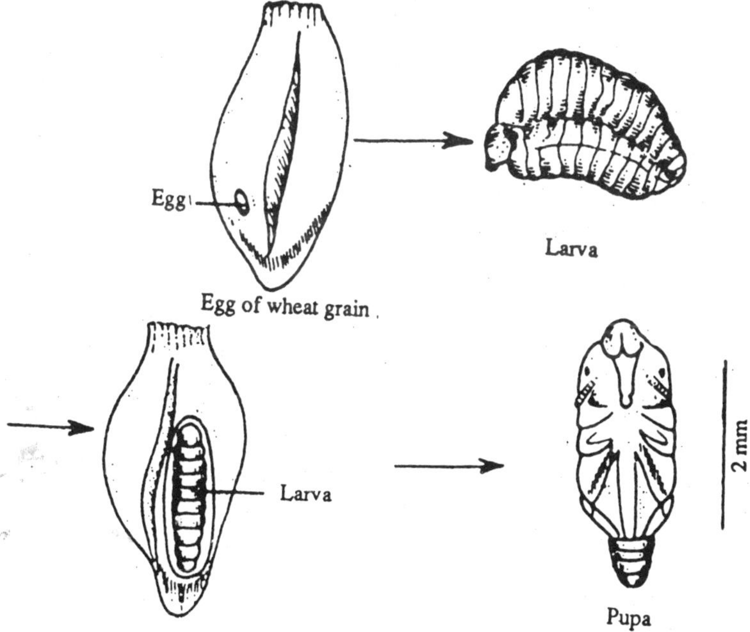

Fig. 6.74 (a) : Life-cycle stages of Rice weevil.

Damage

Tiny circular scales, just visible to the unaided eye, can be seen on the bark of lightly infested trees. If the trees are heavily infested then the bark may be completely covered by overlapping scales. Fruits may also be attacked especially at both stem and flower ends. The bark may crack and exude gum, and the twigs die back; infested trees often die.

Pest Status

A polyphagous pest on many crops and plants, but probably the most serious pest of deciduaus fruit trees; fruit trees are seriously debilitated or killed, and the fruit rendered unsalable.

Life History

This is one of the few viviparous hard scales; each female can produce 100-400 nymphs. The crawlers are minute (0.2 mm long), with well-developed legs and antennae, and yellow in colour. Overwintering occurs in the nymphal stage.

The adult female scale is flattened, nearly circular, with a raised central nipple, grey in colour, 1-2 mm in diameter, and completely covered by the circular yellow scale.

The male scale is oval, about 1.0 by 0.5 mm.

The total life-cycle takes 18-20 weeks in California, and in the warmer parts of the world there may be four or five generations per year.

Distribution

Widely distributed throughout the warmer parts of the world but absent from many tropical countries: recorded from S. Europe, Algeria, S. Africa, Zimbabwe, Zaire, Iraq, Turkey, India. Pakistan, China. Korea, Japan, Australia, New Zealand, USA, W. Indies, Mexico, and much of S. America.

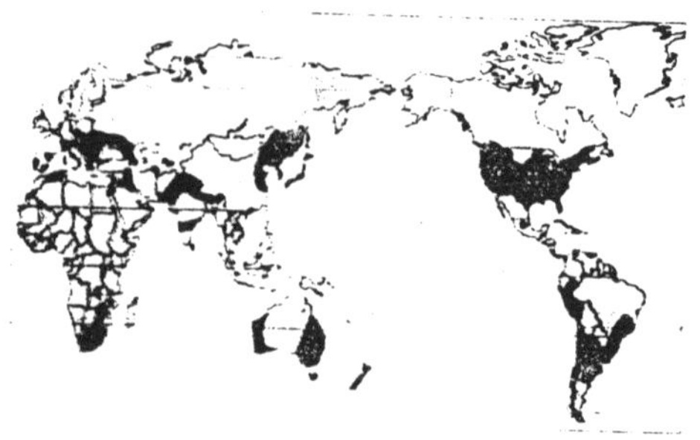

Fig. 6.76

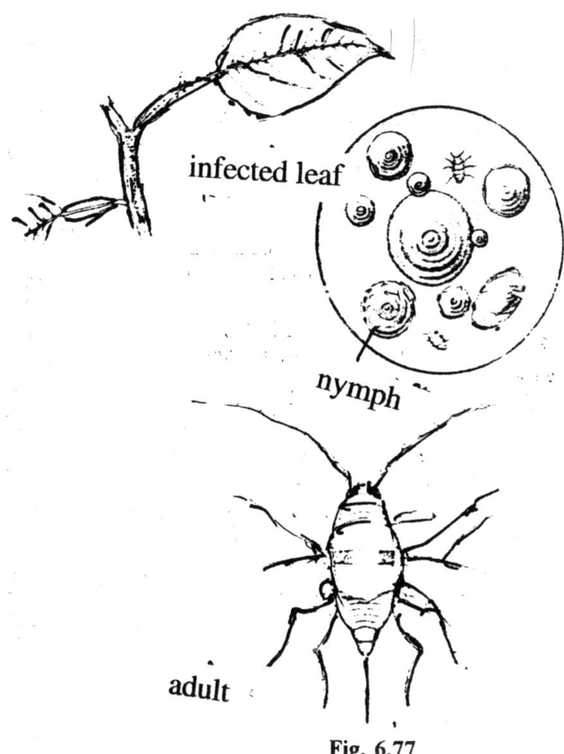

Fig. 6.77

Control

As the most important hosts are deciduous trees this scale can be best controlled by sprays during the dormant period. Recommended sprays are lime-sulphur, parathion, and petroleum oils, fumigation with HCN or CH_3Br.

Biological Contol : Natural enemies

1. *Aphytis diaspidis* in J&K and H.P.
2. *Prospaltella perniociosi* (in U.P.)

Dacus Dorsalis (Hend.)

(= Dacus ferrugineus)

(= Strumeta dorsalis Hend.)

Common name : Oriental Fruit Fly (Mango fruit fly)

Family : Tephritidae

Hosts (main) : Guava, mango. *Citrus*, banana, avocado, papaya, etc.

Alternative : Peach, passion fruit, coffee, melons, pineapple, jackfruit. strawberry; in Hawaii 173 species of plants in 112 genera were recorded.

Damage

Females oviposit through the skin of the fruits and sap may ooze from the punctures. The maggots feed inside the fruits, and their infestation may be associated with fungal and bacterial rots. *Pest status*. A serious pest of all fleshy fruits and vegetables in the general S.E- Asia region, and Hawaii.

Life History

The female fly uses her ovipositor to deposit eggs about 5 mm beneath the surface of ripening fruits. The batches of eggs hatch in about two days, and larval development can be as short as seven days before the mature maggots drop out of the fruits to pupate in the soil. Pupal development takes about ten days.

The adult flies are dark brown with bright yellow markings on the thorax, the scutellum is either white or pale yellow, and the hyaline wings have a fine of infuscation along the leading edge and down the anal veins. Adult wingspan is about 15 mm with body length about 8 mm. The female maturation period is 5-7 days. The entire life-cycle takes only about 25 days in the tropics, where there may be many generations per year, but in cooler regions development is much slower. Temperature limits are 14°C for larval development and 21°C for adult.

Distribution

From Pakistan and India through S.E. Asia to N. Australia, and to China, Taiwan and the Ryukyu Islands and Hawaii.

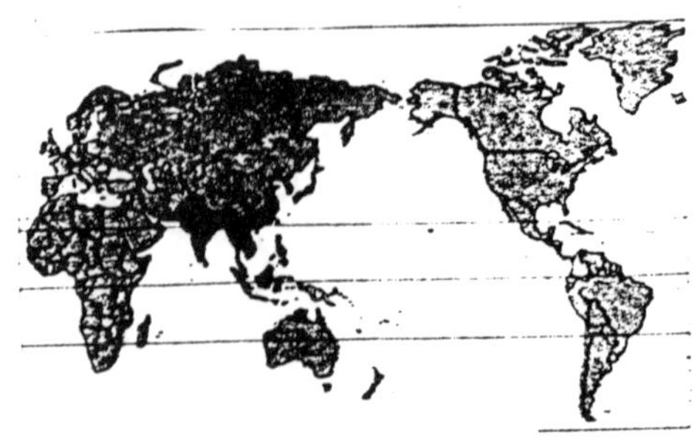

Fig. 6.78

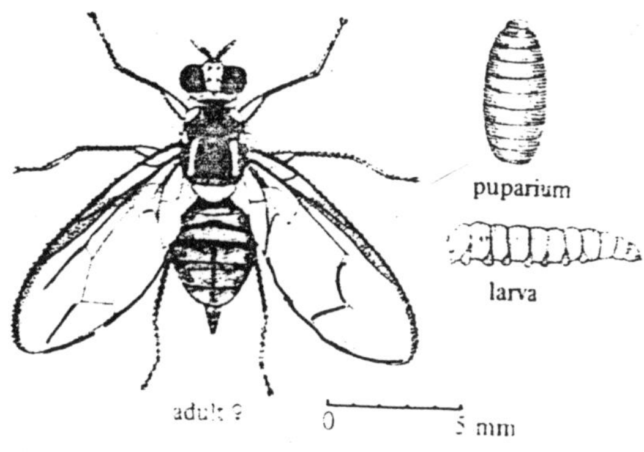

Fig. 6.79

Control

Bagging of fruits is practised in S.E. Asia to deter ovipositing female flies. Other methods of control include destruction of unmarketable fruits.

Pheromone traps are used for fruit fly monitoring purposes in many orchards. Natural parasitism levels are often quite high, and this pest has, to a great extent, been controlled in Hawaii by the parssite Opius (Braconidae).

Chemical control is clearly difficult, and is generally a suppressive procedure aimed at the female flies, usually with a protein bait spray incorporating malathion (or naled), and a chemical attractant.

Some countries have quarantine regulations aimed specifically at this pest.

Sesamia inferens (Wlk.)

Common name : Purple Stem Borer

Family : Noctuidae

Hosts (main) : Rice, and sugarcane.

Alternative : Maize, sorghum, wheat, other cereals, *Eleusine coracana*, and many other grasses.

Damage

Small plants typically show 'dead-hearts', and older plants have extensive parts of the stem hollowed out, with a consequent physical weakening of the stem, and a reduction of crop yield.

Pest Status

One of the major pests of rice and sugarcane; a polyphagous pest, and of importance on several other cereals in the tropics. Sugarcane is not a preferred host for oviposition-rice and grasses being preferred for this.

Life History

The eggs are bead-like, and laid in rows within the leaf-sheath; some 30-100 eggs per batch. Incubation takes about seven days.

The caterpillar is purple-pink dorsally and white ventrally; the head capsule is orange-red. After about 36 days the mature caterpillar is up to 35 mm long and 3 mm broad.

The pupa is dark brown with a purple tinge in the head region, and is about 18 mm by 4 mm. Pupation takes about ten days.

The adult moth is fawn-coloured with dark brown streaks on the forewings and white hindwings. The body length is 14-17 mm and wingspan up to 33 mm. The adults survive in the field for 4-6 days.

The total life-cycle takes 46-83 days.

Distribution

Pakistan, India, Bangladesh, Sri Lanka, S.E. Asia, China, Korea, Japan, Philipines, Indonesia. Papua, New Guinea, West Irian and the Solo Isles

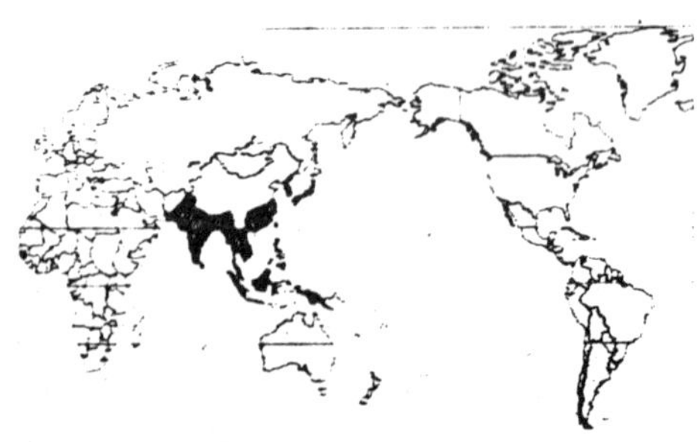

Fig. 6.80

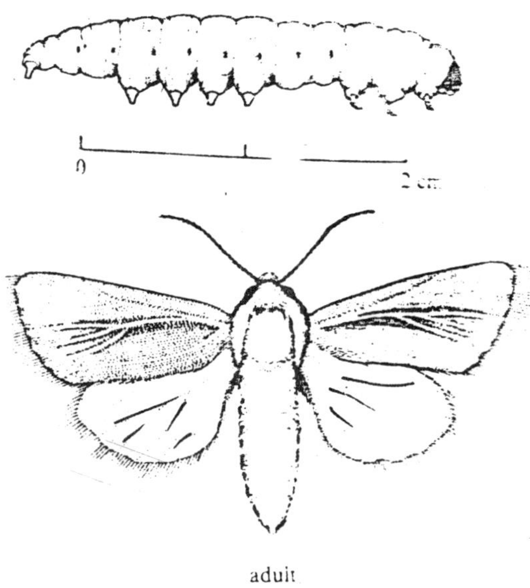

Fig. 6.81

Control

The different ways in which stalk borers can be controlled a enumerated below:

1. Cultural Control

(a) *Resistant varieties* : Some varieties with tight or extensive leaf sheaths are not favoured for oviposition; other varieties have an increased silica content in their tissues and feeding larvae usually die.

(b) Simultaneous sowing over a large area prevent papulation build-up.

(c) Destruction of crop residues-important for killing of pupae left in old stems and tall stubble.

(d) Destruction of thick-stemmed grass weeds which would a as alternative hosts.

(e) Close season of at least two months to prevent population continuity.

2. Biological Control

(a) Trichogramma spp. as egg parasities.

(b) Apanteles spp. as larval parasites.

3. Insecticidal Control

(a) *Dusts* : Contact insecticides applied down

(b) *Sprays* : The funnel of yong plants to kill the emerging and feeding first instar larvae.

(c) *Granules* : Applied either by folier lodging, to the soil, or to the water for paddy rice; usually dystemic in action.

(d) Systemics applied as sprays includes DDT, r-BHc, endrin, a_2 azinphos methyl, diazinon, endosuffan, fenthion, fenitrothion, monocrotophos, phorate, phosphamidon, tetrachlovinphos, triazophos, carbarys and carbofuran, etc.

Resistance is a problem in some areas, and also certain species are less sensitive to some chemicals, so that local advice should be sought for control of any particular stalk borer.

Leptocorisa acuta (Thunb.)

Common name : Rice Seed Bug (Asian Rice Bug)

Family : Coreidae

Hosts (main) : Rice

Alternative : Various species of wild grasses.

Damage

The bugs usually appear in the young crop with the early rains, and both nymphs and adults suck sap from the developing grains at the 'milky' stage. All soft milky grain is susceptible to attack the bugs suck the sap until the grain is emptied. Before the grain is formed the bugs will fed on succulent young shoots and leaves.

Pest Status

Rice Bugs are very destructive in areas where rainfall is evenly distributed throughout the year, and also in irrigated crops. Yield losses of 10-40% are common, and in severe infestations the entire crop may be destroyed.

Six other species of *Leptocorisa* are also pests of rice in different parts of the tropics.

Life History

Eggs are laid in rows along the rice leaves: they are red to black in colour, and flat in shape. The incubation period is 5-8 days. Newly hatched nymphs are green, but they become browner as they grow. There are five nymphal instars distinct colour changes occurring after each moult. The nymphal period lasts 17-27 days.

The adult bugs are slender and some 15 mm in length, greenish-brown in colour, and have been recorded to survive up to 115 days in favourable conditions. In the absence of rice plants the bugs live on wild grasses.

The complete life-cycle takes some 23-34 days, and there are several generations per year.

Distribution

Found in Pakistan, India, Sri Lanka, through S.E. Asia to S. China, Philippines, Indonesia, Papua New Guinea, West Irian and N. Australia.

Fig. 6.82

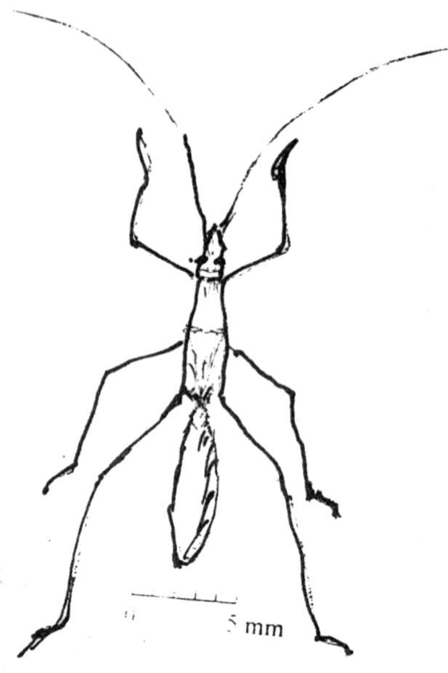

Fig. 6.83

Control

The use of resistant varieties is probably the best method of control to be arived at, but as yet insufficient plant breeding has been carried out. Sprays of carbaryl, dimethoate or demeton-S-methyl can be effective, especially if followed by roguing of diseased plants.

Cultural Method

(i) Keep paddy field clean of all weeds

(ii) Clipping of all egg bearing leaves and burn them to destory the eggs.

(iii) Putting up light traps to altract and destroy the insects.

Biological method : Natural enemies.

(i) *Cicindela sexpunctata.*

(ii) *Ommatius* (Parasite of nymphs).

Tryporyza incertulas (Wlk.)

(= Schoenobius incertulas (Wlk.)

Common name : Yellow Paddy Stem Borer

Family : Pyralidae

Hosts (main) : Rice

Alternative : Wild rice and various wild grasses.

Damage

The caterpillars bore into the rice stems and hollow out the stem completely. Attacked young plants show 'dead-hearts' and older plants often break where the stem is hollowed out causing lodging.

Pest Status

A serious pest of rice throughout India and S.E. Asia.

Life History

Egg masses are laid in batches of 80-150 on the leaf sheath, and covered with the brown anal hairs of the female moth. The incubation period is 4-9 days.

The carerpillars are yellow in colour, about 18-25 mm long when mature; the head capsule is black. Total larval development takes about 40 days. Pupation takes place inside the stem; the pupa is pale and soft; the pupal period is from 7-11 days.

The adults are sexually dimorphic and quite distinct; the female has one dark spot in the centre of the yellow forewing, whereas the male has a series of small dark spots on a brown forewing. Body length is about 13-16 mm and the wingspan 2 2-30 mm, the males being smaller. The adults do not feed and only live for 4-10 days.

The entire life-cycle takes from 35-71 days.

Distribution

Recorded from Pakistan, India, Sri Lanka, through S.E. Asia, Indonesia, Philippines, up to China and S. Japan.

Control

In some areas early or late planting is recommended so as to avoid crop continuity and a build-up of the borer population. Various other types of cultural control are practised in different countries, according to the local condition. A number of rice cultivars show general resistance to stem borers. Foilar sprays of parathion, and parathion-methyl are successful, but their mammalian toxicity is too high for safe use on a large scale. Endrin, dieldrin, trichlorphon, dichlorros and diazinon have also been successful, but several applications at weekly intervals are required.

Fig. 6.84

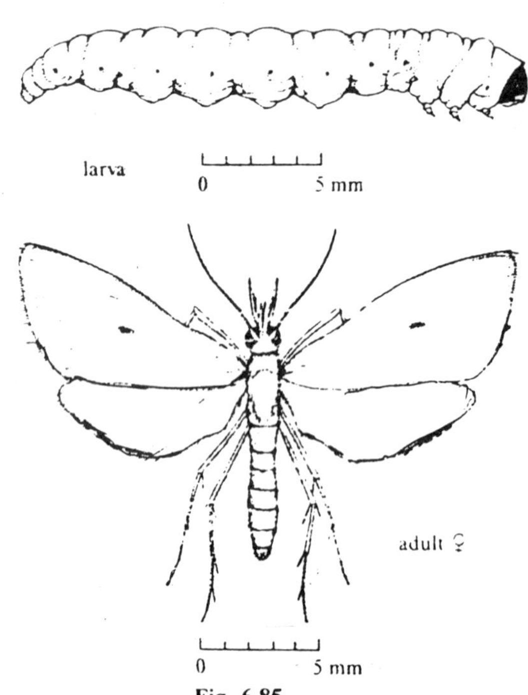

Fig. 6.85

Tribolium castaneum (Herbst)

Common name : Red Flour Beetle

Family : Tenebrioiadae

Host (main) : Maize, wheat and other stored grains.

Alternative : Many types of stored foodstuffs.

Damage

Infestation is apparent by the appearance of adults on the surface of the grain; there is extensive damage to previously holed or broken grains, or grain damaged by other pests. Damage is done by both larvae and adults.

Pest Status

A serious secondary pest throughout the warmer parts of the world in food stores.

Life History

The eggs are small, cylindrical, and white. The female lays the eggs scattered in the produce. The larvae are yellowish-white, about 6 mm long when fully grown. The head is pale brown, and the last segment of the abdomen has two upturned dark pointed structures. The larvae live and develop inside the grain till pupation. The damage to the stored grains is done by the larvae of the Red Flour Beetle.

The pupa is yellowish-white, later becoming brown, the dorsum hairy, and the tip ot the abdomen having two spine-like processes.

The adult is rather flat, oblong, reddish-brown in colour, and about 3-4 mm long. Each female is capable of laying about 400-500 eggs and the adults are long-lived, under some circumstances living for a year or more. The life period from egg to adult is 35 days at 30°C. Adults fly in large numbers in the late afternoon.

Distribution

This pest is cosmopolitan in warmer countries.

Control

The shelled grain should be thoroughly admixed with BHC dust or powder, if locally permitted.

Fig. 6.86

Fumigation should only be carried out by approved operators.

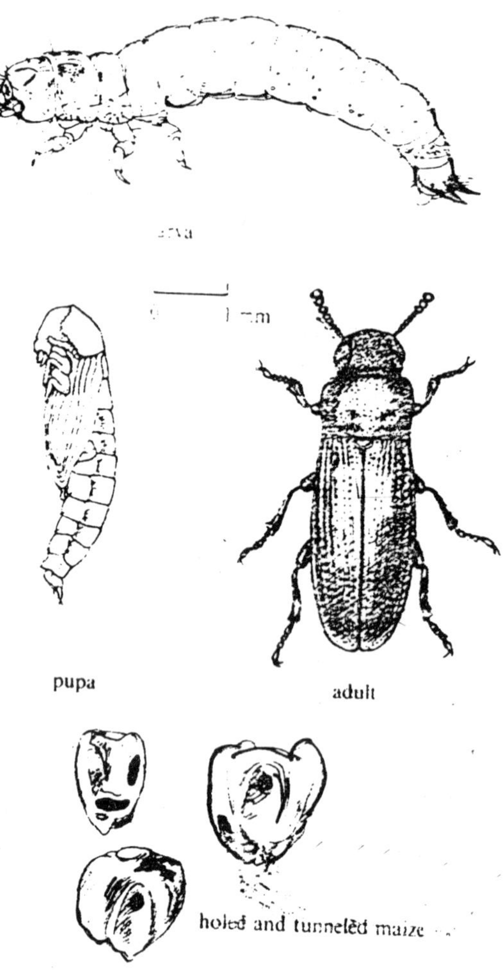

Fig. 6.87

Trogoderma granarium Everts

Common name : Khapra Beetle

Family : Dermestidae

Hosts (main) : Cereals and groundnut.

Alternative : Pulses, spices, and various cereal and pulse products (cakes).

Damage

The larvae bore in the stored cereal grains and pulses, usually hollowing out the grain. Development is rapid in the hot humid tropics and very large populations may build up quickly. The pest is fairly polyphagous and can survive in facultative diapause for a year or longer in the absence of food.

Pest States

Although virtually cosmopolitan, this pest is absent from E. Africa and S. Africa, and these countries have legislation to prevent its importation. A very serious pest of stored grains and pulses in most of the warmer parts of the world. It is the only member of the Dermestidae that is phytophagous.

Life History

Eggs are laid in the stored produce, and the larvae are regarded as primary pests of stored grains in that they can damage intact grains and seeds. The larvae often develop at different rates, some coming to maturity in two weeks and others taking months or even more than a year. In the absence of food the larvae can go into a state of facultative diapause for many months which ensures their survival at times when the store is empty; at these times the larvae congregate in large numbers in crevices and cracks in the buildings or storage containers.

The adults are small dark beetles 3-4 mm in body length; they are wingless, neither do they feed; they live for about 14 days. Thus, this species is spread almost entirely through the agency of man, and quarantine regulations against it may be easily enforced.

Under optimum conditions the life-cycle can be completed in about three weeks (37°C and 25% RH).

Distribution

Almost cosmopolitan in the warmer parts of the world, but as yet kept out of the countries of E. and S. Africa; it prefers a hot dry clinate.

Control

A difficult pest to control in that many of the usual contact insecticides used against Coleoptera have little effect, the effect of its being able to practice facultative diapause results in infestations reappearing after a long storage period. Fumigation with toxic materials such as methyl bromide are most likely to give the best control results, but pirimiphosmethyl is reputed fo be effective.

Cultural method : For safe storoge clean, dry, insect freegrains should be stored.

Biological control : No specific parasite or predator is known for this pest but certain sps. of reduviidae family have been observed feeding on the grub of this pest.

Fig. 6.88

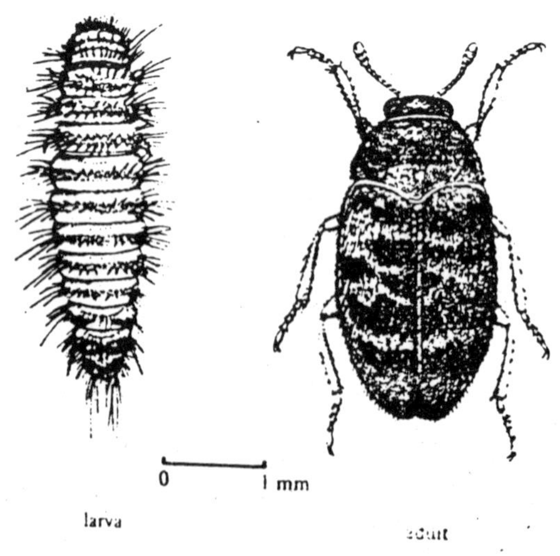

Fig. 6.89

Pectinophora gossypiella (Saunders)

(= Platyedra gossypiella (Saund.)

Common name : Pink Bollworm

Fanrily : Gelechiidae

Hosts (main) : Cotton

Alternative : Hibiscus and other Malvaceae, but only cotton can support a large infestation.

Damage

The entry hole of the caterpillar into a large green boll is almost invisible. If the boll is opened, however, the red and white caterpillar can be found, especially inside the developing seeds. Damaged bolls fail to open completely and often have secondary rots where the caterpillar has been feeding.

Pest Status

Potentially very serious in all cotton-growing areas, but control can be achieved by enforcement of a close season.

Life history

The egg is oval, about 0.55 mm long; laid singly or in small groups, on or near a bud or boll. Hatching takes 4-6 days.

Full-grown larvae are 10-12 mm long, white with a double red band on the upper part of each segment. There are four larval instars, the total larval period lasting 14-23 days. Caterpillars feed on flowr buds or young bolls and cause shedding. This is a late species and does most damage to large green bolls, the small larva enters a boll near the base and bores into a developing seed. The mature larva leaves the boll through a neat circular hole about 2 mm in diameter, drops to the ground and pupates in the litter. Pupation takes place inside a loose cocoon. The pupa is brown, 7-10 mm long and about 2.5 mm broad. Pupation takes 12-14 days.

The adult is a small brown, inconspicuous moth of wingspan 15-20 mm; nocturnal in habits. After a pre-oviposition period of about four days, it may live a further ten days and lay about 300 eggs.

Distribution

Completely pantropical in distribution, including the subtropical regions.

Control

In some areas a mixture of DDT and endrin is recommended, or carbaryl, trichlorphon and azinphos-methyl, and sprays are applied at

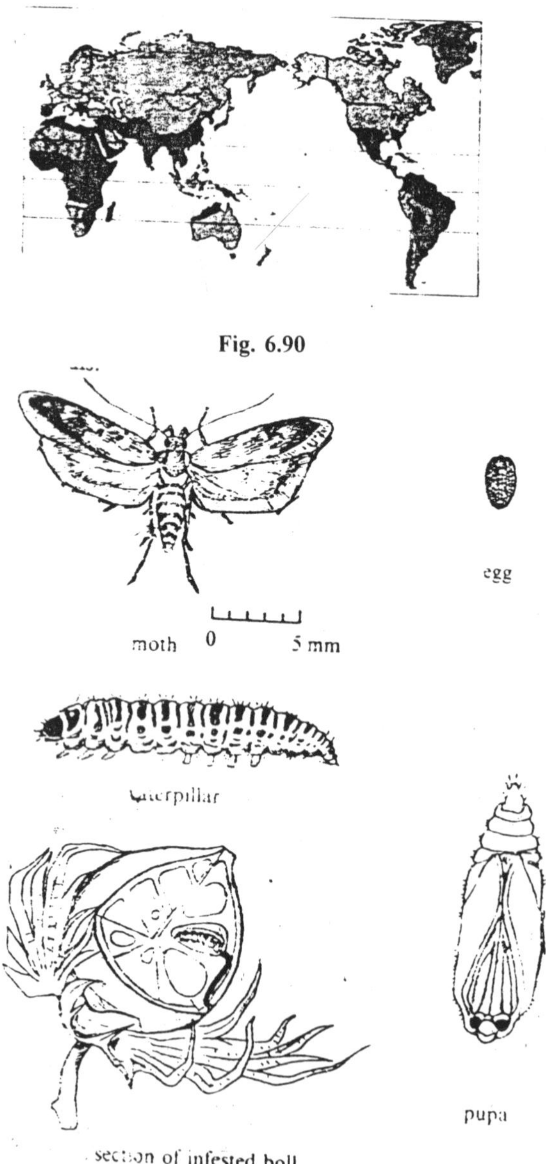

Fig. 6.90

Fig. 6.91

weekly interrals, seed should be fumigated with carbon disulphide at 5ml funigant per 100 kg seed or 70 ml per cubicmetre of space with an exposure of 24 hrs. Pickup and destroy the infested bolls.

Biological method : Natural enemies

(i) *Triphleps pectinophorae* (egg parasite)

(ii) microbracon sps.

(iii) *Apanteles pectinophorae.*

(iv) Elasmus platydrae (larval and pupal parasite).

Heliothis armigera (Hb.)

(= Heliothis obsoleta)

(= Helicoverpa armigers Hb.)

Common name : American Bollworm

Family : Noctuidae

Hosts (main) : Cotton, beans, maize and sorghum.

Alternative : Tobacco, tomato, many legumes, some vegetables, and other plants.

Damage

Clean circular holes are bored in flower buds, cotton bolls of all sizes and some fruits.

Pest Status

A sporadically very serious pest of cotton and beans in many parts of the Old World; completely polyphagous, and a minor pest on many cultivated fruits (in the botanical sense).

Life History

Eggs are spherical. 0.5 mm in diameter, yellow turning brown; hatching takes 2-4 days. Each female moth may lay 1000 or more eggs.

The larva is stout, greenish or brown, but variable in coloration. Body bears long dark and pale bands; fully grown larvae are 40 mm long. There are six larval instars and the total period lasts 14-24 days, but 51 days at 17°C

Pupation takes place in the soil; the shiny brown pupa is about 16mm long; pupal development takes 10-14 days in the tropics.

The adult is a stout-bodied, brown nocturnal wingspan about 40mm.

The complete life-cycle can be as short as about 28 days in the tropics,and there are several generations each season; in Europe there are probably only two generations per year, the second over-wintering as pupae in diapause.

Distribution

Widespread throughout the tropics, subtropics and warmer temperate regions of the Old World, extending as far north as Germany and Japan

Control

Maize during the tasselling period, and flowering legumes, are attractive to this species and will divert egg-laying females from the cotton; little damage is done to the maize and beans because of the high larval mortality on these crops.

For chemical control, apply insecticides when caterpillars are small; the usual insecticides are DDT, BHC (or a mixture of the two), endosulfan, carbaryl, and now the polyhedral virus is available.

Thorough ploghing the infested field after the harrest to expose and destroy the pupae.

Fig. 6.92

Biological method : Natural enemies

(i) Trichogramma minuthm (parasite of eggs).

(ii) Triphleps insidiosus.

(iii) Bracon SAS.

(iv) Aphanteles spe.

(v) Chelonus narayani (parasites of larvae).

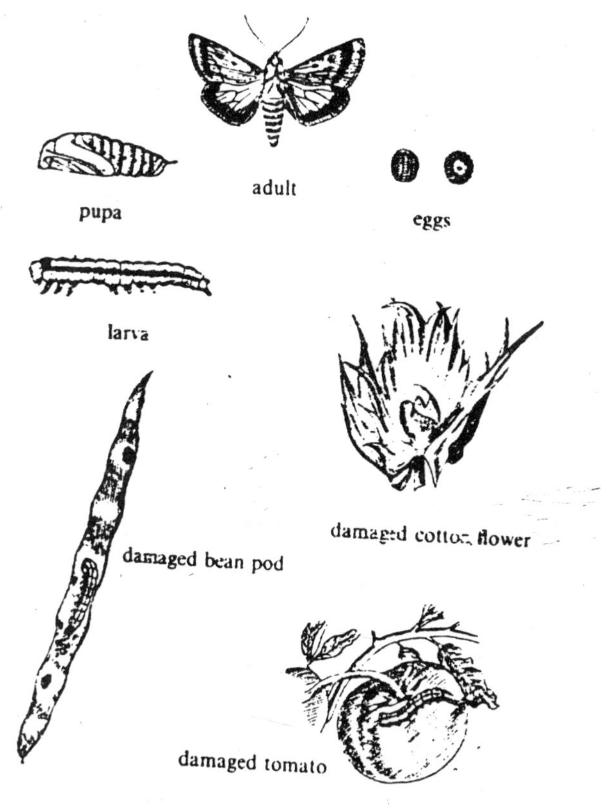

Fig. 6.93

Phthorimaea operculella (Zeller)

(= Gnorimoschema operculella (Zeller)

Common name : Potato Tuber Moth

Family : Gelechiidae

Hosts (main) : Potato

Alternative : Tobacco, tomato, eggplant, and other Solanaceae, and Beta vulgaris.

Damage

The leaves have silver blotches caused by the young larvae mining in the leaves. Leaf veins, petioles and stems are tunnelled followed by wilting of the plants. Eventually the tubers are bored by the larger caterpillars, and they often become infected with fungi or bacteria.

Pest Status

An important pest of potato in warmer countries. Infestations arising initially in the field and continuing during storage of the tubers. There is a serious risk of transportation from country to country through infested tubers.

Life History

The egg is minute and oval, 0.5 × 0.4 mm, and yellow. Eggs are laid singly on the underside of the leaf, or in tubers (usually in storage) near the eye or on a sprout. Each female lays about 150-250 eggs. The eggs on the leaves hatch in 3-15 days and the first instar larvae bore into the leaf, where they make mines. The caterpillars are pale greenish. They gradually eat their way into the leaf veins and into the petioles, then gradually down the stem and sometimes into the tuber. The full-grown caterpillar is 9-11 mm long. The larval period lasts for 9-33 days.

Pupation takes place in a cocoon in the surface litter or in the tuber, and takes 6-26 days.

The adult is a small moth with narrow fringed wings; the forewings are grey-brown with dark spots, and the hindwings are dirt white. The wingspan is about 15 mm. The moths are very short-lived.

One generation takes some 3-4 weeks, and there can be up to 12 generations per year, but development is very dependent upon temperature.

Distribution

Almost completely cosmopolitan in distribution. but with limited records from Asia and none from W. Africa

Control

Effective insecticides are DDT. dicrotophos, dimethoate. and parathion, all as sprays. As a preventative measure sprays should be applied every 14 days after the first mines are found in the leaves.

Fig. 6.94

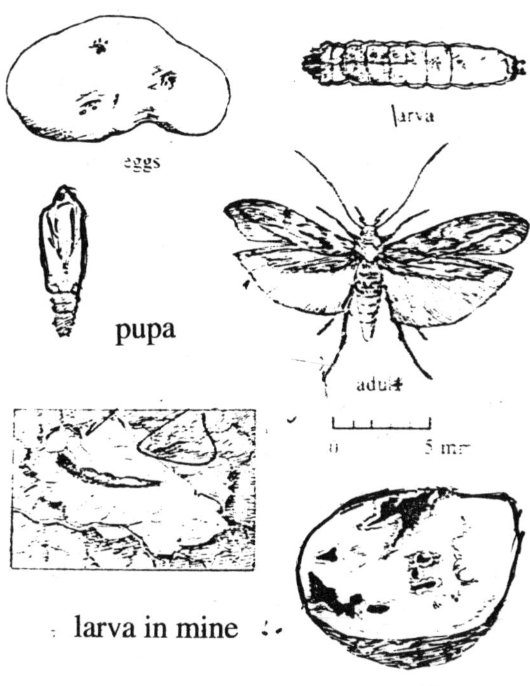

Fig. 6.95

Athalia Spp.

Common name : Cabbage Saw fly

Family : Tenthredinidae

Hosts (main) : Brassicas of all species

Alternative : Other members of the Cruciferae.

Damage

Leaves are eaten by the larvae, often leaving only the midrib. Blackish larvae are present on the plant and they fall to the ground if the plants are shaken.

Pest Status

A sporadically serious pest of all cruciferous crops; turnip, Chinese cabbage, kale and crambe are particularly susceptible to attack.

Life History

Eggs are laid singly in small pockets cut in the leaf by the female sawfly.

The larva closely resembles a lepidopterous caterpillar, the important difference is that these sawfly larvae have six pairs of prolegs on the abdomen instead of the four found in caterpillars. The full-grown larva is about 2.5 cm long, and is oily black or green; black specimens often have yellow spots. The head is shiny black and that part of the body just behind the head is often slightly swollen, giving the Larva a humped appearance.

The full-grown larva burrows into the soil and spins a tough silk cocoon to which particles of soil adhere. The yellowish pupa forms within the cocoon.

The adult remains in the cocoon for a while before it emerges and pushes its way through the soil to the surface. It is about 1.5 cm long and has a dark head and thorax with a bright yellow abdomen. The adults may often be seen flying about slowly, just above the crop.

Distribution

Great Britain, and Europe from Spain through to Siberia and Japan, down to Asia Minor, N. Africa, E. and S. Africa, and S. America.

Control

Destruction of wild crucifers on the head-lands will help to keep the pest population down.

The following pesticides have been found to be effective as foliar sprays: DDT, carbaryl and pyrethrum.

Fig. 6.96

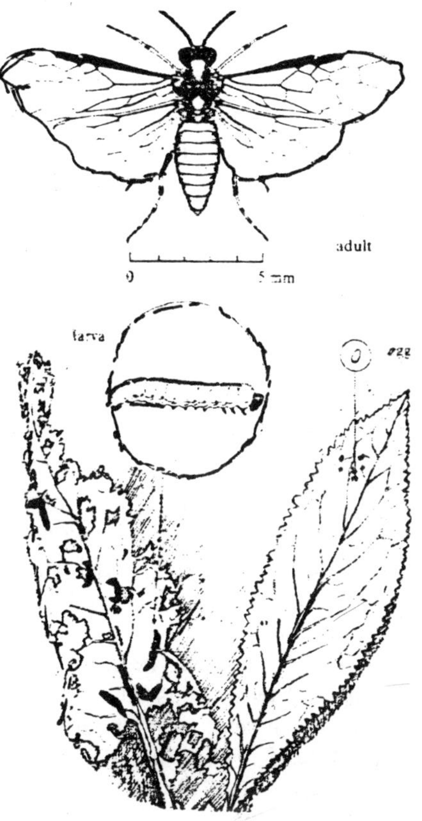

Fig. 6.97

Cultural Control

(i) Collection and destruction of grubs in the morning and evening.

(ii) Use of bittergourd seed oil emulsion as a anti-feedant.

Biological method : Natural enemies

(i) Perilissus cingulator (parasitises the grubs).

Agrotis Ipsilon (Hfn.)

(= Euxoa ypsion Rott.)

(= Scotia ipsilon Rott.)

Common name : Black (Greasy) Cutworm

Family : Noctuidae

Hosts : A polyphagous cutworm attacking the seedlings of most crops, in particular cotton, rice, potato, tobacco, cereals, and crucifers.

Damage

The young larvae feed on the leaves of many crops; the older in caterpillars feed at the base of crop plants or on the roots or stems underground. Seedlings are typically cut through at ground level; one caterpillar may destroy a number of seedlings in this manner in a single night.

Pest Status

A cosmopolitan pest ol sporadic importance on many crops in different parts of the world. It can cause severe damage on rice in S.E. Asia and in Australasia. On other crops the occasional severe infestation usually results in devastating damage.

Life History

The eggs are white, globular, and ribbed; 0.5 mm in diameter; and hatch in 2-9 days. Each female may lay as many as 1800 eggs.

The larvae are brownish above with a broad pale grey band along the mid-line, and with grey-green sides with lateral blackish stripes. The head capsule is brownish-black with two white spots. The general appearance of the caterpilar is blackish. The mature caterpillar is 25-35 mm long; larval development takes 28-34 days. In temperate countries some larvae overwinter as such, and pupate in the late spring. The first two instars feed on the foliage of the plant, the third instar becoming non-gregarious, in fact often cannibalistic, and adopting cutworm habits.

The pupa is dark brown, 20 mm long, with a posterior spine; pupation takes 10-30 days, according to temperature

The adults are large, dark noctuids with wingspan of 40-50 mm, with a grey body, grey forewngs with dark brownish-black markings: the hindwings are almost white basally but with 2 dark terminal-fringe; paler in the males.

The life-cycle from egg to adult takes 32 days at 30°C, 41 days at 26C. and 67 days at 20°C

Distribution

Almost completely cosmopolitan, from northern Europe, Canada, Japan, down to New Zealand, S. Africa, and S. America. It has not been recorded to date from a few areas in the tropics *e.g.*, S. India, N.E. South America.

Fig. 6.98

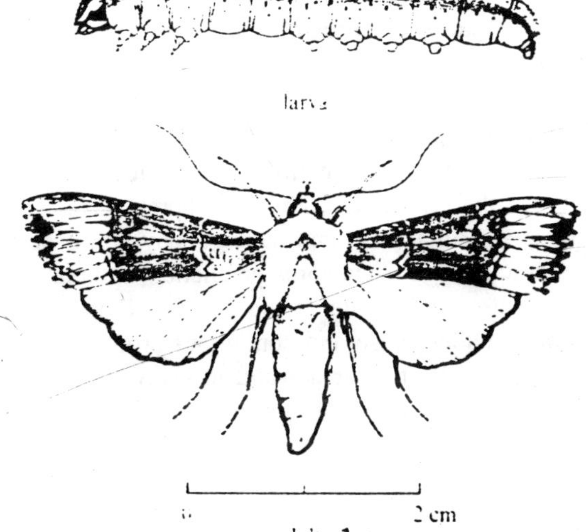

Fig. 6.99

Control

1. Cultural method

(a) *Weed destruction*—these plants are often preffered sites for oviposition and food for the instar larvae.

(b) Hand collection of larvae-often more suistable for gardens and small-holdings.

(c) Flooding of the infested field may be feasible for some crops.

(d) Deep plouging will bring larvae and pupae to soil surface for exposure to predators and sun.

2. Chemical methods

(a) High-volume sprays of insecticide.

DDT (25%) – 4.2 l/ha.
Endrin (w.p.) – 3.75 l/ha.
Chlorpyrifos (48% e.c.) – 2.5 l/ha.
Tria zophos (40% e.c.) – 2 l/ha.

The spray should be directed along the plant rows, aiming at run-off to the soil below. Timing should idedly be aimed at the young caterpillars whilst still feeding on the plant leaves or on the soil surface.

(b) Soil application of bromophos (w.p.) or chlorphyrifos granules.

(c) Baits of moist bran mixed with DDT, r-BHC, or endrin may be effective against older caterpillar.

Generally cutworms are extremely difficult to control, especially in 'boom' years, for by the time infestations are apprent the susceptible stages of the larvae are often past, and damage may be already quite serious. The sporadic nature of cutworm papulation outbreaks makes preventative treatments rather futile in most areas. Finally the soil-dwelling larvae often under dense and continuous crop foliage, make targets difficult to 'hit' with insecticides and in many area such high-volume spraying is not feasible because of water shortage and equipment restrictions.

(B)

PRINCIPLES OF PEST CONTROL AND PRACTICES

Agriculture consists of a single plant species cared and managed by man, so as to yield maximum consumable parts of the plant. The agricultural production depends upon high input of energy and materials in the form of deep ploughing of the soil, inorganic fertilizers, irrigation, as well as use of pesticides to protect yield from being destroyed by different types of Pests. Pest in the widest sense is any animal (or plant) causing harm or damage to man, his animals, his crops or possessions, even if just causing annoyance. From an agricultural point of view, an animal or plant out of context is regarded as a pest (individually) even though it may not belong to a pest species. Thus an elephant in a shamba (a cultivated plot of ground) is a pest, but next-door in a game park it is not, and is infact a valuable national asset there. Similarly, maize plants growing in a field with cotton have to be regarded as 'weed' pests.

Many insects belong to generally accepted pest species, but individual populations are not necessarily always-pests; that is, of course, not necessarily *economic pests*.

Economic Pest

On an agricultural basis, we are concerned when the crop damage caused by insects leads to a loss in yield or quality, resulting in a loss of profits by the farmer. When the yield loss reaches certain proportions (5-10% loss) the pest can be defined as an economic pest. Obviously a loss of 10% of the plant stand in a cereal or sugarcane field is not particularly serious, whereas the loss of a single mature tree of Citrus or Mango is important.

Economic Damage

This is the amount of damage done to a crop that will financially justify the cost of taking artificial control measures, and will clearly vary from crop to crop according to its basic value; the actual market value at the time and other factors.

Economic Injury Level (EIL)

This is the lowest population density that will cause economic damage and will vary between crops, seasons and areas. But it is of basic agricultural importance that it is known for all the major crops in an area.

Economic Threshold

It is desirable that control measures, to be taken to prevent a pest population from actually causing economic injury. So the economic threshold, according to *Stern, Smith, Vanden Bosch* and *Hagan*, 1959, is the population density of an increasing pest population, at which control measures should be Started to prevent the population from reaching the economic injury level.

Pest Load

This is the actual number of different species (and number of individuals) of pests found on either a crop or an-individual plant at any one time; this would usually be a pest complex but could also be a monospecific population.

To meet people effective control of a pest probably means its total elimination from an area, but complete eradication is rarely attainable so that such a definition is inadequate. Effective pest control may be described as reduction or maintenance of a pest pollution below the damage threshold.

Categories of Crop Pests

Individual species of cultivated plants are almost invariably affected by a range of pests of differing importance. Three categories of plant pests may be recognised and are shown diagrammatically in (Fig. 6.100).

1. **Key pests :** In any one local pest complex it is usually possible to single out one or two major pests that are the most important; these are defined as key pests, and are usually perennial and dominate control practices. If these key pests left uncontrolled, population levels remain above the damage threshold, most if not all the time. It is of course necessary to establish economic threshold for these key pests in order to be certain when to apply control measures. For it has been often observed that the only presence of a few individuals of a key pest species in a crop may cause undue alarm and lead to unnecessary pesticide treatment. Key pests owe their status to several factors, including their usually high reproductive potential, and the type of damage they inflict on the host plant. An example is provided by codling

moth which is a serious pest almost everywhere that apples are grown.

2. **Occasional or secondary or sporadic pests :** Defined by *Coaker* as a species whose numbers are .usually controlled by biotic and abiotic factors which ocasionally break down, allowing the pest to exceed its economic injury threshold or in other words for a long periods of time populations remain low and insignificant but every so often, on either a regular or sporadic basis they increase to damaging levels and cause a problem. The reasons for fluctuations in populations of occasional pests are often climatic or due to biological factors such as the incidence of natural enemies, but in many cases the details are as yet unknown.
3. **Potential pests :** This term is used occasionally in the literature and refers to a minor pest species that could become a major pest following some change in the agroecosystem *i.e.,* when their natural control is interfered with and they become elevated to actual pest status. European red mite follows this pattern. Earlier this century it was practically unknown in apple orchards and never reached damaging levels. However in recent years, European red mite has become a major pest in many parts of the world because its natural enemies have been seriously depleted by the widespread use of broad spectrum synthetic insecticides.

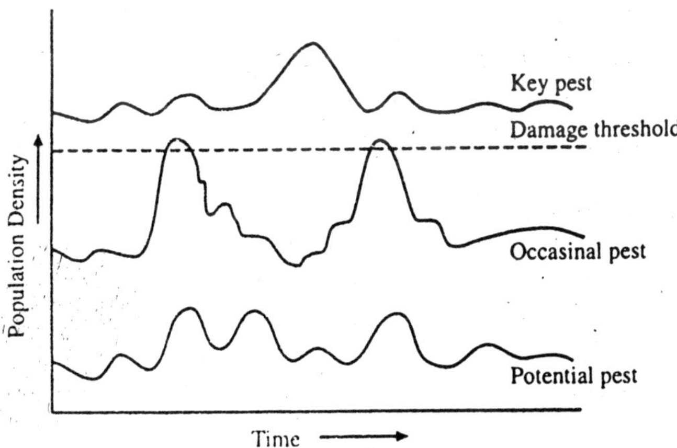

Fig. 6.100 : Categories of crop pests and their population levels with respect to damage threshold.

Applied or Artificial control methods are mostly divided into the preventive and curative or *directive methods. Preventive* or prophylactic methods to ward off pest invasions include measures such as the removal of weeds, grasses and dead branches from fields; through usage of good seeds, .proper cultivation, manuring, irrigation, removal of crop remains and stubble, growing of pest resistant varieties, treatment of seeds with chemicals, rotation of crops, mixed cropping, changing of sowing, planting and harvesting times, cutting and ranking of field bunds etc.

Among the *direct* methods are the physical and mechanical methods. The physical methods involve the use of electricity, sound waves, infra-red and X-rays. The mechanical method is concerned with the operation of machinery such as hopper dozers, fly and maggot traps, light traps, centrifugal force machines for stored grains and manual operations involving the hand picking of egg masses, larvae and adults.

In general pest control procedures fall rather naturally under a number of headings such as chemical control, biological control and cultural control, which are descriptive of the techniques involved and which have until recently been distinct well defined categories. However, some new procedures cut across these established boundaries and furthermore there is now a strong tendency to use two or more approaches together in system of integrated control or pest management. The main pest control procedures are listed in Table 6.1 with an explanation as to what each involves.

Table 6.1 : The main types of pest control procedures and what they involve.

Procedure	Components
Cultural control	Manipulation of cultural practices to the pest's disadvantages by such means as method and time of cultivation, modification of sowing dates and manipulation of irrigation practices.
Plant resistance	The use of species or varieties of plants that can grow and produce despite the presence of the pest.
Biological control	In an narrow sense involves the use of predators and parasites (mostly other insects) to control pest species. In the broader sense also includes disease organisms (pathogens). Use of the latter is sometimes referred to as microbial control.

Chemical control	Principally involves the use of chemicals which are toxic to insects (insecticides) but also includes the use of chemicals which modify insect behaviour, (*e.g.*, attractants and repellents).
Plant and animal quarantine	Involves restrictions on the international movement of plant and animal material to minimise further spread of pests (and of diseases).
Mechanical control	Includes killing or trapping pests by mechanical means or the use of barriers to prevent pests from gaining access to plants, stored produce or other materials.
Physical Control	Involves modification of some physical feature of the environment to render it unsuitable to a pest (*e.g.*, the lowering of temperature of stored grain), or the utilisation of some physical property as an attractant (*e.g.*, light traps for night flying insects).
Pest management (integrated control)	The blending together of two or more of the foregoing procedures into an overall harmonious system of control. Applies particularly to the integration of chemical and biological control.

Note : In addition to the above there are other specialised pest control procedures (such as the sterile male technique) that do not fit any of the described categories).

Chemical Control

Chemical control in the conventional sense involves the use of chemical (insecticides) to kill pests. However, in recent years many discoveries have been made of substances that influence insects behaviour, for example, chemicals which are attractive to pests or which produce some other effect such as sterility, without killing the insect concerned. Where such substances can be exploited to manipulate pests and provide or help to provide solutions to pest problems their use clearly falls within the scope of chemical control. A modern definition of chemical control might therefore be the use of chemicals to kill, deter or in the other ways influence pests for control purposes.

Since World War II insecticides have played a dominant role in pest control and despite increasing attention to alternative means of dealing with pest problems, continue to do so in most situations. There are number of reasons for this. Firstly, insecticides are usually rapid and curative in effect so that action can be taken quickly against a pest problem which threatens a grower's crop. Moreover such action can be undertaken by the farmer or grower himself so that he does not have to rely on the services of a specialist. A further point is that application of insecticides can provide a very high level of pest control and thus enables high quality blemish free produce to be grown. In many branches of horticulture in particular, such as fruit growing, top quality produce is essential to ensure good market prices and profitability.

Inherent in the use of insecticides for pest control are a number of drawbacks, as with other control methods. A major one is that insecticides for the most part are not very selective for the pest to be controlled and to some extent also affect other living organisms. If these are parasites or predators of the pest against which treatment is aimed, control initially may be good but when the pest population recovers (as it inevitably does in time) it may rise to higher levels than before, because some of its natural controls have been destroyed, a phenomenon known as resurgence. Insecticides may also eliminate natural enemies of potential pests so that they are raised to pest status. A further group of beneficial insects that may be harmed by insecticides are pollinators of cultivated plants, bees in particular.

Another aspect of the rather non-selective action insecticides is that many are toxic to higher animals resulting in harmful effects on wild life. There may also be hazards to humans. These take the form of risk of poisoning for those manufacturing and applying pesticides and of traces of chemicals (residues) associated with harvested produce.

A point of great practical concern for continued effectiveness of chemical control is the widespread ability of pest species to develop resistant strains. With many crop and public health pests this is a very serious problem for which there is as yet no entirely satisfactory answer.

Finally, although the great majority of uses of.chemicals for pest control have been profitable up to now, rapidly increasing costs of manufacture are drastically changing the situation and many farmers and growers are questioning whether such expenditure can continue to be justified. For some crops and some situations the answer may well be that it cannot. In any case, chemical pest control inevitably involves continual expenditure in contrast to some other methods, such as biological control by inoculation, where the initial cost may be the only cost.

Most problems associated with chemical control arise from the fact that use of an insecticide against a crop pest does not simply involve action of the chemical on the pest alone. This simple interaction can be represented by the notation:

chemical → pest

In fact the interaction is between the chemical and a complex biological system of which the pest is only one component. This interaction is more truly depicted by

chemical ← ecosystem in which the pest occurs.

Failure to recognise the complexities that may be involved accounts for many problems in the use of insecticides.

Origin of Chemical Control

Insecticides were in use from very early times and as early as 200BC; boiling a mixture of bitumen (mineral pitch, or asphalt) and blowing the fumes through grape leaves was advocated to keep away the insects. Sulphur was considered to be injurious to insects in 100 BC. The toxic nature of arsenic was known about 40 to 90 AD and arsenic sulphide was used by the Chinese before 900 A. D. Arsenic in honey was suggested as an ant bait since 1669. The use of tobacco in the control of the lice bug on pear trees was in vogue in 1690. Pyrethrum was widely used before 1800 in Persia.

The use of modern insecticides commenced in 1867 with the application of Paris Green for the control of the Colorado beetle. Mostly organic chemicals and a few plant products were in use until 1939 when the whole concept of insecticides and of insect control was revolutionised with the discovery of the insecticidal spectrum of DDT. Since then a large number of compounds have been synthesised, examined for their insecticidal activities and developed. In 1941-42 .the insecticidal properties of BHC were discovered by British and French investigators. During the same period the potentialities of phosphorus chemicals were investigated by the Germans, which led to the development of parathion, malathipn, demeton, TEPP, etc. Since then phenomenal progress has been made in the development of insecticides, as a result of which insecticides have now been found among phosphates, phosphorothiolates, phosphorothionates, phosphorothiolothionates (phosphorodithioate), phosphonates, phosphonothiolates, phosphonothionates, phosphonothiolothionates (phosphonodithioate), carbamates and other classes of compounds. Other families of chemicals are also being investigated for possible insecticidal activity.

CLASSIFICATION OF INSECTICIDES

Insecticides may be classified in several ways based on the mode of entry, mode of action and the chemical nature of the toxicant or active ingredient.

Classification Based on Mode of Entry

Based on the manner in which toxicants are administered to the insects and their mode of entry into the body, the insecticides are classified into three groups-stomach poison, contact poison and fumigant.

Stomach Poison

The toxicant, applied to the food, when ingested by the insect kills it primarily by action on or absorption from the digestive system. Though usually it is limited to the control of chewing insects, under certain conditions it may also find use in controlling insects with sucking, siphoning, sponging or lapping type of mouth parts.

The possible ways of ingestion of the stomach poisons are :

1. The insect while feeding on its natural food, such as foliage, feathers of birds, etc., covered thoroughly with the toxicant, ingests the poison also.
2. By feeding on poison baits consisting of the toxicant and an attractant a suitable food material.
3. The toxicant sprinkled on runways is picked by the antennae and legs of the insect and it is likely to ingest the poison while cleaning these appendages with its mouthparts.
4. Sucking insects, while feeding on plants treated with systemic insecticide, suck the plant sap containing the toxicant which acts as a stomach poison in the insect.

A stomach poison should be sufficiently stable, cheap and available in large quantities and should not be distasteful as to repel the insect. It should be able to bring out quick kill of the pest. The chemical should not be soluble in water and phytotoxic to plants. It should have finer particles and spread uniformly and adhere well to the treated surface. It should not leave any harmful residue on treated surface.

Examples : Lead Arsenate to control Apple orchard and shade three pest control; sodium fluride (Naf) to control cutworms, earwings, grasshoppers etc.

Contact Poison

A toxicant which kills the insect by contact and entry into the body through the vulnerable sites found on its body is said to be a contact poison. It may be applied directly on to the body of the insect in a spray or dust. On the other hand, if application is made so as to leave a residue of the toxicant on plant surfaces, animals, habitations, etc. frequented by insects, the toxicant is likely to be picked up by the insect when it comes in contact while crawling or passing over the treated area. This type of poison is particularly effective for the control of sucking insects.

When applied, the toxicant spreads quickly over the entire surface of the body of the insect, possibly in the wax monolayer and apparently from it gets absorbed on to the surface of the cuticle. It penetrates into the body of the insect through sutures, membranes, bases of setae and possibly other parts of entry. Though it was widely held that the insecticide penetrates through the integument of the body wall and gets carried to the target organ, the central nervous system by the haemolymph; recent investigations have shown the toxicant does not penetrate into the haemolymph in significant amounts and reaches the site of action via the integument of the tracheal system.

Contact insecticides are found in:

1. toxicants of plant origin such as rotenone, nicotine, anabasine, ryania, sabadilla and pyrethrum;
2. synthetic organic compounds such as BHC, DDT, toxaphene, chlordane, organic thiocyanates, organic phosphates and dinitrophenols;
3. oils and soaps;
4. inorganic compounds such as lime-sulphur and sulphur, and arsenic trioxide and sodium fluoride to a limited extent.

Many contact insecticides at higher dosage levels may prove to be phytotoxic on plants.

Fumigant

A toxicant applied as a vapour enters the tracheae of the insect through the spiracles in the form of gas and brings about its kill; such gaseous toxicant is called a fumigant. This finds use in the control of all kinds of insects irrespective of the type of mouth-parts they possess. Their application is limited to plants or products in air-tight enclosures, tents or buildings or to soil. A fumigant which generally boils at about room temperature is the most useful one and such chemicals include methyl bromide, hydrogen cyanide, and ethylene oxide. The essential requirement of a soil fumigant is slower release of vapour from chemicals boiling as high as 180°C. Fumigant action is evidenced when naphthalene and paradichlorobenzene having relatively high vapour pressure are used in tight containers, and also in certain contact insecticides like DDVP, lindane, etc. An ideal fumigant is determined by its relative effectiveness, cost, safety to human beings, animals and plants, inflammability, penetrating power, effect on germination of seeds, reactivity with household furnishings, etc..

CLASSIFICATION BASED ON MODE OF ACTION

Based on the ways in which the chemicals act upon the system of an insect to cause its death, insecticides may be classified as follows:

Physical Poison

A physical poison exerts a physical rather than a biochemical effect and brings about kill of insects by asphyxial (exclusion of air) effect as with heavy oil and tar oils, or by effecting a loss of body moisture from the insect by inert dusts. Epicuticle of the insect gets lacerated by abrasive dusts like aluminium oxide and this may cause water loss. Water absorbent materials like charcoal remove water. Certain dusts may combine both the characteristics.

Protoplasmic Poison

A protoplasmic poison is primarily associated with precipitation of protein. The cellular protoplasm of midgut epithelium is destroyed by inorganic stomach poisons like fluorides, arsenites, arsenates, fluosilicates, fluoaluminates and borates, alkaloid reagents—nitrophenols and nitrocresols, fatty and mineral acids, formaldehyde, and ethylene oxide and heavy metals like mercury and copper.

Respiratory Poison

A respiratory poison is associated with blocking of cellular respiration and inactivation of respiratory enzymes as with fumigants like HCN, H_2S and CO (carbon monoxide).

Nerve Poison

The action is primarily associated with the solubility of the toxicant in tissue lipoids and inhibition of acetylcholinesterase in insects and warm-blooded animals.

A chemical substance acetylcholine is formed in the nervous system of insects and mammals mediating in the conduction of the nervous impulses over the microscopic gap between nerves or between nerve and gland or muscle. Acetylcholine, after it has served its function, is destroyed by an enzyme, acetylcholinesterase present in the tissues so that the end-organ (nerve, gland or muscle) may return to its resting state preparatory to repeating its function. It acetylcholine is not destroyed by acetylcholinesterase, it will continue to cause impulses to move along the nerve and cause increased excitation which induces the production of a new coactive substance by the central nervous system. When produced in large quantities this natural substance becomes a toxicant and disrupts the normal nerve functions resulting in tremors, convulsions, muscle paralysis and finally death.

The acetylcholine is hydrolysed by the enzyme acetylcholinesterase to form a choline and an acetate-enzyme, the latter reacts instantaneously with water and splits into acetic acid and the free enzyme. When required the enzyme is used again to hydrolyse more acetylcholine whereas the acetic acid and choline are used in other physiological processes.

Organic insecticides are active inhibitors of acetylcolinesterase enzyme in insects and mammals. The nerve poisons include the organochlorine compounds (DDT, lindane, paradichlorobenzene, carbon tetrachloride, ethylene dichloride), aromatic and olefinic hydrocarbons (kerosene, gasolene, naphthalene), botanic insecticides (pyrethrin, nicotine), organic phosphates (parathion), N-methyl carbamates and miscellaneous chemicals like aniline and carbon disulphide. DDT does not inhibit cholinesterase but initiates high nervous excitation. In the case of poisoning with organophosphatic compounds the phosphorylated enzyme is irreversibly inhibited. On the other hand, carbamates are irreversible chlorinesterase inhibitors.

CLASSIFICATION BASED ON CHEMICAL NATURE OF INSECTICIDES

With the advent of more and more newer groups of insecticides, many of them entering the insect body is more than one way, it becomes difficult to classify them based on mode of entry. This overlapping is avoided if insecticides are classified based on their chemical nature. Two major divisions, viz. inorganic and organic insecticides, are recognised. The following is a broad classification of insecticides (includes acaricides, rodenticides and some nematocides also) based on chemical nature.

CHART

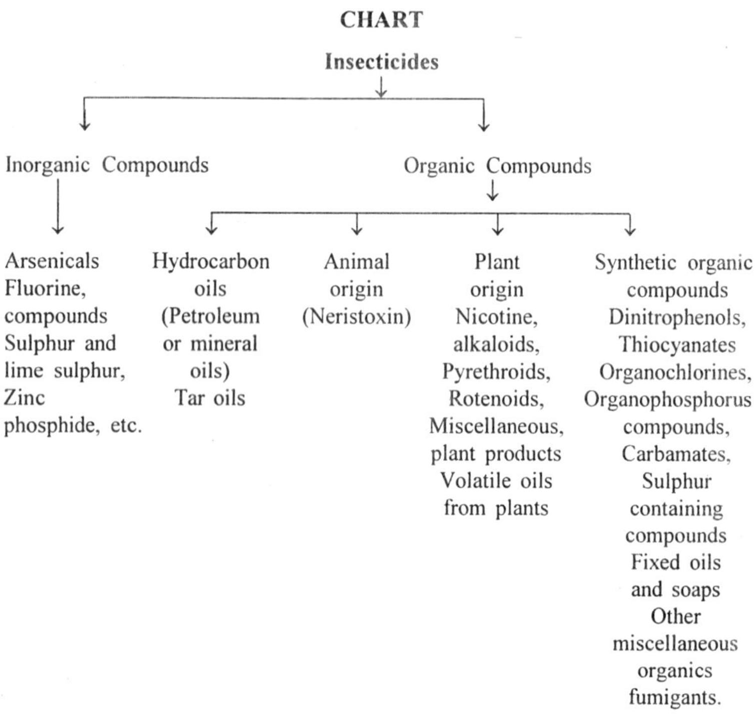

Inorganic Insecticides

Inorganic insecticides are compounds of antimony, arsenic, barium, boron, copper, fluorine, mercury, selenium, thallium and zinc which are of mineral origin and elemental phosphorus and sulphur.

Lead arsenate is commonly used in the form of acid lead arsenate ($PbHAsO_4$) and less so as basic lead arsenate, a mixture of Pb_4 (PbOH) $(ASO_4)_3$. H_2O and $Pb_5(PbOH)_2$ $(ASO_4)_4$. Acid lead arsenate is used primarily in fruit orchards, on forest and shade trees, and on shrubs to control chewing insects. Although not as toxic to insects, basic lead arsenate is safer to use on plants in foggy coastal regions than is the more soluble acid lead arsenate.

Sodium fluoride (Naf) was once a common insecticide used in cockroach as antibait. A related compound, sodium, fluosilicate (Na_2SiFe_6) is used now in baits for ants, cockroaches and grasshoppers. Cryolite (Na_3 Alf_6), which is mined in Greenland as well as manufactured in the United States, has proven effective against a number of truck-crop insects and is used on plants that are generally sensitive to chemical injury.

Sulfur is the only element that has been found valuable as an insecticide. Finely ground sulfur has been applied widely as a dust for control of mites and of certain fungi. The addition of a wetting agent to sulfur makes it possible to use sulfur as a spray. Boiling together sulfur and freshly slaked or hydrated lime produces a mixture of compounds, known as liquid lime-sulfur. The active materials in the mixture are probably the calcium polysulfides (CaS_4, CaS_5 and others). This material has been used as a fungicide, and on fruit trees against scales, aphids and mites. Liquid lime-sulfur may be mixed with a stabiliser and evaporated to dryness to form dry lime-sulfur. Dry lime-sulfur is less effective than the liquid but is easier to handle.

Many other inorganic compounds are now or have been employed as insecticides.

Sodium selenate (Na_2SeO_4) applied in the soil, acts as a systemic and was used in the post primarily to control aphids and mites on greenhouse ornamentals. Cabbage maggots were controlled with mercurous chloride (Colonnel), Hg_2Cl_2 solutions. Phosphorus has been used in cockroach bite and thallous sulfate (Tl_2SO_4) in ant bait.

Organic Insecticides

Hydrocarbon Oils

The mineral oils are the petroleum oils derived from sedimentary rocks; complex solutions of hundreds of hydrocarbons are present in the petroleum oils and kerosene, gasoline, lubricating oils, asphalt, tar and many other mixtures are derived from them.

Oils in their natural state are highly phytotoxic but when used in an emulsion they may under certain conditions be safely applied to plants. Mineral oil is a heterogeneous mixture of saturated and unsaturated chain and cyclic hydro carbons. Certain parts of this mixture are much more useful as insecticides than others. To designate the quality of oil, data commonly supplied on the label are viscosity, boiling or distillation range, and sulfonation rating (purity or degree of refinement).

Viscosity is usually stated in terms of the time in second for 60cc to flow through a standard orifice. In general oils of low viscosity are safer to use than are those of higher viscosity; however, the viscosity rating may be misleading, because oils from different parts of the country having the same viscosity rating may react quite differently when sprayed on foliage.

Boiling or *distillation range* is a more important character of oil and is an indirect indication of volatility. Phytotoxicity increases with increase in distillation range. However, it is also true that heavier spray oil fractions have a greater effectiveness against insects that do lighter oils. In both cases this probably has to do with the period of time the oils remain in contact with the plant or animal surface. The lighter oils, being more volatile, soon escape into the air, whereas the heavier oils remain in contact with the surface for a longer time.

Oils are composed of both saturated and unsaturated hydrocarbons. The unsaturated hydrocarbons are unstable and readily form compounds that are toxic to plants. Consequently if the amount of unsaturated hydrocarbons present in an oil is lower, that oil is safer for use on plants. The usual testing procedure of determining the amount of unsaturated hydrocarbons present is by means of the sulfonation test. The oil is to be tested is reacted with strong sulfuric acid and the unsaturated hydrocarbons that react with the acid sink to the bottom. The unreacted part, the *unsulfonated residue* (U.R.) is measured as a percentage and is used as a measure of purity. Dormant oils have a U.R. rating of 50 to 90 per cent and the more highly refined summer oils a rating of 90 to 96 per cent.

Oils are employed in a number of ways. They may be used as solvents or carriers for insecticides. Diesel fuel is often used as a carrier for insecticide in airplane application. Oil may also serve to carry an insecticide over water being treated to control mosquitoes.

Oils by themselves are insecticidal and based on time of usage, are classified as either summer or dormant oils. *Summer oils*, which are highly refined and less phytotoxic are applied to trees in foliage. For example, citrus trees may be treated with summer oils to control mites and scale insects. The *dormant oils*, which are less refined are applied when no foliage is present. To increase plant safety as well as insecticidal action, new types of oils have been developed; these are called *superior* (or *supreme*) *spray oils*. From the time buds show green upto the time leaves are one-half inch long; apple trees may be given a delayed dormant spray with these oils to control aphids, scale insects and mites.

Oil is not applied full strength to trees but is diluted with water and applied as an emulsion containing around 1 to 4 or more per cent oil. Several types of agricultural spray oil stocks are formulated. One type, called *emulsible oils*, contains 95 to 99 per cent oil pulse emulsifier. Some formulations of these produce an emulsion instantly when poured into a tank of water, whereas others require preliminary agitation with a small amount of water. The former are ofen referred to as miscible oils and the latter as emulsible oils.

A second type of spray oil stock is the *concentrated emulsion*. These oils are performed emulsions in a concentrated state and contain about 83 per cent oil plus emulsifier and water. They have the appearance of a whitish plate; some are flowable, others are thick like mayonnaise.

A third type, *tank mix oil*, has the oil, the emulsifier, and the water added separately to the spray tank. Violent agitation forms the emulsion. Such an emulsion is called 'quick breaking' because it separates quickly into its component parts upon coming in contact with the plant surface.

The use of oils has several advantages. They are relatively cheap, have a good spreading capacity, are easy to mix and are reasonably safe to animal life. An important consideration is the fact that insects have not developed any resistance to them. They have several disadvantages such as injury to rubber hose of sprayers, instability in storage, phytotoxicity, and low toxicity to insects. The last characteristic may be corrected by preparation of combination sprays. A highly efficient pesticidal spray is a fungicide-insecticide-oil combination that can be prepared in the spray tank just before application.

Organic Compounds of Animal Origin

Nereistoxin

It is a toxin, chemically known as 4-(N, N-dimethylamino)—1, 2-dithiolano, isolated from the marine annelids Lymbrineris (Lumbri conereis) heteropoda and *L. brevicirra*. It was found to possess insecticidal activity and in insects caused paralysis due to competitive synaptic blocking action in the central nervous system. Among various related compounds synthesized and tested, the compounds belonging to the group of 1, 3 dithiol esters showed insecticidal activity. A product known as cartap, having the chemical name 1, 3-bis (Carbamoylthio)-2-(N, N-dimethylamino) propane hydrochloride, has been proved to be a practical insecticide against rice stem borer and cabbage diamond back moth. It is available under the trade name Padan.

Botanical Insecticides

Toxicants derived from plants have been and are still used as arrow tip poisons and fish poisons. Plants products are used in many ways in insect control and among these nicotine, pyrethrum and rotenone are well known.

Nicotine ($C_{10}H_{14}N_2$)

Though tobacco was used in insect control as early as 1763, its principle alkaloid was discovered only in 1828. At least 15 species of

Nicotiana, Duboisia hopwoodii and *Asclepias syriaca* are known sources of nicotine; the chief sources being *Nicotiana tabacum* and *N. rustica.* The alkaloid nicotine, 1-1 methyl-2-(3-pyridyl)-pyrrolidine, in its pure form is a colourless liquid soluble in water in most organic solvents; when exposed to air it darkens and becomes more viscous. As a nerve poison it is highly toxic to a great number of insects. It is toxic when absorbed through the body wall or taken in through the tracheae or when ingested. It is also highly toxic to mammals and is absorbed through the skin. In India it was mainly in use for the control of cardamom thrips. The dust formulation consists of nicotine sulphate absorbed on active bentonite or inactive gypsum carrier and in presence of moisture nicotine sulphate changes into free nicotine. When used as a fumigant, dried tobacco may be burnt to produce the smoke. Vapourisation of extracts by heating or burning the extracts absorbed on a combustible material releases the nicotine.

Pyrethrum

derives its toxicity from five esters, pyrethrin I, pyrethrin II, cinerin I and cinerin II and jasmoline II are the active toxicants. The esters or toxic principles are derived from the two acids, chrysanthemic acid (leading to pyrethrin I) and pyrethric acid (leading to pyrethrin II) and the alcohols cinerolone, jasmololone and pyrethrolone.

The insecticidal activity of pyrethrum was discovered in Iran (then called Persia) around 1800. Great secrecy surrounded the source of the material, which was called Persian Powder and was sold at extravagant prices for louse and flea control. The discovery of the source of the powder, according to some, was made by an Armenian merchant who has been travelling in the Caucasus Mountains. Another version of the story is that the secret was given to Russian soldiers by prisoners of war during the Russo-Persian War in 1827 or the Russo-Turkish war of 1828-29. The original powder came from the daisies *Chrysanthemum coccineum* and *C. carneum*. In 1840 it was found that *C. cinerariaefolium*, produced in Dalmatia, had a higher insecticidal activity and this species of daisy soon replaced the previous sources.

Pyrethrum was introduced into the United States about 1858 and found wide application. From 1876 to about 1914 an attempt was made to grow pyrethrum commercially in California. The flower has also been experimentally grown around Fort Collins, Colorado. But because neither of these two ventures was a commercial success, the United States still imports all of its pyrethrum.

Pyrethrum is produced by grinding the flowers and mixing them with a dust diluent or more commonly by extracting the active materials with solvents and formulating the extracts into sprays and dusts. The active materials are esters that are rapidly hydrolyzed by alkali and decomposed by sunlight.

Pyrethrum is toxic to most insects with which it comes in contact. Its primary target is the ganglia of the insect central nervous system; because it acts rapidly and affected insects fall quickly. Such action is called 'knock down'. Insects knocked down do not necessarily die; some will recover if allowed to do so. Pyrethrum lacks persistence in the field because of its breakdown by sunlight; thus it leaves no harmful residue. It acts almost entirely as a contact poison. Pyrethrum is relatively harmless to mammals by ingestion, although it is poisonous if injected into the bloodstream, and the dust may cause allergic reactions in some people. It is harmless to plants. Two disadvantages of pyrethrum are that it breaks down rather rapidly and it is relatively expensive.

Pyrethrum is formulated as dusts, sprays and aerosols. Usually a synergist such as sulfoxide or piperonyl butoxide is added to increase toxicity and also to reduce the amount of pyrethrum necessary for a satisfactory kill.

A material with such low mammalian toxicity would obviously find many uses. It can be applied to edible plants shortly before harvest because it leaves no harmful residue. It is effective against a wide range of insects and is one of the insecticides preferred by the backyard gardener. Pyrethrum has been used on livestock as a toxicant for external parasites and as an insects repellent. One of its largest uses is in household sprays. Pyrethrum mixed with stored grain protects the grain from insect injury.

Rotenone ($C_2H_{22}O_6$)

Rotenone is found in 68 species of leguminous plants. The important commercial ones are two species of *Derris*, which grow in the Far East and contain 5 to 9 per cent rotenone and several species of *Lonchocarpus*, which grow in the Amazon Valley of South America and contain 8 to 11 per cent rotenone.

To obtain rotenone the roots of the plants are dried, powdered, and either mixed with a dilutent powder to be used as a dust or extracted and the extract used in making sprays or dusts, Pure rotenone is white crystalline and insoluble in water but soluble in many organic solvents. It is rapidly oxidized in the presence of light or alkali.

Rotenone is a selective insecticide; it is very effective against some insects and inactive against others. It kills insects by inhibiting respiratory metabolism and blocking nerve conduction. Rotenone acts as both a stomach and a contact poison. Although, mammalian oral toxicity is high—the LD_{50} ranges from 10 to 30 mg/kg—rotenone may be used with relative safety on most mammals except swine, which are highly susceptible. It causes no phytotoxicity.

The usual formulations of rotenone are dusts, which must be mixed with nonalkaline carriers to prevent breakdown wettable powders; emulsifiable powders; emulsifiable concentrates and aerosols. Rotenone finds many applications. It is widely used on vegetable and ornamental and is used for control of external pests of livestock. The material has several advantages, such as safety to both mammals and plants, but it also has disadvantages, such as breakdown in storage, slow action and low toxicity to some insects.

There are several other botanical insecticides of minor importance. *Sabadilla* comes from the seeds of a lily grown in Venezuela. It has been used against human lice and for control of Homoptera and Hemiptera. *Ryania* is obtained from stems and roots of *Ryania speciosa* family Flacourtiaceae. It is less toxic to mammals than rotenone; it is more stable and possesses a longer residual action. It has been used primarily against the European corn borer, codling moth, citrus thrips, and some other lepidopterous insects. *Hellebore*, which comes from a lily, was in wide use for many years ago for control of pests of vegetables, fruit and livestock.

Lobelia excelsa plant is found on the western ghats in South India. The leaves are curved in shade and chopped and 1 kg of the leaves are soaked in water for about one day. The infusion is filtered and the filtrate made upto 20 litres and 60g of soap is added. It has been found effective against aphids on snakegourd and cowpea, tingid bugs on brinjal and mites on castor and lady's finger.

Volatile Oils, from Plants

The volatile or ethereal oils are obtained from special glands of plants and their pungent odour is characteristic of the plant source. The oils are not saponified and are not greasy or viscous. The volatile oils are chiefly used as attractants in baits or as repellants. The attractants include eugenol-and geraniol. Citronella and oil of cedar are repellants. Other common examples of volatile-oils from plants sources are menthol, comphor, oil of peppermint and wintergreen.

Synthetic Organic

Synthetic organic compounds dominate the field of insecticidal control today. Although several synthetic organic insecticides, were known previous to world war II, not until the discovery of DDT did interest in these chemicals become widespread. Following the release of DDT, many companies became active in searching for new insecticides. The majority of synthetic organic insecticides now is use were developed after 1947, although several were marketted before. Dinitrocresol was sold as an insectiside in Germany as early as 1892 and by 1932 several thiocyanates were commercially available in the United States.

The development of the organophosphorus insecticides began in Germany. During second world war *Gerhand Schrader* and other German chemists, searched for synthetic compounds to replace insecticides that had become unavailable in Germany because of war conditions. Several organophosphorus compounds proved highly toxic to insects. From these chemicals came a number of effective and practical insecticides, such as parathion, tepp, schradan, and demeton. Rapid developments make an up-to-date classification difficult. One generally applicable insecticides are described below:

Dinitrophenols

These are derivatives of 4-6-dinitro 2-alkylphenols and their salts or esters.

Examples:

(a) DNOC (4, 6-dinitro-o-cresol). It is used as herbicide. It is available as 20 to 33% water paste of sodium salt and as a 40% wettable powder. It is used as dominant ovicide applied for the control of aphid, mite and scale insects on fruit trees and in poison bait for grasshoppers.

(b) DINOCAP ($C_{18}H_{24}N_2O_6$): The structural formula is 4, 6- dinitro-2-capryl-phenyl crotonate. It is an acaricide and available as 25 % wettable powder.

Organic Thiocyanates

Several insecticides of organic thiocyanates have appeared since 1932. Loro is a commercial product containing technical louryl thiocyanate and a wetting agent and is used for the control of aphids, thrips, spider, mites etc. The technical product containing 80% isoborynyl thiocyanoacetate and the remainder closely related bornyl and fenchyl thiocyanpacetates, as a 20% solution in deodorised. Kerosene is available as Thanite and is used as fly and cattle sprays and as household insecticide. They are contact poisons and act on the nervous system of insects.

Organochlorine Compounds

The chlorinated hydrocarbon insecticides are molecules composed of chlorine, hydrogen and carbon and occasionally oxygen or sulfur. They are also known as organochlorine insecticides.

DDT (Dichloro Diphenyl Trichloroethane)

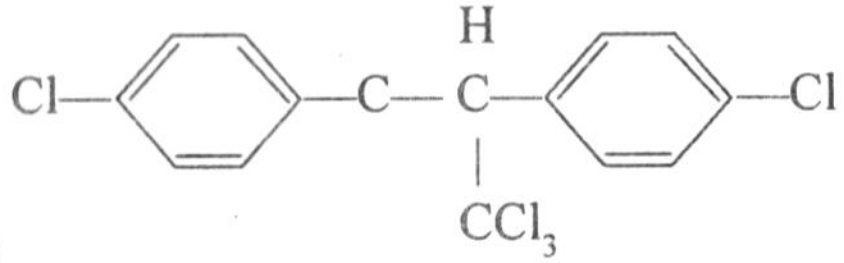

H

Cl— —C— C— —Cl

CCl_3

DDT is one of the most important insecticides ever devised by man. It has saved millions of lives directly through control of insect vectors of human diseases and indirectly through control of crop pests with resulting increase in food. Though condemned by environmentalists, it has saved millions of *acres* of forest from destruction by insects.

In 1932 chemists of the Geigy Company in Switzerland began a search for an improved mothproofing compound. Seven years later *Dr. Paul Muller* found a material that showed great promise. The company was given a patent on it by the Swiss government on March 7,1940, and later it became known as DDT. DDT soon proved itself as an agricultural insecticide by stopping an outbreak of the Colorado potato beetle during 1941 in Switzerland. At the same time it was found highly effective in controlling fleas, lice, mosquitoes, and flies. In 1942 samples of the compunds were sent to the British branch of Geigy Company and two hundred pounds were sent to the United States. Production of DDT in the United States began after a pilot plant was established at Norwood, Ohio, in May 1943. All of these developments were kept secret at the time.

The first wide use of DDT same in 1944, when typhus broke out among the civilian population of Naples. Mass delousing with the new insecticide was started and within three weeks the typhus outbreak was brought under control. Never before during a war had an epidemic of typhus been halted. In World War I, 25 per cent of the Serbian soldiers perished from typhus and Russia lost several millions to the disease.

In Naples 1.3 million civilians were treated with DDT during January, 1944. It became more and more difficult to keep secret such mass medication and such spectacular control of treaded disease. On August 2, 1944, the British government disclosed the identity of the chemical. The world recognised the major advance made in medical science through the development of DDT and in 1948 Dr. *Muller* was awarded the Noble Prize for medicine.

DDT is made by the condensation of chloral and monochlorobenzene in the presence of sulfuric acid. In its pure state it is crystalline powder with a melting point of 108°C. The crude form is a waxy solid with a melting point near 89°C is insoluble in water, but soluble in most organic solvents. DDT is broken down by alkali and by organic bases, otherwise it is stable and inert. It is not attacked by ordinary acids.

DDT is effective against a great number of insect pests. However, it has little toxicity to most Orthoptera, the boll weevil, Mexican bean beetle and most aphids. Also, it has relatively little acute toxicity to mammals. But there is concern about possible chronic effects in mammals exposed to low levels of DDT over a long period of time. Human deaths attributed to DDT have always included solvent, and in all probability it was the solvent which caused the death and not the DDT. This chemical has been widely used in insect control, yet no increase in illness of applicators has been detected. Workers exposed to DDT in its manufacture have shown no ill effects. DDT is known to be stored up to certain levels in the fatty tissues of animals and to be excreted in the milk.

Although DDT is safe, to use on most plants, the crude form causes phytotoxic symptoms on cucurbits, tomatoes several other plants. Pure DDT (mp 108°C) does not cause this reaction. One surprising effect of DDT is that at very-low concentrations stimulates growth in potatoes and cabbage much as plant hormones do. In fact, symptoms of plant injury by DDT are often 'hormone like' in nature.

DDT is formulated in several ways; solution, emulsifiable concentrates, dusts, wettable powders, aerosol bombs and granules. Its greatest boon to mankind has been its effectiveness in controlling the insects that carry human diseases.

DDT is a stable product that does not break down readily and thus tends to remain in the environment. It is highly insoluble in water and does not leach to any extent from the soil. It accumulates in fat and thus becomes stored in the bodies of animals.

DDT is virtually banned in the United States and its demise is a sad example of hasty research, biased reporting, unfounded claims and emotional response. Despite much experimentation, there still remain many unanswered questions concerning DDT. Unfortunately emotionalism has so clouded the issues that any research done now is so violently attacked or supported by the pros and cons that the factual data become lost in the fray.

Methoxychlor, another development of the Geigy Company is closely related to DDT but is safe to use where environmental contamination is a concern. In general it is useful against the same insect pests as DDT; however, it is more effective against plum curculio and the Mexican bean beetle and less effective against the European corn borer and corn earthworm. It is only 1/25 to 1/50 as toxic to mammals as DDT and accumulates less in the fat and is excreted less into milk. It is usually safer to apply on plants than DDT. Another valuable insecticide related chemically to DDT is *perthane*.

Toxaphene ($C_{10}H_{10}Cl_8$), an American development, was discovered by George Buntin of Hercules Incorporated. It is a yellow, waxy solid, insoluble in water but soluble in many organic solvents. Toxaphene is made by chlorinating camphene until it contains 69 per cent chlorine. It is broken down by strong alkali and sunlight. It is effective against a large number of insects and is about four times more toxic to mammals than DDT. It has low chronic toxicity and accumulations in fat dissipate rather quickly in the living animal. It is safe on most plants, but is highly toxic to cucurbits. It is formulated much like DDT. It has been used against grasshoppers, cutworms, webworms, cotton insects, legume insects and livestock insects. It is very toxic to fish and care must be taken to see that none of its gets into a water system.

BHC

1, 2, 3, 4, 5, 6-Hexachlorocyclohexane ($C_6H_6Cl_6$) is known as BHC or HCH. Its common name benzene heachloride is incorrect from a chemical standpoint as HCH could be confused with hexachlorobenzene.

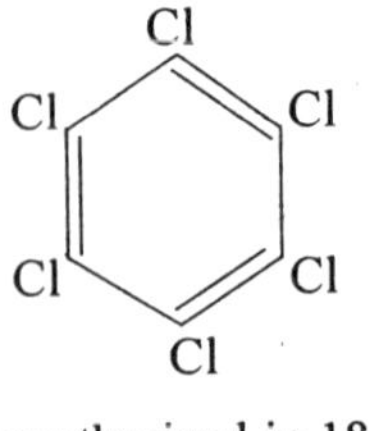

Cl
Cl Cl
Cl Cl
Cl

This compound was first synthesized in 1825 by *Michael Faraday*. It was independently discovered to be an active insecticide by *A P. W. Dupire* in France in 1941 and by *F. D. Leicester* in Endand in 1942.

BHC is produced by the chlorination of benzene in the presence of sunlight. Among the 13 possible isomers found in varying proportions in BHC, the gamma isomer is insecticidally most active. The gamma isomer comprises about 12.5% of crude BHC. The crude product consists of 10 to 18% of the active gamma isomer four other nearly inactive terioisomers (α, β, δ, ε isomer), upto 4% heptachlorocylohexene and a trace of octachlorocyclohexene. It acts as a stomach and contact poison and also to some extent as a fumigant. It is more toxic to insects than DDT. It is also safe to use on a variety of crops but phytotoxic on some especially curubits. The BHC poisoning increases respiration rate and symptoms of poisoning are tremors, ataxia, convulsions and prostration.

It is generally formulated as a 50 % wettable powder or as 5 to 10% dust and is used for the control of a wide variety of pests of crops. BHC in oil solutions are readily absorbed through the skin. When applied as a dust or spray it is very irritating to mucous membranes and eyes.

Cyclodiene Insecticides

These are highly chlorinated cyclic hydrocarbons with "endomethylene-bridged' structures. The toxicity of the compounds is attributed to their high lipoid solubility. Though they are known to act on the ganglia the precise biochemical lesion responsible for toxic action is not understood. Aldrin ($C_{12}H_8Cl_6$), dieldrin ($C_{12}H_8Cl_6O$), chlordane ($C_{10}H_6Cl_8$), heptachlor ($C_{10}H_5Cl_7$), endrin (isomer of dieldrin), and endosulan ($C_9H_6Cl_6O_3S$) are chemically related cyclodiene insecticides. All are insoluble in water and soluble in most organic solvents. They are more toxic to mammals than DDT and endrin is highly toxic. Most are residual insecticides, effective against wide range of insects, for example, chlordane (lice, ticks, fleas, biting flies, household pests etc.); heptachlor (grasshopper, soil insects like white grubs, wire worms, termite etc.). Aldrin (termite, white grubs); Dieldrin (termites, grubs); endrin (Lepidopterous borers infecting sorghum, rice, sugarcane etc.)

In general symptoms of poisoning are hyperactivity, convulsions and prostration.

Organophosphorus Insecticides

The synthetic compounds known as organophosphorus insecticides are organic molecules containing phosphorus. Thousands of these compounds have been synthesized, and many more are possible at least in theory. The group represents the largest'number (about the hundred) of synthetic insecticides under intensive experimentation and in commercial use. Most organophosphorus insecticides have the structure

```
RO    A
  \   ||
   \  ||
      P—X
     /
    /
RO
```

where R is an ethyl or methyl group, A is either a sulfur atom or an oxygen atom, and X is variable and more complex than the other substituents. Compounds such as malathion and parathion contain a sulfur atom in place of A in the above structural formula. Such compounds must be activated or converted to their oxygen analogue (= O substituted for = S) before they are insecticide. This conversion actually occurs in the animal's body. The oxygen analogue of malathion is called malaoxon; that of parathion is called paraoxon.

Some organophosphorus insecticides have the ability to act systemically that is they are absorbed by plants, rendering the sap , toxic to insects, or they are taken in by animals, rendering the blood toxic. Examples of nonsystemic or only slightly systemic compounds are parathion tepp. malathion, and diazinon. Systemic compounds include crufomate, demeton coumaphos, famphur, dimethoate, disulfoton, phorate, mevinphos, and trichlorofoe.

Parathion

Pure parathion is a pale yellow liquid, but in its usual crude form it is a brown liquid with an odour of garlic. It is slightly soluble in water and hydrolyzes rapidly in alkali solution. It is highly toxic to both insects and mammals. Symptoms of parathion poisoning are headache, nausea, and constriction of pupils. Atropine is an antidote.

Parathion is effective against a great many insects and mites. It is non-cumulative in mammals, but highly toxic. It is safe for use on most plants, except for a few varieties of plums and pear and on McIntosh apples. It is used widely in production of friut vegetables ornamentals and field crops. Its high mammalian toxicity precludes its application on livestock, in the household, or where it will come in contact with man. To avoid the danger of dermal toxicity of parathion a method has been developed to enclose it in micro-capsules. Under field conditions the capsules break down and release the parathion. Methyl parathion, the dimethyl analogue of parathion, is less hazardous to mammals than parathion, apparently because of lower dermal toxicity. It has the same toxicity to insects. Among the organophosphorus insecticides synthesized in the United States, methyl parathion is produced in the greatest quantity.

Production in 1974 reached a record high of 51.4 million pounds. Much is used to control insects on cotton and as a substitute for DDT.

Naled is a moderately toxic insecticide that can be used on numerous crops close to harvest'and for control of insects that threaten public health. *Chlorpyrifos* (Lorsban) and *fenthion* (Baytex) are registered for use against mosquitoes and flies and are employed by pest control operators against household and premise pests. A formulation of fenthion (Queletox) is effective in bird control.

A number of organophosphorus insecticides have found specialised uses : the insecticide *dichlorvos* (Vapona) is only moderately toxic to humans and animals, yet highly effective in controlling insects and threaten public health and those that threaten livestock. Both dichlorvos and *crotoxyphos* (Ciodrin), another moderately toxic phosphate, provide effective control of livestock ectoparasites with no problems of residues in milk. A unique characteristic of dichlorvos is its very high vapour pressure, which allows it to act as a fumigant as well as a contact insecticide. *Ternophos* (Abate) is effective against the larvae of mosquitoes, black flies, and midges, yet it is only slightly toxic or has no effect on mammals, wildlife and many aquatic organisms. Animal-systemic insecticides, now commonly employed by livestock growers, belong to the organophosphorus group of insecticides. Examples of compounds effective against cattle lice and cattle grubs are *coumaphos* (Co-Ral), *crufomate* (Ruelene), *ronnel* (Korlan), *famphur* (Warbex), *phosmet* (Prolate), and *trichlorfon* (Neguvon). Trichlorfon is also registered for use against insects on forages; grains, vegetables, ornamentals and in the home as well as in fly baits.

A number of organophosphorus insecticides are plant-systemic and are effective against certain pests of a wide variety of crops. Example are *demeton* (Systox), *dicrotophos* (Bidrin), *dimethoate* (Cygon), *disulfoton* (Di-syston), *mevinphos* (Phosdrin), *monocrotophos* (Azodrin), *oxydemetonmethyl* (MetalSystox R), *phorate* (Thimet) and *phosphamidon* (Dimecron). Most of the plant-systemic phosphates are highly toxic to mammals, with the exception of oxydemetonmethyl and dimethoate, which are moderately toxic. Plant systemic insecticides are valuable where injurious forms such as aphids and spider mites must be controlled and beneficial forms such as pollinators and predators must be protected.

Carbamates

Carbamates in use and activity resemble organophosphorus insecticides. Some have low mammalian toxicity, whereas others are very toxic. These compounds break down readily and leave no harmful residues. Carbamates are derivates of carbamic acid and dithiocarbamic acid and have an –OCON = group in the molecule.

Carbaryl

($C_{12}H_{11}NO_2$): This compound is well known by its trade name Sevin. It is extremely toxic to honey bees and should not be applied where there is danger of killing these valuable insects. It is recommended in home gardens and on commercial crops against such insects as blister beetles, japnese beetle, grasshoppers, scale insects, and colorado potato beetle.

Other carbamates have proved to be effective in insect control. *Carbofuran* (Furadan) is recommended for the control of soil insects and nematodes and of foliar insects pests of a variety of crops. Applied to the soil *metalkamate* (Bux) is used in large quantities to control rootworms infesting corn *Methomyl* (Lannate and Nudrin) control broad spectrum of insects in many commercial vegetable and ornamental crops, field crops, and certain fruit crops. *Propoxur* (Baygon), a moderately toxic carbamate with fast knockdown and long residual, is recommended against household and turf insects, *Aldicarb* (Temik), a highly toxic carbamate, is a systemic insectiside, acaricide and nematicide for soil application as granules. It has good residual activity of three to ten weeks or more Several other carbamates have systemic activity including carbofuran and methomyl. *Formetanate hydrochloride* (Carzol), both an acaricide and an insecticide, is effective against mites, thrips, and lygus bugs. It is registered for use on citrus, deciduous fruit, and alfalfa.

Organic Sulphur Compounds

These are compounds which are effective aganist mites on crops and ticks on animals and are called *Acaricides. A* number of them are inactive against insects. Certain compounds are exclusively oricides and may kill the newly emerged nymphs, the others kill stages of mites. These are in general stable and possess long residual action and low mammalian toxicity. Some examples of organic-sulphur compounds are Tetradifon (Tedion); Tetrasul (Animert); fenson (Murvesco); Genite (Genitol); Sulphenone; Omite; Aramite etc.

Fixed Oils and Soaps

Fixed oils are derived are from both plants and animals and include oils like castor oil, linseed oil, soyabean oil, fish oil etc. They are glycerides and with alkaline bases they saponify or form soaps, setting free glycerin. Fish oil is used in making insecticidal soaps. Soyabean oil is also used in insecticides.

Soaps dissolved in water at sufficient quantities act as contact insecticides and have been used for pest control since 1787. A soap is a salt of a fatty acid and an alkali-metal base such as potassium or sodium hydroxide; the former base forms soft soap and the latter, hard soaps. The fatty acids are derived from animal and vegetable oils and may include oleic, palmitic or steric acids. Rosin fish oil soap is available as a commercial insecticide soap and is effective against hairy caterpillars and scale insects. In the preparation of emulsions and as spreaders, wetting agents and stabilisers for pyrethrum and nicotine sprays, soap is used. Soft soaps are used in solution in oils to form the miscible oils.

Miscellaneous Compounds

Insecticides have been discovered among a miscellaneous group of compounds. Two of these, *chlordimeform* and *chlordimeform hydro-chloride* have proved to be both insecticidal and acaricidal. As acaricides they control eggs, nymphs, and adults of plant-feeding mites, including strains that are resistant to organophosphorus compounds. Because of their possible carcinogenicity the manufacturer has withdrawn them from the market in 1976 as a precaution until all studies are completed on their safety.

Allethrin, was synthesized ty USDA chemists. Its characteristics of slight toxicity to mammals, high toxicity to insects, and quick knockdown are similar to those of the pyrithrins. Resmethrin discovered by scientists at the Rothamsted Experiment station in England; it is less toxic to mammals and more toxic to insects; it does not require synergists, has greater residual activity and is lower priced. Rosmethrin is particularly effective against the greenhouse whitefly. It is also useful outdoors to control flying insects causing annoyance to people.

Fumigants

Fumigants are chemicals that exist as gases at required temperatures and pressures and are insecticidal at specific concentrations. The gas enters the insects body through spiracle and brings about death. A fumigant that vaporises readily at room temperature is the most useful. An essential requirement of a soil fumigant is that the vapour should emerge slowly. An ideal fumigant is determined by its relative effectiveness, cost, penetrating power, safety to human beings, living animals, plants and germinating seeds, reactivity with household furnishings, flammability etc. Fumigants are employed to control great varieties of stored product pests household articles, etc. Soil fumigation is done to eradicate soil dwelling insects and nematodes. Live plants are fumigated for controlling subterranean pests.

In any fumigation operation the first principle to be observed is to safeguard human lives and only trained personnel with necessary protective appliances should be entrusted with the work enclosure to be fumigated must be a nearly air tight as possible and it is better to avoid windy or cold weather. There should be provision for ventilation after fumigation. If living plants are fumigated, accurate dosage must be used in addition to scrupulously following the proper time for exposure and ventilation and the temperature should be optimum between 40 and 80°F. It is desirable to have the treatment done at night or in darkness. If no live plants are involved, the fumigant is used slightly in excess and the materials are exposed as long as convenient. The temperature is to be between 70 and 100°F. Food substances containing moisture may get poisoned after fumigation.

The dosage for fumigation is expressed in terms of lb/1000cu.ft and an effective fumigant should bring about 99 per cent kill of the insect population which is referred to as the critical (Ct) value. The Ct value is the product of the concentration of the gas per unit volume × the duration of the exposure.

If the exposure period is shorter the concentration of the gas will be more. The sorption of the gas by the material fumigated affects the dosage and the Ct value of the fumigant, and further the rate of sorption is dependent on the moisture content of the material. The dosage recommended is intended for a standard period of exposure to achieve useful results, and in live plants it is limited to the amount that could be tolerated without any apparent phytotoxic effects. Tent fumigation is adopted for fumigating trees, tents being relatively air tight into which a hose introduces the gas. Vacuum fumigation is done using gas in a partial vacuum to enhance penetration and to save time.

Hydrogen Cyanide or Hydrocyanic Acid

HCN This is the most extensively used fumigant, originally employed for the control of the cottony cushion scale in the USA. It is a volatile, colourless liquid with a bitter almond odour and a sp. gr. 0.699 at 20°C. The gas has sp. gr. 0.9483 and is highly inflammable and explosive in mixtures above 5.6 per cent or above 4 lb/1000 cu. ft. Under normal atmospheric pressure the gas is not able to penetrate and so partial vacuum fumigation is employed. Metals like gold, brass, nickel, etc. are tarnished by HCN which could be removed by prompt rub with polishing cloth or prevented by prior application of grease. HCN is liberated when sodium or potassium cyanide is treated with sulphuric acid or on exposure of calcium cyanide to moist atmosphere. In the market two forms of calcium cyanide are available. Cyanogas contains 40 to 50 per cent CaCN and the gas is liberated slowly. It is used in fumigation of burrows of rats. The Calcyanide dust contains 88 per cent CaCN and the gas is liberated rapidly when exposed to moist air. The plants/trees to be fumigated should not be watered for some hours previous to fumigation. The gas is not compatible with bordeaux mixture and hence should not be applied on plants before or after fumigation. HCN is one of the most deadly gases and at higher dosages or if exposed too long it may kill plants. The gas becomes toxic to insects and warm blooded animals as it combines with iron atoms of cytochrome oxidase. Only experienced person should be allowed to handle fumigation. Gas masks should be used. For general fumigation the dosage is 1 lb sodium cyanide (98 per cent pure), 1½ pints sulphuric acid and 3 pints water per 1000 to 1500 cu. ft. For fumigating live plants the dosage is 3.5 to 7 g sodium cyanide, 5.5 to 11 ml of sulphuric acid and 11 to 22 ml of water or 1000 cu. ft.

Carbon Disulfide

(CS_2) is colourless to yellow liquid with an unpleasant odour. It is highly flammable and is usually diluted with four parts of carbon tetrachloride to reduce fire hazard. Carbon disulfide is highly absorptive and penetrating; it has been used in buildings and in the soils. Besides its high flammability it also has the advantage of reacting with many surfaces, leaving yellow stains.

Methyl Bromide

(CH_3Br) has in recent years become a widely utilized fumigant. It is a colourless gas with a faintly sweetish odour. It is stable and nonflammable and has high insect toxicity and some acaricidal properties. It has very high penetrating properties and is rapidly desorbed from treated materials. It is quite safe for use on dormant plants. Methyl bromide is used in the fumigation of plant products imported into this country. It is useful in mills, warehouses and granaries. Methyl bromide is highly volatile and needs special equipment for application.

Ethylene Dichloride

($C_2H_4Cl_2$) is a sweet-smelling, noncorrosive liquid. Because it is flammable, it is usually mixed with carbon tetrachloride. This mixture is used mainly for fumigation of stored products, but has also been used in emulsion form to control the peachtree borer. It has relatively low mammalian toxicity.

Carbon Tetrachloride

(CCl_4) is much less toxic to insects than the usual insect fumigants. It is neither flammable nor explosive and is mixed with other fumigants to reduce the hazards of fire and explosion.

Paradichlorobenzene

($C_6H_4Cl_2$) and *naphthalene* ($C_{10}H_8$) are solids which slowly give off gas. They have been used as soil fumigants and are popular in the form of moth balls or flakes for clothes moth control.

Dichloropropene : ($C_3H_4Cl_2$) and dichloroporpane ($C_3H_6Cl_2$) mixture (D-D mixture) is a common soil fumigant. The mixture is toxic to nematodes as well as to all soil insects. Because this fumigant is phytotoxic and may cause off-flavour in potatoes, the soil must be treated well in advance of planting date. Other fumigants applied to soil include *ethylene dibromide* (EDB), *dibromochloropropane* (DBCP), and *dichloropropenes mixture* (DCP).

Biological Control

As long as the 17th century man first conceived the idea of taking advantage of the food preferences of natural enemies of insects to destroy or supress pest species. Biological control may be defined as the destruction or supression of undesirable insects, other animals, or plants by the introduction, encouragement, or artificial increase of their natural enemies.

Among the natural enemies of insects which may be used in this way are:

1. Predaceous and parasitic insects
2. Predatory vertebrates
3. Nematode parasites
4. Protozoan diseases
5. Parasitic fungi
6. Bacterial diseases
7. Viral diseases.

The basic principles to be considered in the selection of a natural enemy are the similarity of natural conditions which would help in the establishment of the natural enemy in the new environment; the degree of host specificity and genetic races of a given natural enemy may also play an important role in view of the possibly better bio-ecological adaptations of particular races. It is also essential to develop a genetic strain of the parasites superior to the wild stock to have better effects.

The main advantages of biological control are:

1. The absence of toxic effects.
2. No development of resistance by the pests.
3. No residues of poison in the solids and rivers, etc.
4. No build-up of toxins in food chains.
5. No killing of pollinators, or development of secondary pests through the destruction of their natural enemies.
6. The permanence of successful biological control programmes where repeated application of chemicals would be required.
7. The fact that biological control is self-adjusting and does not require the careful timing and organisation which should be given to pesticide applications, and which often make it impracticable or small peasant holdings in underdeveloped areas.

The principal methods employed in biological control include :

1. Collecting parasites and predators from places of their origin and releasing them in places where they are absent.
2. Collecting and storing the host insects in such a way as to kill them, but permitting the parasites to escape.
3. Rearing under favourable conditions great numbers of parasites and predators and releasing them whenever needed.
4. Importing parasites, predators or disease producing organisms from a foreign country. A recent development is control of insect behaviour by natural products, employing products of insects themselves (like pheromones and hormones) to combat them when they assume proportions of pests.

Parasitic and Predaceous Insects

1. The cottony cushion scale or fluted scale *Icerya purchasi* is a scale insect which is a polyphagous pest on a variety of fruit tress in western countries. In Tamil Nadu it spread to an alarming degree on a variety of wild vegetation on the Nilgiris by about 1928. It had over 100 host plants but the only important crop affected was the wattle of commerce *Acacia decurrens*. The only successful way of controlling the pest is by its beetle predator *Rodolia cardinalis*. The beetles were initially got from California in May 1929, multiplied in the laboratory and released in infested areas. By 1931, the incidence of pest was practically reduced to negligible limits.

2. The coconut black headed caterpillar *Nephantis serinopa* is one of the most serious pests of the coconut palm. The natural enemy searched in 1926 was *perisierola nephantidis* (ant like) (Bethylidae) parasiting the grown up caterpillars and *Trichospilus pupivora* (wasp) (Eulophidae), a pupal parasite.
3. *Spoggossia bezziana* the exotic tachinid fly parasite was obtained recently from the Institute of Biological Control, Banglore for mass multiplication and liberation in the coconut gardens.
4. The apple woolly aphis *Eriosoma lanigera*. This is a serious pest of the apples and the recognised method of control is by a systematic colonisation of its specific parasite *Aphelinusmali*. Consignments of the parasite were obtained from Punjab in 1940 and liberated in the Pomological Station, Coonoor. The work was intensified from 1944 onwards and appreciable control has been effected.
5. *Other pests :* Weekly releases of *Trichogramma* at 10,000 per acre from flowering season till the ripening of the bolls showed a progressive decline in infestation of the pink bollworm *Pectinophora gossypiella* on cotton. Attempts to control the castor semilooper (*Achaea janata*) with its larval parasite *Microplitis ephisuae* (Braconidae) and the paddy stem borer (*Tryporyza incertulas*) and *Epilachna vigintioctopunctata* on brinjal with their respective parasites proved ineffective. The ichneumonid parasite Isotima javensis has been found promising in the control of the top borer on sugarcane in Pugalur area.

Borers which cause heavy damage to sugarcane in India have been subjected to attack by the cuban fly *Lixophaga diatreae* imported from Taiwan in 1962. The stalk borer *Chilo auricilius* has been examined in experimental trials and has yielded promising results. The fly has been employed with advantage in Taiwan and hence the import of the strain of India from that stock. It has been found that in Uttar Pradesh, the fly has been able to survive the North Indian winter. Another fly *Diatraeophaga striatalis* (tachinid) has been received from Malagasy Republic in 1965 and the release of gravid males was made in Tamil Nadu. As yet no information has been available on its utility in biological control.

The other common predaceous insects in South India include, the coccinellid beetles *Menochilus sexmaculatus* on aphids on a variety of crops and *Chilocorus nigritus* predominantly on scale insects attacking coconut, betelvine, neem, tapioca, etc.

Predatory Vertebrates

Among the vertebrates, the birds are in fact most effective as a proportion of food of most birds is made up of insects. Yet, there has been no spectucular cases of transportation of birds from one country to another to combat insects. In South India, occasionally ducks are allowed into paddy fields for feeding on the striped bug *Tetrodahisteroides*. The ducks feed on hundreds of the bugs and in two or three days the pest is controlled appreciably. For control of mosquito larvae certain species of fishes have been employed. Introduction of the giant toad of Mexico, Central America and South America in the Hawaii islands, Philippines and West Indies is said to have brought under control the white grub pest of sugarcane.

Nematode Parasites

Many species of nematodes are parasitic on beetles, grasshoppers, cockroaches, moths and other insects. The utilisation of nematodes in biological control of insects is in its primary stage. However, recently a species of bacteria carrying nematode, at present known as DD-136, has been imported from Japan by the Indian station of the Common wealth Institute of Biological control with a view of trying it for the control of paddy and sugarcane borers and codling moth in India.

Microbial Control

Another relatively modern aspect of biological control involves the application of insect's own pathogens for control. Like many animals, insects are also subject to many diseases—bacterial, fungal and viral. Many of these pathogens may be macerated or dried and distributed in water suspensions or as dry spores over large areas. Bacillus thuringiensis is a form which produces toxic protein crystals during spore formations. The active principle in the crystal is responsible for paralysis of the midgut within 5 to 20 minutes of imbibition of the sporulated bacillus, followed in 1 to 7 hr by the paralysis of the body. Lepidoptera and Coleoptera are most susceptible to this. Other bacteria are the deadly milky spore- producing bacteria—*Bacillus lentimorbus* and *Bacillus popilliae*—which attack beetle grubs and caterpillars. Their action is to block the haemolymph and the whole blood becomes milky white. A new method for multiplying these spores in a liquid fermentation medium has been undertaken on an industrial scale. A talc-cum-limestone powder is chosen as the carrier of the commercially made milky dust. The dust is incorporated in the soil and the insects feeding on the roots consume the spores, catch the infection and die.

Fungi were the first microorganisms recognised to produce diseases in insects, particularly the fungal diseases of silk worms produced by *Beauveria bassiana*. Subsequently, a more virulent fungus *Metarrhizium anisopliae* was isolated from diseased grubs and even today used in the control of the coconut beetle. These fungal pathogens are known from all fungal groups—Asco-, Basidio-, Phyco- and Deutero-mycets. Of the phycomycetes, infections caused by Blastocladiales and Entomophthorales are very common. The Ascomycets include a very common genus called *Cordyceps*. The Basidiomycetes are of little importance while the fungi *imperfecti* play useful role in insect control. They infect their hosts not so much by ingestion but more by penetration through the integument, softening the hard chitin by releasing some enzymes. Once within the body cavity the fungus proliferates, invades all tissues and fills up the body cavity of the insect with thickly grown hyphae. Owing to heavy growth of the mycelium in the insect body, it becomes hard, stiff and mummified. Death is caused by the liberation of toxic substances—Mycotoxins—having tetanic reactions. Death may also be caused by exhaustion or by mycosis or by paralysis or by asphyxiation through the plugging of tracheae. About 250 virus infections have been recognised in about 175 insects. Of these, two are most outstanding, affecting most insects—the polyhedroses viruses and granuloses viruses. The polyhedroses viruses form polyhedra shaped bodies containing the viruses in the infected tissues. Two types of polyhedra are recognised—the nuclear polyhedroses and cytoplasmic polyhedroses. The incubation period is 5 to 20 days and an infected larva stops feeding, becomes sluggish and dies. Shortly before and after the death, the integument becomes fragile and easily ruptures, emitting the liquified contents filled with disintegrating tissues and polyhedra. The granuloses viruses form small, granular, inclusion bodies called the capsules, each containing a virus particle. They are commonly used in the control of Lepidoptera and they attack the tracheae and blood cells.

Biological Control of Weeds

Started with the introduction into Hawaii of *Lantana* for ornamental purposes, the accidental introduction of a bug *Orthezia insignis* proved as a check. Insects should not be considered for noxious weed control, if they have been recorded as attacking plants of economic value. Insects recorded as attacking only the genus to which the noxious weed belongs or allies of it, having no economic value, should be subjects of future study.

Cactus and prickly pear have been well known to encroach and render useless thousands of acres of arable and pasture lands. Artificial methods by burning, hand-picking, etc. were of no avail and insects had to be used to control these plants. Cactus in America, and Aastialia is controlled by the tunnelling caterpillar. *Cactoblastis cactorum* which is native of South America.

Introduction of plants of similar nature in India includes the prickly pear, *Lantana, Cuscuta, Xanthium* and *Alternanthera*, causing economic losses. Both lantana and prickly pear are natives of the new world and in South India the prickly pear enjoys a very wide distribution growing luxuriantly covering vast areas of arable and pasture lands and levying a heavy toll of the former every year to keep them within limits. Ordinary methods of control proved ineffective and only a single natural enemy was known to exist in its native home Mexico. This is also a scale insect called the Cochineal insect *Dactylopius tomentosus*. They are minute forms feeding in clusters and covered by a profuse secretion of white cotton material. This was imported by the Madras Agricultural Department and distributed over vast areas. The insect sucks up the sap, big clumps of prickly pear die out and today the country sides show a varity of prickly pear. Similarly *Lantana* is an equally troublesome weed but efforts to prevent its growth have not yielded any results. Only for sometime it was checked by the lantana bug. But recently the lantana seedfly *Agromyza lantanae* has been introduced from Hawaii but again its efficiency is also low so that we find lantana still spreading rapidly.

Autocidal Control or Sterilisation

This usually refers to the sterilisation of males by X-rays or γ-rays and is called the *sterile-male-technique*-control of a pest by this technique is called autocide. Sterilization can be effected by exposure to various chemicals and this practice is called *chemosterilisation*. The fundamental basis behind this method is that male sterilisation is effective in species where females only mate once and are unable to distinguish or discriminate against sterilized males. Chemosterilisation has now advanced from a theoretical technique to a practical one; a variety of chemicals have been demonstrated to interrupt the reproductive cycles of a large number of insect species.

The classical case of autocidal control was in about 1940 on the island of Curacoa against Screw-worm (*Callitroga*) on goats-the male flies were sterilized by exposure to γ-rays and dropped from planes at a rate of 400/square mile/week. The whole pest population was eradicated in twelve months. The life-cycle took only about four weeks to complete, and the females only mated once in their lifetime. Generally, autocide is most effective when applied to restricted population (islands, etc.), but can be effective on parts of continents. The screw worm eradication compaign was extended to the Southern part of the USA where the pest is very harmful to cattle. In texas 99.9% control was achieved in only three years. Male sterilisation trials were effective against Mediterranean fruit-fly (*Ceratitis capitala*) on part of the island of Hawaii in 1959 and 1960, but immigration from untreated parts of the island prevented control from being long-lived.

This method of control could quite possibly be effective against. *Orcytes spp.* attacking the coconuts along the coastal strips of Kenya and Tanzania, and work is in progress on the feasibility of autocide as a method of controlling Tsetse.

Insects are irradicated by exposure to cobalt 60 irradiation, where large populations could be handled. Radioactive isotopes are also applied on insects by spraying the insects with a solution of a radioactive compound or applying it with the aid of the brush on a particular area of the body or by feeding the insects on diets containing the isotope. This method also serves in the identification of the treated insects by their radioactivity. The common radioactive isotopes used for labelling studies with insects are $carbon_{14}$, $phosphorus_{32}$, $iodine_{131}$, $bromine_{82}$ $arsenic_{76}$, $sulphur_{35}$ etc.

When the extension of sterility techniques to other insects is considered, several points are looked into, especially the effect on survival and reproductive behaviour, methods to be adopted for mass rearing and proper methods of dispersal of released sterile insects so that they could mix with natural popultions, further the sterile insects should also not cause undue loss of crops and livestock. Through repeated release of sterile males, the ratio of sterile males to normal males becomes increased.

Insect Pheromones In relation to pest control (Insect Attractants)

The number and variety of methods are used by insects to communicate among themselves are amazing for such unreasoning creatures. Although many insects commonly communicate by visual signals or by calling through their chirps, clicks, songs and sounds of wing beat, many others communicate by release of chemical messengers, special compounds produced by insects. A receiving insect may make a behavioural response upon contact or at very close range with the released chemical, or in cases of sex attraction, the response may be elicited by following a stream of attractive vapour for a mile or more. Such responses to chemical clues are divided basically into two types:

1. *pheromones*, the groups of chemicals that provide messages between individuals of the same species (intraspecific),
2. *allelochemicals*, the substances that affect individuals or populations of a species different from the source (interspecific).

Two types of pheromones have been discovered among insects, the fast-acting releasers and the slow-acting *primers*. An important group of releasers are the *sex pheromones* that are secreted by several orders of insects, notably moths, beetles, flies and the Hymenoptera. Much is now known about the sex pheromone of the codling moth. The female of this species secretes the pheromone in a gland (Fig. 6.101) that opens dorsally in the intersegmental fold between the last two segments of the abdomen. Normally these two segments lie telescoped within the abdomen, but when a virgin female seeks a mate she will protrude them and release the pheromone. Codling moth males of the right age detect the pheromone in the air, become sexually excited, and respond immediately by flying to the source and mating with the 'calling' female. The sex pheromone of the codling moth has been identified as (E, E)-8, 10-dodecadien -1 -ol.

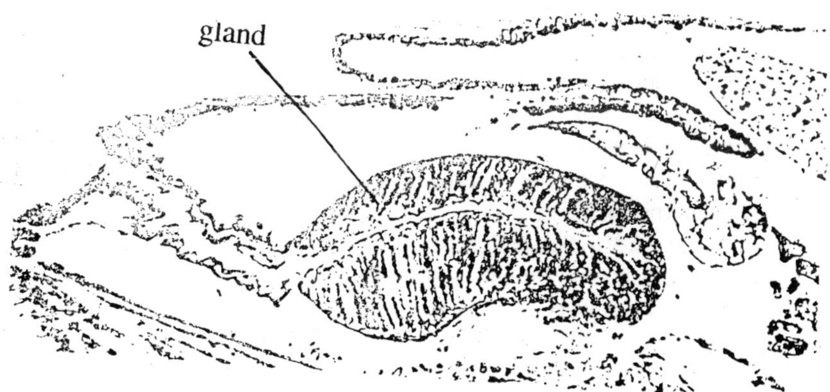

Fig. 6.101 : Longitudinal section of the pheromone gland and duct of the codling moth.

A second group of pheromones brings about aggregation or assembly of individuals of a species, a type of behaviour that has survival value. Assembly may be temporary, as in the overwintering of lady bird beetles, or permanent, as for honey bees in a colony. Bark beetles produce *aggregation pheromones*. The first bark beetle that finds a susceptible tree releases a pheromone that attracts others to join in the attack.

A third group of pheromones is the *trail-marking secretions*. Ants and termites leave a chemical trail which others of the same species follow. Honey bee workers release an aerial odour that allows other workers to locate a new food source.

Alarm pheromones, a fourth group, are chemicals that alter the behaviour of social insects to retreat or to attack. In the well-known behaviour of the honey bee, a sting from one bee elicits aggressive behaviour of other bees. This aggression of guard bees starts when the sting apparatus of one bee is torn from its body and embedded in the victim's skin. A special gland that remains attached to the sting releases the sting pheromone sensed by the other bees. Immediately it stimulates additional stinging of the invader.

The slow acting *primer pheromones* are secreted by social insects to maintain the organisation of the colony. The mandibular glands of the queen honey bee produce a chemical called queen substance that inhibits the workers, which ingest it from producing eggs and building queen cells. Termites produce, and pass from one individual to another, primer pheromones that foster a favourable caste ratio of soldiers, workers and reproductives.

Pheromones, originally reffered to as ectohormones, are complex chemical compounds, basically long-chain hydrocarbons such as alcohols, esters, ketones, aldehydes and sometime ethers. Many have now been successfully identified, and some synthesised a number are now available commercially from some chemical companies for pest monitoring or control purposes.

During the fifties and sixties entomologists discovered several synthetic attractants through a tedious programme of screening and testing thousands of compounds prepared by chemists. A powerful lure for the male of the melon fly was found in the compound 4-(p-hydroxyphenyl)-2-butanone acetate. This substance, now called *cuelure*, is used in traps to detect infestations of the pest. Other synthetic insect attractants in current use are medlu-re and *trimedlure* for males of the Mediterranean fruit fly.

Medlure

Many insects produce pheromones that attract the opposite sex. Entomologists have been able to locate the glands and to extract the attractive substances from insects and chemists have succeeded in identifying and synthesizing several of these natural compounds. Pheromones are effective at very low concentration, 10^{-7} to 10^{-12}μg and can attract insects from a distance of several miles. Most are specific in their action, attracting only one sex of one species of insect. Synthetic pheromones are available for attracting the males of several moths, including *disparlure* for the gypsy moth, *gossyplure* for the pink bollworm, *codlelure* for the codling moth, *looplure* for the cabbage looper, and *virelure* for the tobacco budworm. *Grandlure*, a mixture of four synthetic compounds that copy the natural phenomone produced by the male boll weevil, acts as a sex attractant for the female and as an assembly attactant for both males and females of the species.

There are three main ways in which they may be used to assist with control of pests :

1. *For incorporation in traps or poison baits :* If traps or baits incorporating an insect sex attractant are to provide control of a pest species, two conditions must be satisfied. Firstly, the trap or bait must compete successfully with natural pheromone emitting females of the species. Secondly, a very high proportion of males must be eliminated from the popultion before they are able to mate with female if there is to be effective suppression of the next generation. There are few examples so far where success has been achieved with sex attractants used in this manner.
2. *To disrupt mating :* The idea here is that, if a sex attractant is released into the atmosphere and maintained at sufficient concentration over the period of mating of a species, the natural scent of females will be swamped and males will thus be unable to locate them. The procedure is sometimes referred to as the "male confusion technique". Success has been achieved experimentally with the technique for several species of moths but how practical it will prove to be for general use is as yet uncetain. There are many difficulties involved such as providing for adequate distribution of the pheromone and maintaining high enough concentrations for sufficient periods of time. Cost is also likely to be a major factor. Despite such problems research into this use of sex attractants is likely to continue because it offers the possibility of very selective pest control with minimum disturbance of other organisms.
3. *For survey and monitoring :* The specific and potent properties of sex attractants make them ideal for use in insect traps to

survey for the presence of particular pests on an area basis and to monitor for seasonal occurrence. They are already being extensively used for these purposes and commercial supplies of the sex attractant, together with suitable traps, are available for several important pests such as codlingmoth.

Hormonal Control

Many chemical compounds have been synthesized, having actions similar to those of juvenile hormone and moulting hormones of insects. To distinguish them from the natural hormones, they are often called "*mimics*' or '*hormone mimics*'. In recent years moulting hormone mimics have been extracted in abundance from plants and are known to contain *ecdysterone* chemically identical with the moulting hormone of crustaceans and insects. They are also called *phytoecdysones*. Some of the plant sources are mulberry leaves, some ferns, and several gymnosperms, conifers, roots of *Cyathula capitata*, etc. Investigations on the activity of these mimics on other insects and their efficiencies as compared with natural ecdysone showed that in many cases , these plant steroids proved to be a little more active. Experiments on these ecdysterone are not much successful due to their apperent inability to pass through the cuticle or to be absorbed by the gut.

Inhibition of yolk deposition in *Musca* and *Tribolium* after feeding on plant steroids for five days and subsequent failure to oviposit after a return to normal diet was noted. In inhibiting ovarian development as high doses of ecdysterone, it has been known to act as a chemosterilant.

Synthesising other compounds related to ecdysone, results in compounds such as ecdysterone, rubrosterone I, rubrosterone, ponasterone A, etc. Several others have also been synthesised varying in the degree of hydroxylation producing an array of compounds which are yet to be fully tested.

Entire extracts of male Cercopia moths, have been known to produce an oil with juvenile hormone activity and its chemical composition is known to be methyl *trans, trans, cis*- 10-epoxy-7-ethyl-3, 11-dimethyl-2, 6 tridecadienoate. *Wigglesworth* tested 42 compounds with juvenile hormone activity. These also included farnesol derivatives. When applied to the cuticle in a small drop of octane or injected in arachis oil quite some differences in potency were observed. The trans-trans-cis form was seen to be much more active than the *cis-cis-trans* and all *cisforms*. In some cases a thirty-fold activity occurred in the degree of juvenilisation.

Best results were always obtained when cercopia oil (than in olive oil) in injecting the compounds.

An interesting discovery of far reaching effects is their usage as fumigants of pests of stored products. Some insecticide synergists like sesoxane have been observed to have juvenile hormone activity. Cuticular application of sexoxane in acetone, resulted in adults possessing varying larval and pupal feaures and these adults did not breed. A wide range of compounds have been discovered such as piperonyl butoxide, piperonyl farnsyl, sexone etc. Their peculiar action appears to be the prevention of the breakdown of moulting hormone so that the insects suffer from an overdose of moulting hormone. When these compounds were mixed in cercopia oil and topically applied in acetone, even low doses produced all kinds of abnormalities. A promising insecticide appeared to render 80 per cent of the treated insects sterile even at such low doses as 0.001. Even the vapour of these active compounds have been known to produce abnromalities.

The active principles, juvabione and dehydrojuvabione or paper factors as they were called, have been known to prɔduce abnormalities of growth and several analogues of these have been synthesised.

Recent investigations have also shown that juvenile hormone mimics pass through the cuticile of insects, when dissolved in acetone but still the topical application of moulting hormone and mimics is not solved. It is a healthy trend to observe the inequality in effectiveness of mimics on all insects, in view of the prospects of these pesticides being 'tailor made' to particular species. Further, alteration of part of the natural hormone in the synthesised compounds appear for more effective in developments than those caused by an overdose of the natural hormone. However, the persistence of these compounds and their effects on nonpest species has not been assessed at all.

Plant and Animal Quarantine

Although human activities have greatly assisted the spread of pest species around the world this process is fortunately by no means complete and for most countries there is a long list of plant and animal pests (and diseases) which have not yet gained entry. The aim of plant and animal quarantine is to prevent the introduction and further spread of pest and disease organisms. It cannot be undertaken by individuals but must be the responsibility of governments to establish suitable legislation and to provide for its implementation.

A basic problem in attempting to prevent further introduction of pests and diseases into a country is that international trade and travel cannot be completely shut down. It is not even practical to totally prohibit the importation of plant material. Any quarantine action therefore must be to some extent a compromise with restrictions on trade and imports being commensurate with the risk of introduction and likely importance of a new pest or disease problem. Such measures might include a total trade ban on certain items and would certainly include total prohibition on import of live animals which could act as carriers. On the other hand the risk posed by a minor pest of a little grown ornamental plant could not justify such drastic measures.

Another problem in drawing up quarantine regulations is that plant and animal pests (and diseases) may be able to survive for some time away from their hosts. Quarantine restrictions therefore must also take into account likely non-host sources of infestation.

There are four methods of plant quarantine viz., inspection at point of destination, inspection and certification at point of origin, complete embargoes and controlled introduction of plants.

(i) **Inspection at point of destination :** This inspection is necessary because many insect pests can be discovered and their accidental introduction thereby eliminated. Often it is physically not possible to examine all materials likely to carry insect pests with sufficient care to detect them. Also because some insects and plant diseases have inconspicuous stages they are extremely difficult to locate.

(ii) **Inspection at the point of origin :** Most countries and many states follow this plant *i.e.,* inspection and certification at the point of origin. In this case, materials are allowed to enter the state or the country as the case may be, provided they bear a certificate issued by the plant quarantine officer which mentions that the materials have been inspected and found to be free from insect infestation or plant disease infection.

(iii) **Embargoe :** By this we mean the exclusion of all plant materials or commodities which are classed as hosts of the insects. This is also not possible in practice. Because the area where the pest exists cannot always be definitely determined and the material listed in embargo may not include all of those which might carry or harbour the pests.

(iv) **Controlled introduction :** This consists of allowing the importation of only a very limited amount of material for propagating purposes, obtaining it from pest free or disease free areas and from plants believed to be insect or disease free, inspecting the shipment very carefully then planting in an isolated place for a preliminary period.

Pre-requisites of Quarantine

There are four fundamental pre-requisites desired are establishing quarantine in any country:

(i) The pest concerned must be a threat of substantial nature.

(ii) The pest must not be susceptible to control by other measures involving less interference with normal activities.

(iii) The objectives of the quarantine in preventing the introduction or in limiting the spread of the pest must be reasonably possible of accomplishment.

(iv) The economic gain from quarantine must exceed the cost of enforcement.

Several types of measures may be employed in quarantine programmes:

(a) Total prohibition on imports of certain materials : This can only be justified where a total ban imposes no great inconvenience and/or where the pest or disease is potentially so serious that risks cannot be taken. For example, import of unginned cotton, Mexican jumping beans and potato tubers is completely prohibited.

(b) Goods provided with a Certificate of Health : In the case of plant material, a Certificate of Health, phytosanitary certificate issued by authorities in the originating country may be required before entry is permitted. Phytosanitary certificates must state specifically freedom from pathogens and pests as required under provisions of the Quarantine Act. Unfortunately certificates from some countries are not reliable and other measures need to be taken.

(c) Importation by licensed organisation only : Many kinds of fresh fruit and plants cannot be brought into India by individual travellers because of the risk of introducing pests and pathogens. However, imports are permitted by certain licensed importers under strictly controlled conditions.

(d) Importation from certain designated areas only : Where the overseas distribution of a pest or disease is limited, imports of plant material may be allowed from some countries or states but not from others. Sugarcane cuttings are not imported from Australia, New Guinea, Fiji and the West Indies.

(e) **Preferred types of plant material :** Seeds are less likely to harbour pests than vegetative parts of plants and thus, where a choice is possible, these are the preferred type of plant material for importation. If propagative material such as cuttings or bulbs is to be introduced greater care must be taken and more restrictions enforced to ensure that it is pest (and pathogen) free.

(f) **Chemical treatment :** Fumigation or other chemical treatment of plants or plant material, either before or immediately after entry, may be undertaken as an added precaution.

(g) **Isolation and inspection after entry :** It may be required that imported plants be grown initially in isolation and be inspected before distribution is permitted. This may cover a period of twelve months or more. If diseased or infested plants are found they may have to be destroyed.

Insecticidal Laws

The Fedral Government of many of the states have the laws regulating the sales and usage of insecticides. Labels of economic poisons are essentially required to indicate clearly the following:

(i) Name and address of manufacturer.

(ii) Trade mark.

(iii) Ingredients with percentage of active and inert substances along with their common or chemical names.

The Government of India have enforced several Acts like sea customs Act of 1876 to stop the entry of Mexican cotton boll weevil, destruction Insect Pests Act of 1914, Poison Act of 1919, Drug Act of 1940 etc. for preventing the entry of infested seeds and not allowing any material without proper inspection and certification that the material is free from all sorts of insect infestation as well as fungal, viral and bacterial diseases. As a rule all the food stuffs are fumigated.

(C)

INTEGRATED PEST MANAGEMENT (IPM)

Insects and land plants have coexisted for nearly 400 million years. During this period, plants developed a variety of physical and chemical defences to prevent colonization by insects and other herbivores. Simultaneously, insects have also developed the ability to overcome these defences. Thus, in natural ecosystems, insects and plants existed in a dynamic equilibrium. As soon as the land was cleared of natural vegetation and replaced by a few species of food plants, humans came into conflict with phytophagous insects. These insects, and other organisms feeding on the valuable crops which humans planted, were called *pests.* Broadly speaking, a pest is any plant or animal species that is in competition with humans for some resource. In agriculture, a phytophagous insect is labelled as a pest when its population is high enough to cause significant damage in yield or quality to any of the economic plants grown by man.

During ancient times, humans had to live with and tolerate the damages of insects and other pests but gradually learned to improve their condition through trial and error experiences. Over the centruries, farmers developed a number of mechamical, cultural, physical and biological control measures to minimize the damage caused by phytophagous insects. Synthetic organic insecticides developed during the mid-twentieth century initially provided spectacular control of these insects and resulted in the abandonment of traditional pest control practice. The intensive cultivation of these varieties, together with the application of increasing amounts of fertilizers and pesticides, has resulted in a manifold increase in productivity. However, this technology package has also resulted in aggravation of pest problems in agricultural crops.

The era of pesticides (1939-1975), however, began with the discovary of the insecticidal properties of DDT by *paul multer* in 1939 and he was awarded Noble prize for this work in 1948. Later on it is soon followed by number of insecticides in different groups like organo phasphorus, organo chlorine, carbamates and synthetic pyrethroids etc. Due to their efficacy, convenience, flexibility and economy, these pesticides played a major role in increasing coop production. The success of high yielding varieties of wheat and rice that ushered in the "green revolution" was partially due to protection umbrella of pesticides. The spectacular success of these pesticides masked their limitations. The intensive and extensive use, misuse and abuse of pesticides during ensuing decades caused widespread damage to the environment. In addition, insect pest problems in some crops increased following the continuous application of pesticides. This in turn, further increased the consumption of pesticides resulting in the phenomenon of the pesticide treadmill. The combined impact of all these problems together with the rising cost of pesticides provided the necessary feedback for limiting the use of chemical control strategy and led to the development of the *Integrated pest management* (IPM) concept.

To overcome increasing problems associated with the strategy of exclusive and indiscriminate use of pesticides, *stren et.al.* (1959) propounded the concept of integrated control. The use of term 'pest management was advocated by Geier and Clarke (1961). But the concept of IPM really came of age at the XV International congress of Entomology during 1976. IPM refers to an ecologial approach in pest management in which all available techniques are consolidated in a unified programme so that pest populations can be managed in such a manner that economic damage is avoided and adverse side effects are minimised. In its early stages, IPM was a technical approach designed to reduce the number of pesticide applications. It subsequently developed into a methodology in which farmers were encouraged to develop IPM interventions themselves, in the process of coming to a better understanding of their agro-ecosystem.

In fact, IPM is a dynamic and constantly evolving approach to minimise crop losses in which all the suitable management tactics and available surveillance and forecasting information are utilised to develop a holistic management programme as part of a sustainable crop production technology. Here it needs to be emphasized that the aim of IPM programmes should not be restricted to a mere efficient use of pesticides and product substitution, within an agricultural system that essentially remains unchanged, rather, these programmes should aim at fundamental structural changes through a better understanding of ecolocial processess and synergy between crops.

Scope

When scope of IPM is discussed many actions and reactions come to our mind. Though IPM concept is gaining popularity, it is only theoretical since actual figures do not support these beliefs. Despite the potential of IPM, the use of pesticides continues to grow. The agro chemical industry is concentrating its efforts on promoting conventional pesticides in Latin America and Asia. In 1996, pesticide use increased by 6% in Latin America and countries such as Brazil, Chiana and India have become important producers of conventional pesticides. In a number of developing countries these can now be obtained very cheaply at the local merket, lowering the economic threshold for pesticide use and allowing more frequent application. Basically, if scope has to be thought of, there is sky is the limit. But it is also necessary that project design has to be appropriate, there has to be proper transparency and participation of farmers. If all parameters are taken care properly, the project bound to succeed thus anticipated scope can be effectively enhanced.

COMPONENTS OF IPM

The major components of IPM include pest forecasting management tactics, decision making and implementation.

Forecasting System

The forecasting system is based on surveillance and monitoring combined with a computer data base regarding weather predictions, costs and efficiencies of various management tactics, prices of inputs and farm produce, etc.

Monitoring phytophagous insects and their natural enemies is a fundamental tool in IPM for taking management decisions. Monitoring requires estimation of changes in insect distribution and abundance, information about insects life history and the influence of important biotic and abiotic factors on pest populations. A number of sampling tools have been developed to make inferences regarding the total insect population as based on smaller collections (samples). A number of specialized sampling techniques becoming increasingly popular in IPM programmes like sequential sampling, variable intensity sampling, double sampling and binomial sampling.

Sampling techniques need to be modified to overcome some of the limitations of sampling as a component of IPM programmes. Because IPM is a continuous process, temporal changes in insect abundance should be explicity considered in sampling programmes. Secondly, two or more pest species causing the same type of damage may be simultaneously present in the field and make sampling difficult. Thirdly, few IPM programmes at the farmers' level attempt to estimate the abundance and impact of natural enemies on insect pest populations. There is an urgent need to develop efficient and cost reliable estimation procedures and forecasting models which incorporate the role of natural enemies in the decision making process (Wilson, 1985).

Management Tactics

Pest management tactics are either preventive or therapeutic. Preventive practice utilizes tactics to lower environmental carrying capacity (reduce the general equilibrium position) or increase tolerance of the host to pest injury. Prevention relies on an intimate understanding of the pest's life cycle, behaviour and ecology. The preventive tactics involve natural enemies, host resistance and cultural practices. In addition, quarantines are also an important component of preventive tactics. Therapeutic tactics are applied as a correction to the system when necessary. The objective of therapy is to decrease pest populations below the economic injury level. The only widely used therapeutic tactic is the use of conventional insecticides but other approaches like microbial agents, augmentation of natural enemies, use of growth regulators, etc. may also play a vital role.

Host plant resistance and cultural practices combined with natural biological control should form the core around which IPM systems are developed. A number of varieties, each having different bases of resistance, should simultaneously be in the field to reduce the chances of development of new insect biotypes. Multiple pest resistance has the potential to play a much more important role in crop pest management systems. Many of the traditional cultural practices may not be useful under modern intensive agriculture. Cultural and mechanical control techniques need to be modified and refined to make them compatible with intensive agriculture. A thorough understanding of the crop stress-insect interactions may enable us to manipulate the crop, environment to the detriment of the pest. The conservation and augmentation of natural enemies must receive priority. Genotypes favouring colonization by the natural enemies due to the presence of allelochemicals should be preferably utilized in IPM programmes. Transgenic plants developed through biotechnological approaches are likely to occupy an increasingly important place in future pest management programmes.

Pesticides will continue to play an important role in IPM programmes. But their use must be made compatible with other components of pest management and with society's requirement for maintaining environmental quality. This requires the development of selective insecticides exploiting weak links in insect physiology and behaviour. Efficacy of insecticides can also be increased by improvements in formulation and application technology.

In addition to conventional insecticides, therapeutic control may be achieved by use of parasitoids, predators, pathogens, behaviour modifying chemicals and selective toxicants. Methods for mass multiplication of parasitoids, predators and pathogens need further refinement, so that periodic releases of natural enemies become competitive with the cost of insecticides. Another important consideration in such releases is the difficulty in maintaining the quality of mass produced natural enemies. The global IOBC working group on 'Quality Control of Mass Reared Arthropods' is making efforts to develop quality control criteria and tests. Genetic improvement may help to enhance the efficiency of natural enemies by increasing host range and stability. Artificial selection, hybridisation and recombinant DNA techniques may be used for this purpose. Contrary to popular perception, biocontrol agents may also damage the environment, result in pest resurgence and secondary pest outbreaks. Therefore, ecological studies and environmental impact assessment of biocontrol agents must be undertaken before they are released in the field.

Chemicals disrupting normal patterns of insect development, growth and behaviour, offer immense opportunities for suppressing insect populations in a selective manner. Many of these chemicals are essentially non-toxic to humans and domestic animals. These compounds are non-persistent and usually active at very low concentrations. Further improvements are needed in controlled release formulations, application technology and quality control in order to increase their utilization in IPM programmes. Plant derived products which act as repellents, antifeedants, oviposition deterrents, development disruptors and toxicants against phytophagous insects, may also serve as substitutes for synthetic pesticides.

Decision Making Systems

Pest management is a combination of processes that include decision making, taking action against a pest and obtaining the information to be used in reaching these decisions. In assessing, evaluating and choosing

eventually are the end users, and who select to adopt or reject the technology based on its appropriateness. Institutional barriers to research scientists in national programmes conducting on-farm research in developing countries are real, and need to be addressed.

(i) *Informational constraints* : The lack of IPM information which could lie used by the farmers and by extension workers is a major constraint in implementation. While the individual control techniques are well known, little knowledge is available on using these in an integrated fashion under farm conditions. The lack of training materials, curricula and experienced teachers on the principles and practice of IPM is another major constraint.

(ii) *Sociological constraints* : The conditioning of most farmers and farm level extension workers by the pesticide industry has created a situation where chemicals are presented as highly effective and simple to apply. This acts as a major constraint in IPM implementation. There is direct conflict between industry's objective of more sales and the IPM message of rational pesticide use. There is a need for private industry and the public sector extension agencies to work in a more complementary manner.

(iii) *Economic constraints* : A major constraint, even if IPM is adopted in principle, is the funding for research, extension and farmer training needed for an accelerated programme. IPM must be viewed as an investment and as with other forms of investment, requires an outlay. In the long run, IPM programmes may become self-generating due to savings on resource inputs for production.

(b) *Political constraints* : The relatively low status of plant protection workers in the administrative hierarchy is a constraint to general improvement in plant protection. Associated with the above are the morale and financial standing of these workers.

The continuance of pesticide subsidies by the government for political reasons and its tie up with the government-provided credit for crop production act as major constraints to farmers' acceptance of IPM. Various vested interests associated with the pesticide trade also act as a political constraint on the implementation of IPM. Indeed, the illicit financial rewards for licensing the import, formulation or distribution of agricultural chemicals are one of the major obstacles limiting the expansion of IPM activities.

(B) Measures for Improving IPM Implementation

Apiculture of IPM implementation in developing countries requires farmers' participation, increased government support, legislative measures, improved institutional infrastructure and a favourable environment.

(i) *Farmers' participation :* It is not an exaggeration to say that the dawn of civilization started with farmer innovation. Ever since that day farmers have improved ways of growing crops through successive innovations. Prior to the emergence of crop protection sciences and even before the broad outlines of the biology of pests were understood, farmers evolved many cultural, mechanical and physical control practices for the protection of their crops from insect and non-insect pests. Farmers' innovations were the only source of improvements in crop production and protection technology until formal research by scientists started complementing it during the late eighteenth and nineteenth century. Unfortunately, with the advent of modern high-tech agriculture comprised of HYVs, fertilizers and pesticides, farmers have been completely displaced from the research and development process. Instead this role has been usurped by the private industry and the government agencies. The technology generated by scientists is being transferred through the extension agencies to the farmer. The new technology package has created a number of ecological and environmental problems. The alternative path of sustainable agriculture requires farmers' participation at every step of the research and development process in order to draw on his understanding of the local conditions and constraints, his innovativeness and his skills at making the best possible use of limited resources.

Placing the farmer at the centre of development process is wholly consistent with the IPM goal of making the farmer a confident manager and decision maker, free from dependence on a constant stream of pest control instructions from outside. The role of researchers, extension and non-government organizations (NGOs) and field workers is to act as consultants, facilitators and collaborators, stimulating and empowering the farmers to analyze their own situation, to experiment and to make constructive choices. A number of terms have been proposed for the new approach. These include: 'Farmer-first-and-last', 'farmer participatory research', 'people-centered technology development (PCTD)' and 'participatory technology development (PTD)'. PTD serves to improve the experimental capacity of farmers and helps in development of locally-adapted improved technologies.

The approach has been used in the development of Farmers' Field Schools (FFSs) in IPM in Indonesia and the Philippines. The Field Schools are designed to build up farmers' understanding of agro-ecological relationships and to improve their capacities to systematically observe, document and interpret these. Experimental learning is based on the principles of adult learning, and follows a cycle of focused experience, systematic observation and measurement, critical reflection followed by planning and taking action. The action, in turn begins a new cycle of experience-based learning and critical reflection. Thousands of farmers have been trained utilizing this approach and it is being tried on a pilot scale.

The evident advantages of this approach are a marked reduction in the use of chemical pesticides, with measurable benefits to the environment. Households save on cash, while crop yields tend to increase owing to better crop management.

(ii) *Government support :* Both the national programmes of developing countries and the donor agencies must have a policy commitment to IPM in the context of national economic planning and agricultural development. The costs to developing countries of not bringing their policies in line with the objectives of IPM are relatively greater than the costs of developed countries. National policies to promote IPM require close regulation at all stages related to the import and/or manufacture, distribution, use and disposal of pesticides. In the case of pesticides that do not meet prescribed standards for safety, persistence, etc., import and manufacturing bans should be enacted. At a minimum, the conditions laid out by the FAO Code of Conduct on the regulation, distribution and use ot pesticides should be adopted. Pesticide subsidies need to be eliminated in order to make IPM an attractive alternative.

The funds so saved may be utilized for the implementation of IPM. Funds may also be diverted from some of the current research programmes to IPM-oriented plant protection programmes. Additional monetary resources may be generated through cooperation with bilateral/multilateral agencies willing to support such programmes.

(iii) *Legislative measures :* IPM is an information system and its idoption reduces pest control costs. The alternative to IPM is the indiscriminate use of broad spectrum synthetic organic pesticides. Unfortunately, while pesticide manufacturers and users (farmers) derive the full benefits from the use of these chemicals, they pass on the environmental and ecological costs of their use to society as a whole. If they are made to bear the full cost of the use of these toxicants, they may find IPM a more economical and attractive alternative. This could be achieved by enforcing suitable legislative measures.

Secondly, the success of an IPM programme in any geographical region depends upon its implementation by all farmers in area. Ideally, farmers may voluntarily adopt an IPM programme but some farmers may hold out. Such farmers called 'spoiler holdouts' may impair the success of a programme by failing to adopt a necessary practice thus causing damage to adjacent areas. Besides these, some fanners may free-ride and thus shift the costs of implementing and managing a programme to a group of participating farmers. To overcome 'spoiler holdouts' and Tree riders', it may be necessary to impose a programme upon an unwilling minority through suitable legislative measures.

(iv) *Improved institutional infrastructure :* IPM cannot be implemented unless there is a basic infrastructure for plant protection in a country. There is a need to develop and support national programme capabilities for on-farm testing and technology extrapolation. At the international level establishment of the Global IPM Facility to coordinate and monitor funding of IPM projects was needed to provide impetus to the implementation of IPM.

IPM is predominantly knowledge technology, the use of which requires training of the many groups involved. There is currently little training material for most of these groups including farmers, extension personnel and researchers. If IPM is to become the major approach for pest management in the developing world, this deficiency must be remedied.

Another aspect requiring greater attention is coordination of efforts within and between national research, training and implementation, institutes/programmes and amongst international development agencies.

Lack of a reliable database has also hampered progress of IPM programmes. A reliable source of accurate information on the status of crops and pests in farmers' fields is necessary for many IPM activities. Most of the successful IPM programmes, both in developed and developing countries, have a reasonably accurate system of monitoring and evaluating various biological and environmental parameters in the agroecosystem. A reliable data base on crop yield and pest losses is required for planning and resource allocation at the national and international level. Systems analysis has been used as a problem diagnosis tool for IPM in developed-country cropping systems and may be used in the developing countries as well.

(v) *Improved awareness :* Increased education and awareness regarding the objectives, techniques and impact of IPM programmes is required at all levels including policy makers, planners, farmers, consumers and the general public. The importance and benefits of pesticides are being overemphasized by a multibillion dollar industry utilizing the services of not only their salesmen but also agricultural scientists, administrators and planners. There is not yet a strong market for IPM information. Policy makers and planners need to be convinced that without IPM current agricultural production systems are not sustainable. Similarly, much important information that might induce a farmer to adopt IPM is not immediately observable and is, therefore, not sought by him. A manufacturer has no incentive to recommend a programme that uses less pesticides, or even selective pesticides that kill a limited range of pests.

Consumer groups and the general public may also be able to support the implementation of the IPM programmes by demanding residue-free commodities. There is now a distinct market for organically produced food and other products. Non-Government Organizations (NGOs) and consumer groups need to be strengthened especially in developing countries, so that there is a public-oriented movement for implementation of IPM.

ACHIEVEMENTS IN IPM IMPLEMENTATION

Numerous attempts have been made during the last two decades to implement IPM in different regions of the world. As is to be expected, maximum efforts have been directed towards cotton, a crop which receives a disproportionately high amount of pesticides around the globe. There are many successful examples of IPM implementation on cotton in South and North America and Australia. In the Asian region. excluding China, there has been little progress and cotton ecosystems are in crisis phase due to excessive usage of pesticides. However, in the same Asian region, success of IPM in rice provides one of the best examples of its implementation in the tropical developing countries. Both these crops provide excellent examples of understanding the progress and problems in IPM implementation.

Role of IPM in Crop Production

IMP is acclaimed as a means for the development of ecologically sound and sustainable agricultural practices to intigate crop production losses and for strengthening farming communitties participation in that development. IPM is a key component in the battle for increased food production and has the potential to play a much greater role in the future.

Inspite of significant progress the need for increased production of food, food and fibre continues unabated. The situation is especially acute in Africa where population growth will continue to outstrip growth in food production for many year unless more is done to increase agricultural growth. If current trends continue, by 2025, Africa could have a food gap of 185 million tons, which will be 205 times production.

The green revolution created a delemma in which it was believed that mass starvation could only be avoided by the widespread adoption of techonology that was in some cases harmful to the natural environment. Higher yields were achieved but increased land productivity was accompanied by increased use of chemical impact and the health of farmers. It is now recognised that the real costs of pesticides associated with packaged technoligies are high and increasing. What is required is another Greeen Revolution that is greeen in terms of conserving natural resources and the environment, or a "Super-Green" revolution. Super-Green Revolotion will require a repeat of the successes of the Green Revolution on global scale in many diverse localities and be equistable, sustainable and environmentally friendly.

There has been significant progress in both, the development of IPM tactics, and in the development of more effective implementation strategies. The development of botanicals, behaviour modifying chemicals and bioregulatrors to replace environmentally harmful pesticides will continue to a major research thrust into the next century. The integration of genetically resistant cultivars with natural biological control agents and cultural controls will continue to be a sought after strategy. Although significant progress has been achieved in the development and commercial utilization of insect resistant rice cultivars they have the potential to play an even much greater role in rice production systems in the future. Biotechology techniques such as wide hybridization, transformation, DNA markers and DNA finger printing, are being used to overcome some of the constraints to breeding of insect resistant cutivars. The cultivars developed will contribute to an economically more efficient and an environmentally benign crop protection technology.

Global IPM Implementation Strategies

Implementation of IPM programmes has proven to be highly complex, requiring specialized technical expertise, dedicated and skilled field trainers, detailed analysis of policy, social and economic issues at the national and local levels, political commitment, and careful attention to details in project design, operationalization and monitoring.

Global IPM field implementation gained momentum when FAO Intercountry Programme for IPM in Asia (ICP) organised a Global IPM meeting in 1993. Participants from various regions of the world familiarised themselves with the achievements and approach of the collaborating national IPM Programmes in Asia.

In 1994 an Inter-Agency Task Force was established by FAO, World Bank, UNDP and UNEP for discussing, expanding and intensifying IPM implementation. Because of the dynamic nature of the FFS (Farmer Field Schools) approcah, the 21st century will see an expansion of IPM implementation to cover more crops and more regions of the world. IPM will continue to serve as an entry point for the training of farmers in a holistic approach to sustainble crop production.

(D)

PESTICIDE APPLIANCES

Recent advances in chemical control of insects have come not only from the discovary of more effective insecticides but also from the development of new and better machines for applying these chemicals. Equipment used to apply insecticides must be suited to the crop, to the scale of the operation, and to the formulation of chemical to be applied. There are three major cotegories of plant protection equipments (1) *Dusters :* for applying solid formulations (2) *sprayers :* for applying liquid formulations and (3) *fumigators :* for applying fumigation. Each of these categories are again broadly grouped into further two certegories viz.

(1) Hand or manually operated, and

(2) *Power operated* : For the past few years a new equipment has come into vogue viz. motorised knapsack mistblower. This can be used for spraying as well as dusting (sprayer come duster) only by changing a few attachments.

Dusters

Dustes are used for less than sprays as a means of applying insecticides to plants and so there has been less development of dusting equipment. Most dusters incorporate a metering device of some sort to regulate dust output and a blower to carry dust to the target. Attempts have been made to charge dust particles electrostatically so that they adhere better to plant surfaces but this has not been very successful in practice. All dusting appliances operate on the principle of emitting a blast of air, in which the dust particles are air borne. It consists of a container or hopper, an agitator to keep the dust disturbed and enabling it to be passed by a feeding mechanism into a current of air which carries dust to throw out through an outlet in the form of a cloud. The air current is produced by manual power or by a fan or by bellows. Some common dusting appliances are an follow:

Hand Operated Duster

(a) *Plunger type duster :* It is a simple pump. The pump generate an air blast that passed on the the dust chamber,and the dust blow out through a delivary tube. It is more or less a flit pump.

It consists of a metal chamber or cylinder, a part of which contains a movable plungeı with a handle. The chamber serve the purpose of a recepticle or hopper for dusting material. The dust is filled in the glass bottle or hopper which comes out with air. It is useful on small scale operations such as dusting of pots or small kitchen gardens, cattlesheds, poultry house etc.

(b) *Bellows duster* : It is provided with small container to hold the dust. This equipment works on compressing bellows to produce air blast for ejections the dust. It is also very useful for domestic purpose or for kitchen gardens.

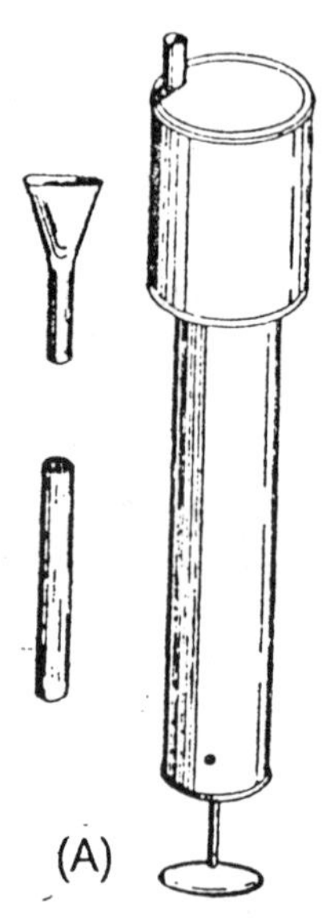

Fig. 6.102

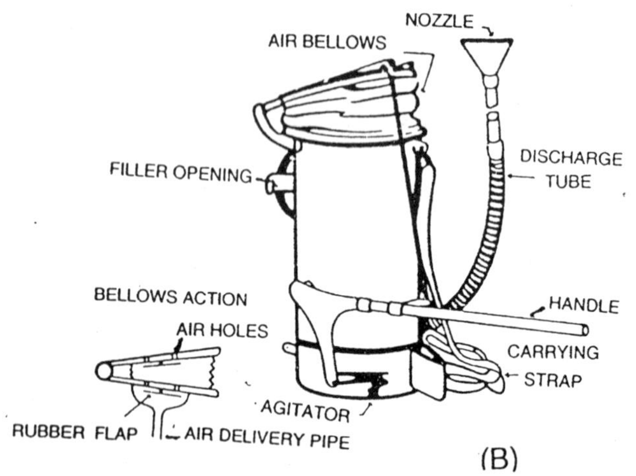

Fig. 6.103

(c) *Hand rotary duster* : It is a more complicated equipment used for dusting agricultural crops and in ware houses. There are two kinds of hand rotary dusters available in market (a) *round* in appearance, and for operation, it is fitted on the chest of the operator and (b) *long and large,* which is operated by fitting to the leftside of the operator. The working of both the dusters is similar.

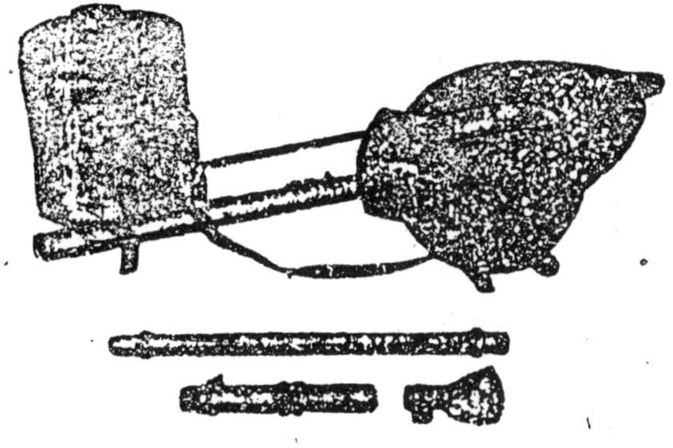

Fig. 6.104

Hand rotary dusters consists of four parts:

(a) *Hopper or container* : It is just like a drum in shape having about 5-6 kg capacity covered by a lid. A chest plate is provided on the back side of the hopper. The crank shaft passes through centre of the hopper and carries two banded iron bars on both sides which agitate the dust contained inside regularly so that it may not block the passage of dust to the blower. They are known as agitators. On the lower portion of the hopper just below the crank-shaft is fan shaft which rotates the fan. When crank is operated both crank and fan shaft moves faster due to gear arrangement. There is an exit hole in the hopper on the fan side provided with a feeder. The feeder supplies dust into the blower through the exit hole. There is a feed regulator on the outer side of the hole for regulation of its size and also the supply of dust.

(b) *Gear box* : The gear box is located on the right hand side of the hopper and contain four gears.

(c) *Blower* : It is a small closed box opening to the exterior through a small projected opening which is about 25 mm in diameter. There is a fan with five blades in the middle of this box. This fan sends the air from the centre to the outside thus creating a vacuum occupied by the dust from the hopper and air from wire netting provided on the left side of the blower.

(d) *Discharge system* : Blower ends are connected with rubber joints called couplings, with a long tube known as lance. The lance is fitted with the nozzle covered at the top and opened at the bottom so that the dust should come out from the bottom.

(1) Power Operated Duster

It is operated by petrol-driver motor. There is a regulator for dust discharge, a fan and a flexible pipe with nozzles. This power duster is mounted on trolleys. The dust comes out from it at the speed of 300-320 km/hr. These are used for much larger areas.

(2) Sprayers and Spraying

No agricultural operation require greater care and throughness than spraying. When stomach poisons are being applied, the aim should be to cover every part of leaf and fruit surface and succulent buds and stems. If a contact insecticide is being used, it should be remembered that only those insects will be killed that are actually hit by the spray. While throughness is required, the application should be stopped as soon as dripping begins.

The application of sprays must ordinarily be very accurately timed to coincide with the period in the development of the insects when they are most easily killed or with the stage in seasonal development of the plant when it will best withstand the treatment. Since these times will vary widely in different sections of the country, the grower should secure the advice of a trained entomologist, who knows the local conditions, regarding spray programmes for different crops and insects. In general, the earlier the spray is given after the insects appear, the easier it is to destroy them.

Spraying should not be done in rainy weather, although many sprays will adhere satisfactorily to plants if the spray has time to dry before rain falls. Winter sprays should not be applied when the temperature is below freezing or when the trees are wet with snow or rain.

Good pressure is essential to the best results in spraying, and the spray should be sufficiently powerful to maintain a pressure of at least 125 pounds. For orchard and vineyard spraying, best results are secured at pressures of at least 250 to 800 pounds per sequre inch.

When using poisons which do not go into solution, good agitation of the spray material is very important to maintain an even mixture.

The best pump can easily be ruined by failure to give it proper care. Clean water should always be pumped through the sprayer after using, to wash put all the corrosive spray materials. Wooden tanks are best stored in damp places. The metal parts should be kept oiled to prevent rust.

Since most spraying materials are violent poisons, they should be plainly labelled and together with mixing vessels, kept out of reach of children. Animals must be kept away from liquid sprays and not allowed to pasture under sprayed trees. Fruits and leafy vegetables must not be sprayed immediately before marketing and should always be cleaned of any spray residue whether or not it is considered poisonous. It is not advisible to apply insecticides with spraying equipment which has previously been used for the application of plant growth regulators such as 2-4 dichlorophenoxyacetic acid (2, 4 D), as very small traces of such materials may severely affect certain plants.

Sprayers vary greatly in size from small hand operated garden syringes to huge orchard air blast machines (mist blowers) that require a large tractor to low them. However, all sprayers must provide some means of breaking up the spray liquid into droplets (atomising) and of conveying the droplets onto the object to be sprayed (target). In most smaller machines hydraulic pressure provides the means both of atomising the spray liquid and of imparting momentum to the droplets so that they carry to and impinge on the target. Many larger machines however, especially those designed for low volume application (see below), employ a fan generated stream of air to carry the droplets. They are referred to as *air assisted* or *air blast sprayers*. Atomisation in such sprayers may be through normal hydraulic nozzles or by power operated spinning discs or spinning cages, which provide more uniform droplet size than conventional nozzles. Until about 25 years ago all sprayers were of the hydraulic type and applied pesticides in very dilute form. The objective of spraying with them was to completely wet plants (or other target surfaces) normally to the point of run off. This is still the procedure with small hydraulic sprayers but the development of air blast machines has permitted the use of chemicals in more concentrated form. The spray pattern is then one of small discrete droplets which do not give complete cover of the target or immediately coalesce but nevertheless provide effective control of most pests. This is *low volume spraying* compared to earlier high volume application. Most large modern sprayers are to the low volume type.

Further refinement of equipment has enabled spray volumes to be progressively reduced and the extreme is reached in *ultra low volume* (ULV) spraying where the insecticide is applied in extremely concentrated form (50% or more) or even as the undiluted technical material if this is a liquid (eg malalhion). With ultra low volume application very small droplets and precise application are essential for effective results.

There is no general agreement as to where dividing lines should be drawn between high volume, low volume and ultra low volume but as a general guide high volume spraying involves the application of 1120 litres or more per hectare (100 gals + per acre), low volume 225-560 litres per hectare (20-50 gals per acre) and ULV 10 litres per hectare (1 gal per acre) or less.

Emulsifiable concentrate or wettable powder formulations (diluted with water) arc used for high volume or low volume spraying but ULV usually requires special formulations with a light oil as diluent.

Low volume (and ultra low volume) spray techniques require less frequent filling of the spray tank compared to high volume spraying. This is very important in aerial spraying and can materially reduce application costs. Also, compared to high volume spraying, less pesticide is normally used to achieve the same result as more is deposited on target. On the other hand more precision is required with low volume spraying in setting up and operating the equipment. This applies even more to ULV. Even with the best available equipment much spray is washed when treating growing plants (as much as 70% commonly does not reach the target.) so there is still great scope for imporvement in the efficiency of spray equipment.

Following are some of the important sprayers commonly used:

(i) Hand Atomiser (Flit Pump)

It is used for the control of mosquitoes and houseflies. It is also used in small kitchen gardens, flower beds etc. The compression is not constant. In each compression stroke the air piston draws the liquid up the aspirator tube and atomises at the top which is simply cut at right angles. It consists of the following parts :

(a) *Barrel :* It is a cylindrical tube 250 mm in length, 37.5 mm in diameter and made of tin. It is connected to the cap of the tank on the lower side of the interior end. This end of the tube is closed with a small hole in the middle known as nozzle. A tube 3.12 mm in diameter projects from the tank through the centre of the filler-cap and end near just half of the nozzle. This tube goes upto the bottom of the tank and throws the liquid in spray form. The barrel is provided with 4.16 mm diameter air-hole on the posterior end of the incoming and out going air.

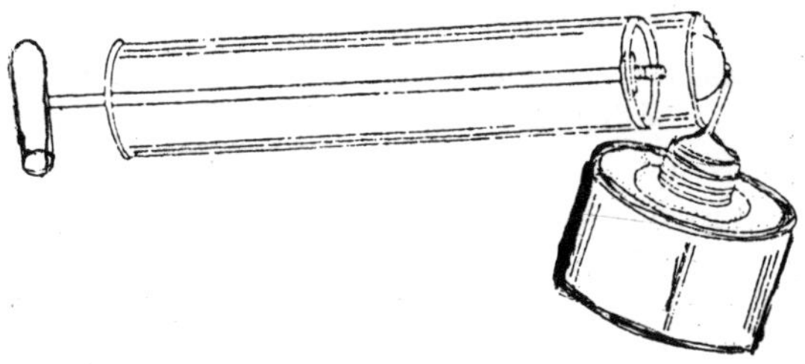

Fig. 6.105 : Flit pump.

(b) *Barrel cap :* Posterior end of the barrel is covered with a lid known as barrel cap which has a hole in its centre through which passes the plunger.

(c) *Plunger :* It is made up of a wire 4.16 mm in diameter having a wooden handle on the posterior end which helps in the operation of plunger. A washer and a nut is attached on the anterior end of the plunger. There is one leather valve in between the washer and the nut which closely fits within the walls of the barrel.

(d) *Tank :* This is round in appearance with 227 ml capacity and provided with a cap known as the filler cap.

(ii) Hand Compression Sprayer

It is a medium sized manually operated machine used to spray the field crops and trees. It works on the principles of stoves and petromax which gives out the spray due to air pressure. It is available in 9, 13.5 and 18 litre capacities. It is made up of three components.

(a) *The pump :* It is just like a foot ball pump and works on its principle. The pump is fitted in between the upper end of the tank and hangs inside it. The only parts of the pump visible outside the tank are handle, plunger and plunger spring while the plump barrel, leather and ball valve checks the incoming of the air which is filled in the tank back into the pump.

(b) *The tank :* It is made up of brass or copper and cylindrical in shape with upper convex end. Its length is determined as per volume and diameter of the tank. The pump is fitted in the middle of the tank. The spray liquid is filled in through a small opening 25 mm in diameter covered by a filler cap which is situated on the upper end of the tank. Just near to it, is the pressure gauge which indicates the pressure of the air inside the tank. There is a hole on the bottom of the tank which is fitted with a tube. The first tube runs parellel to the bottom of the tank and is connected with other tube. The first tube is known as delivery tube and other one is lance. There is one cut off cock between the joint of tubes which checks the flow of the spray liquid.

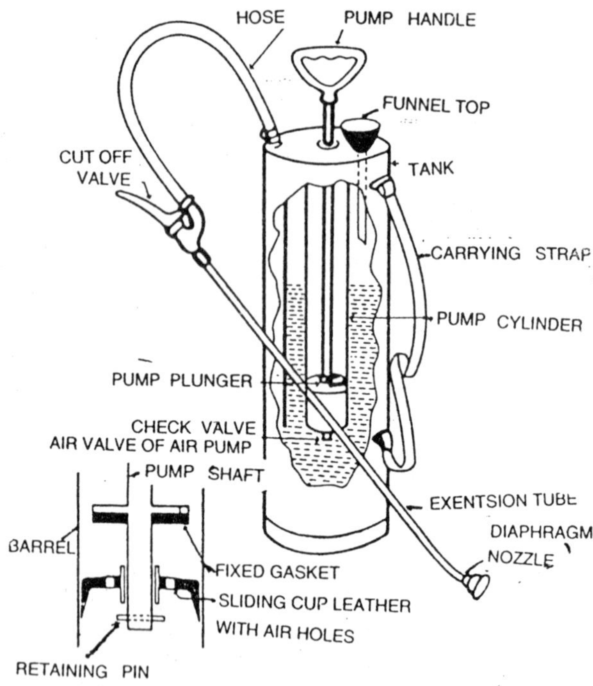

Fig. 6.106

(c) *Spraying parts :* One rubber in plastic tube measuring about 1.25 metre in length and 12 mm in diameter, known as hose, is connected with lance. There is a thin rod on the anterior end of the hose equipped with a cut off cock by which the flow of spray liquid is controlled. The thin rod, held in hand during operation, is connected with the exterior rod 6 mm in diameter. Extension rod at its end possess the nozzle which form the mist.

(iii) Knapsack Sprayer

It is useful for small trees, shrubs, vegetables, low crops or nursery stocks. This has a flat or beam shaped copper or brass tank with 9-22.5 litre capacity to fit comfortably on the back with an agitator inside. The pump handle extends over the shoulder or under the arm and is worked constantly with one hand while the lance is held with the other hand for spraying. Its spraying capacity is 112 litre per day and a man can cover about 1 hectare per day.

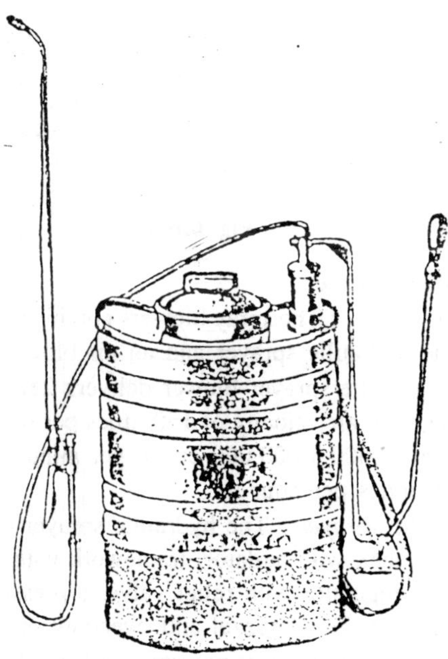

Fig. 6.107 : Knapsack sprayer

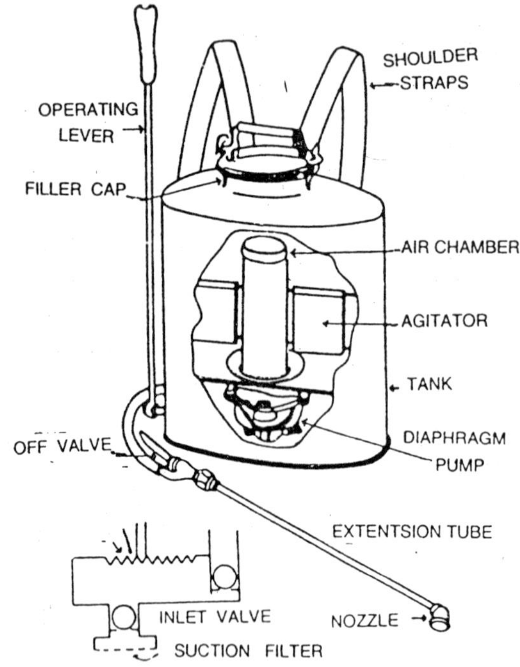

Fig. 6.108

Power-operated Sprayer

(a) *Spray blowers* : The spray blowers combine a low-pressure low volume hydraulic sprayer and an air blower.

A low volume sprayer blower delivers 250-30,000 cu. ft. per minute at an air velocity of 150 miles per hour. A high volume delivers 20,000-1,000,000 cu. ft. per minute at an air velocity of 100 miles per hour.

(b) *Hydraulic sprayer* : The hydraulic sprayer consists of engine, pump, a tank with agilator, framework with or without wheel, pressure regulator with relief valve, a pressure gauge, suction and delivery hose, nozzel and overflow pipe.

The hydraulic sprayer is powered by 1-5 H.P. air cooled engine fitted with single or double or three discharge limes, having discharge capacity of 7-18 litres per minute, at a pressure upto

250-500 pounds per square inch. Three types of sprayer are available in the market :

(i) skid type

(ii) wheel type, and

(iii) wheelbarrow type.

They are suitable for spraying orchards where the spray jet has to reach upto 10-15 metres or more.

ADVANTAGES OF SPRAYING

(i) More efficient than dust

(ii) Spray are not subject to wind drift

(iii) Rate of application can be controlled by droplet size.

DISADVANTAGES OF SPRAYING

(i) Greater cost and smaller out turn of work.

(ii) Spray formulation requires careful storage to prevent deterioration.

CARE AND MAINTENANCE OF SPRAY AND DUST

If the sprayer and duster are given a proper care and maintenance, they remain serviceable for a long time.

1. Common for sprayers and dusters

(a) power engine tends to get heated. Over heating should be avoided.

(b) recommended fuel-oil mixture should be added.

(c) specified lubricant should be used.

(d) clean water should be used in preparing spray.

(e) avoid bending of hose pipes.

2. For sprayer

(a) clean the pump after operation with water and rinse it.

(b) lubricate the washer to keep it soft.

(c) never plunge the piston rashly.

(d) do not clean the nozzles by any hard substances.

(e) nozzles should never be blown through the mouth, a cycle pump can be used.

Fumigators

Insecticides, in gaseous form are known as fumigants, and are used in storage bins, buildings, ship holds and even in soil where the gas can be confined. Fumigants are most often formulated as liquid under pressure and quite often they are mixtures of two or more gases. The liquid fumigants are held in cans or tanks. In some cases the gas is required to be produced at the place to be fumigated; example being dropping of calcium cyanide into earthenware cracks filled with sulphuric acid to produce HCN (hydrogen cyanide). Phosphine or hydrogen phosphide gas is released from tablets containing aluminium phosphide and ammonium carbonated in the presence of moisture.

Fumigation should be attempted only by persons fully acquainted with the type of fumigant to be used and the method of its application. Fumigation of dust or gas is done by cyanogas pump. Cyanogas pump is used to dust calcium cyanide into rodent burrows where, an exposure, emerges a poisonous fume. The duster is of the plunger type which consists of a chamber for the dust, a cylinder with a plunger or air pump, a rod and handle with a delivery tube and nozzle. The smaller type of pump is needed for household pest and the larger one for kitchen garden and in nurseries. Flow of the dust is intermittent and not continuous.

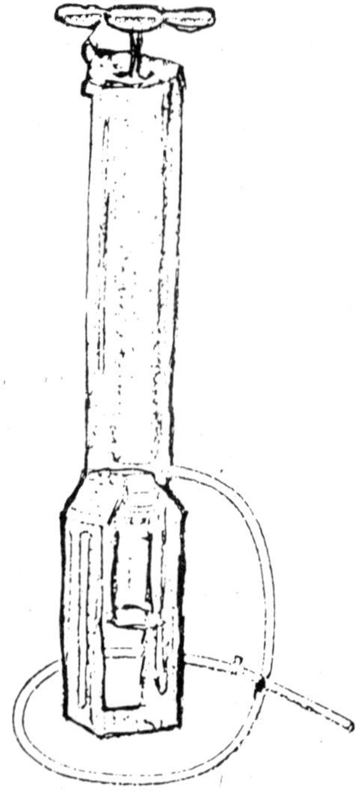

Fig. 6.109 : Cynogas foot pump.

The cyanogas foot pump : Consists of an *air pump* dust chamber and disctarge tube.

(a) *Pump cylinder :* It is a brass tube measuring 450 mm long and 75 mm in diameter closed at the upper end by a pump cap. The posterior end of the brass tube opens into a glass bottle. There is a valve between the pump cylinder and bottle which allows air into the bottle and stop it going back. There is a delivery tube at the base of the pump cylinder connected with the bottle. Power mixed air comes out after passing through this tube. There is foot rest below the pump cylinder for operation by foot. A plastic delivery hose is connected with delivery tube which is connected with 300 mm long brass tube known as lance.

(b) *Pluger rod :* It consists of a handle at the upper end and leather valve on the other end connected with a washer like a cycle pump.

INSECTICIDAL FORMULATIONS

The insecticides are used in various formulations to achieve quick control of the pests. The common insecticidal formulations are the dusts, wettable powders, granules, emulsions, solutions, aerosols and fumigants.

1. Dusts

It is a mixture of 0.1 to 20% toxicant and the rest of the inert carrier such as talc, gypsum, clays, flours etc.. Dusts are used in dry areas against the pests infesting nonirrigated crops.

2. Wettable Powders

Powders are prepared by mixing higher percentage of toxicant with little quantity of wetting agents. The powders contain 15 to 95% of insecticide. Due to presence of wetting agents, the insecticide spreads uniformly on the plants and adheres well providing their longer persistance. The sulphonated aliphatic esters and amides are used as the wetting agents.

3. Granules

The toxicant is impregnated into or on to the absorptive carriers such as clays, bentonite, tobacco waste, etc. The particle size ranges from 0.3 to 1.5 mm which facilitates the entry of insecticide through the leaf whorls. This formulation is very useful for the control of tissue borers.

4. Emulsions

The insecticide is first dissolved in an organic solvent and at the time of application; it is mixed with the water along some emulsifiers to ensure the stability of emulsion. Sometime oil-soluble emulsifiers are used. These formulations are suitable for the low volume spraying. An emulsifiable concentrate (EC) dissolves in water only after adding the emulsifying agent. Some of the commonly used emulsifiers are, glycerol monoleate, manital monoleate, aliphatic esters of sorbitol and polyalkaline ether alcohol etc. Soaps are also used as the emulsifiers.

5. Aerosols

The toxicant dissolved in an organic solvent is available for immediate use without any help of pumps or blowers. The insecticide dissolved in liquified gas is kept under pressure in a metallic container. On opening of the valve, the insecticide forcely comes out as a very fine spray and the fog is composed of liquid while the smoke, of the solid particles. The size of the particles measures 1.30 micra. Aerosol bombs and pins are usually prepared for the control of flying insects inside the closed areas.